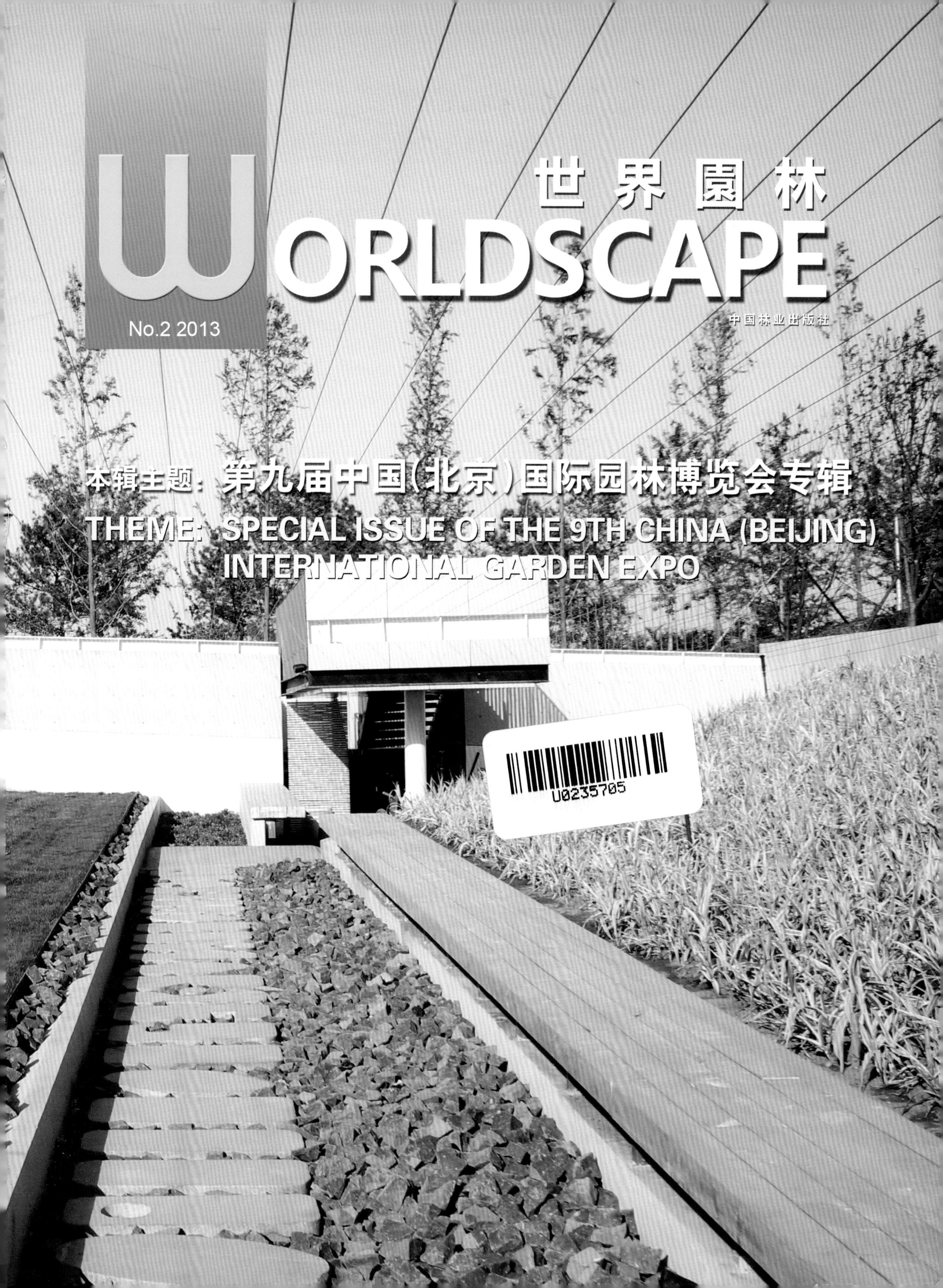

世界園林
WORLDSCAPE
No.2 2013
中国林业出版社
本辑主题：第九届中国(北京)国际园林博览会专辑
THEME: SPECIAL ISSUE OF THE 9TH CHINA (BEIJING) INTERNATIONAL GARDEN EXPO
U0235705

"园冶杯"风景园林(

The 2014 "Yuan Ye Award" International Lan

"园冶杯"风景园林(毕业作品、论文)国际竞赛是由中国建设教育协会和中国花卉园艺与园林绿化行业协会主办，中国风景园林网和《世界园林》杂志社承办，在风景园林院校毕业生中开展的一项评选活动。

时间安排

报名截止日期：2014年4月30日

资料提交截止日期：2014年6月10日

参赛资格

应届毕业生(本科、硕士、博士)

参赛范围

风景园林及相关专业的毕业作品、论文均可报名参赛

竞赛分组

竞赛设置四类：风景园林设计作品类、风景园林规划作品类、园林规划设计论文类、园林植物研究论文类。每类设置两组：本科组和硕博组

Time schedule

The application deadline is April 30.2014:Submission deadline on June 10, 2014

Qualification

This year's graduates (Bachelor,Master,Doctor)

Competition content:

Landscape architecture and the related specialized graduation work, or paper

Types

There are four types:The design works of landscape architecture,the planning works of landscape architecture, the design and planning papers of landscape architecture,and the papers of garden plants. Every type is set into two groups: the undergraduate group,the master and doctor group.

毕业作品、论文）国际竞赛

ape Architecture Graduation Project/Thesis Competition

The 2014 "Yuan Ye Award" International Landscape Architecture Graduation Project is hosted by China Construction Education Association, Chinese Flowers Gardening and Landscaping Industry Association, and is undertaken by China Landscape Architecture network and *Worldscape*. It is a competition among graduates in landscape architecture schools .

地址：北京市海淀区三里河路17号甘家口大厦1409

邮编：100037 电话：（86）010-88364851

传真：（86）010-88365357

学生咨询邮箱：yyb@chla.com.cn

院校咨询邮箱：messagefj@126.com

官方网站：http://www.chla.com.cn

Address: 1409#, Ganjiakou Building,No.17,Sanli River Road,Haidian District,Beijing.

Post Code: 100037 Tel: 010-88364851

Fax: 010-88365357

Student Advisory E-mail: yyb@chla.com.cn

College Advisory E-mail: messagefj@126.com

Website: http://www.chla.com.cn

世界園林

WORLDSCAPE

图书在版编目（CIP）数据

世界园林：第九届中国（北京）国际园林博览会专辑：汉英对照 / 中国花卉园艺与园林绿化行业协会
主编 . -- 北京：中国林业出版社，2013.9
ISBN 978-7-5038-7180-1
Ⅰ . ①世… Ⅱ . ①中… Ⅲ . ①园林 - 介绍 - 中国 - 汉、英 Ⅳ . ① TU986.61
中国版本图书馆 CIP 数据核字 (2013) 第 210718 号

中国林业出版社
责任编辑： 李　顺　成海沛
出版咨询： (010) 83223051

出 版：中国林业出版社（100009 北京西城区德内大街刘海胡同 7 号）
网 址：http: //lycb.forestry.gov.cn/
印 刷：北京卡乐富印刷有限公司
发 行：中国林业出版社发行中心
电 话：（010）83224477
版 次：2013 年 6 月第 1 版
印 次：2013 年 6 月第 1 次
开 本：889mm × 1194mm 1 / 16
印 张：19
字 数：300 千字
定 价：80.00RMB（30USD，150HKD）

WORLDSCAPE

目录

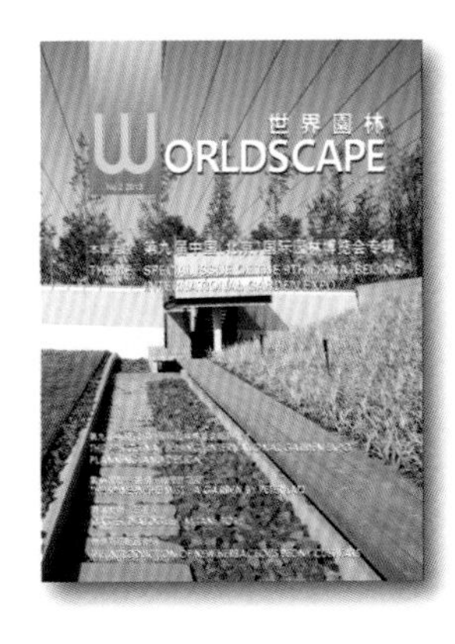

世界園林 第三辑
主　　题 第九届中国（北京）
国际园林博览会专辑

WORLDSCAPE
No.2 2013
THEME: SPECIAL ISSUE OF
THE 9TH CHINA (BEIJING)
INTERNATIONAL GARDEN EXPO

CONTENTS

WORLDSCAPE

源树景观（R-land）是国内顶级的专业 环境设计机构。自2004年成立以来，通过不懈的努力，在景观规划、公共空间、旅游度假、主题设计等领域都获得了傲人的成绩，特别是在高端地产景观的咨询及设计方面，处于绝对的领先地位。

源树景观（R-land）的设计团队中汇集了大量的国内外景观设计精英，其主要设计人员都曾在国内外高水平设计单位中担任重要职务，严格的设计流程确保了每一项设计作品的完美呈现。

源树景观（R-land）历经数年，已完成了数百项设计任务，其中：河北省邯郸市赵王城遗址公园、中关村创新园、山东荣成国家湿地、西安大唐不夜城、北京汽车博物馆、龙湖“滟澜山”、天津团泊湖庭院、招商嘉铭珑原、远洋傲北、中建红杉溪谷、西山壹号院等若干项目均已建成并得到各界的广泛认可。

源树景观致力于最高品质的景观营造，力求为合作方提供最高水准的设计保障。

Add：北京市 朝阳区朝外大街怡景园 5-9B（100020） Tel：(86) 10-85626992/3 85625520/30 Fax：(86) 10-85626992/3 85625520/30 - 5555 Http://www.r-land.com

常务理事单位

无锡绿洲景观规划设计院有限公司
EDSA Orient
北京大元盛泰景观规划设计研究有限公司
北京夏岩园林文化艺术集团有限公司
棕榈园林股份有限公司
岭南园林股份有限公司
北京源树景观规划设计事务所
北京欧亚联合城市规划设计院
重庆金点园林股份有限公司
重庆天开园林景观工程有限公司
杭州天香园林有限公司
北京市园林古建设计研究院有限公司
盛世绿源科技有限公司

理事单位

北京东方园林股份有限公司
重庆华宇园林股份有限公司设计分公司
枫彩农业科技有限公司
江苏山水建设集团有限公司
苏州新城园林发展有限公司
广东四季景山园林建设有限公司
北京乾景园林股份有限公司
广州市林华园林建设工程有限公司

王小璘
Xiaolin Wang

1851 年，英国在伦敦海德公园内建造了一座划时代的旷世巨作“水晶宫”。这座钢结构、玻璃装饰的温室里，展示来自 10 个国家的千余件珍品。在 160 天的展期中，吸引了世界各地约 630 万人次的访客前来观赏，是为近代世界博览会的滥觞！随着时间的推移，世界各国纷纷起而效尤，每年在不同国家、不同城市举办各具特色的博览会，并以主题展现当代人们关注的议题。这项国际性的大型活动，不仅成为城市营销和国际交流的最佳舞台，同时也带动都市化、现代化、科技化的加速成长，促进观光旅游及绿色产业的发展和生活环境质量的提升。

第九届中国国际园林博览会将于 2013 年 5–10 月在北京举行，本刊以专辑介绍园博会的精彩园林作品，由负责规划的北京建筑设计院说明园博会的规划设计构思，使读者得以一窥北京国际园博会的全貌。

本届园博会的最大亮点是，由三位国际知名景观大师设计建造的三座“大师园”，分别有来自德国彼得拉茨先生的“雾中叶脊－彼得•拉茨的花园”、美国彼得沃克先生的“有限／无限”和日本三谷徹先生的“祈祷之园—倾斜土地，创造空间”。透过本刊的介绍，读者可以欣赏到世界一流的园林作品和大师们独特的风采！

为更好地体现园博会的主题和宗旨，园博会组委会举办了“第九届中国北京园博会设计师创作国际竞赛”，由国内外的著名设计师提供设计方案，园博会组委会负责实施，并邀请了当前顶尖的园林专家学者组成评审团，经过审慎严谨的评审后，选出 16 件杰出设计作品。其中，获奖的前六名，将建造于园博会“国际风景园林师创意展园”的“设计师广场”中；其余 10 件为竞赛佳作入围作品。

本刊作为协办单位之一，很荣幸邀请到三位大师，出席 2012 年 2 月和 4 月在北京举办的“国际知名园林大师学术报告会”，并进行“与大师对谈”；其中，与两位彼得先生对话内容已分别刊载于本刊前 2 期的“对话大师”。本期则实录三谷先生与嘉宾和现场观众之间互动的谈话内容。

此外，本刊也介绍了 8 种世界芍药新品种；随着这些品种的引进与推广，相信在吾人生活的周遭环境中，未来将可看到更多美丽绚烂的传统名花—芍药。□

王小璘

In 1851, Great Britain built a masterpiece-the *Crystal Palace* in Hyde Park of London. This ornate steel and glass greenhouse displayed more than 1,000 treasures from ten countries. In the 160 days of the exhibition, it attracted around 6.3 million visitors from all over the world. This was known as the first International Expo. Since then, cities of different countries have organised their own characteristic expos, accompanied by a theme based around relevant contemporary issues. This international activity has not only become the best stage for city promotion and international exchange, but has also accelerated the growth of urbanization, modernization and technology, as well as the development of sightseeing and green industry, and the improvement of the quality of the living environment.

The Ninth China International Garden Expo will be held in Beijing from May- October 2013, and is the theme of this issue. Through the introduction by the Beijing Institute of Architectural Design, which is responsible for the planning and design of the Expo Park, the reader will gain a glimpse of the picture of the whole area.

The highlight of the Park is the three "master Gardens", which are designed by internationally well-known landscape architects. They are "The Spine in The Mist" by Mr. Peter Latz", "Finite/Infinite" by Mr. Peter Walker, and "A Beggining of The Garden - Tilting The Earth and Marking The Place" by Mr. Toru Mitani. Through the introduction of the articles, the reader will gain a glimpse into the unique style of world-class garden works and celebrated landscape design gurus!

As part of the search for designs which reflected the theme and purpose of the Park, the Organizing Committee of the Expo held the "International Competition of the 9th "Beijing Expo designer creation". The designed projects were provided by local and international designers, and implemented by the Committee on site. An invited jury of renowned landscape architects and scholars was responsible for the competition assessment. Sixteen outstanding design works were selected. Among them the works of the top Six have been built at *Designer Square*, an area of the *International landscape architects creative exhibition area* of the Park. The rest of the projects were awarded Honorable Mention.

Our magazine, as one of the co-organizer, was honored to invite the three masters who attended the *Academic report of internationally renowned garden masters* and *the Dialogue with a Master* lecture series held in Beijing in February and April, 2012. The contents of the dialogue with Mr. Peter Walker and Mr. Peter Latz were published respectively in the previous two issues, while the conversation with Mr. Mitani is recorded in this issue.

In addition, new varieties of peony are introduced in this issue. With the introduction and promotion of these cultivars, we believe that more of the gorgeous and traditional *Paeonia lactiflora* will be seen in our surrounding environment in the future. ■

美国加州卡尔斯班花海

这是一座人口不到八万的幽静小城，地处洛杉矶与圣地亚哥之间。因为南加州的气候得天独厚，受惠于常年的温暖日照，一旦春回大地，万物更新，一时间无处不飞花。其中最著名的莫过于令人叹为观止的卡尔斯班花田。

每年春天，由3月中旬开始一直持续到5月初，卡尔斯班农场内50英亩的坡地，都会被五彩缤纷的鲜花所覆盖，由于不同颜色的花朵是分区栽种的，在排列上也间隔有致、颇具匠心，使得整体效果壮观美丽，看上去赏心悦目。

http://www.chla.com.cn/htm/2013/0417/165765.html
来源：中新网

"闯入"异乡的东方古典园林

在美国洛杉矶近郊圣玛利诺市的亨廷顿图书馆内，坐落着一座原汁原味的中国古典园林——流芳园。流芳园以苏州园林为蓝本建设，规划面积约4.9hm^2(12英亩)，该园规划面积接近于苏州的拙政园，其水榭、亭台、

拱桥、长廊、漏窗、怪石等传统苏州园林建筑元素应有尽有，故又有"海外拙政园"的美誉。该园的二期工程正在进行当中。

"流芳园"自2008年2月23日正式开放以来，令当地的美国游客得到最直观的中国文化印像，而踏入"流芳园"的当地华侨华人们，在熟悉的景物中，回家的感觉油然而生。

http://www.chla.com.cn/htm/2013/0419/165969.html
来源：中国风景园林网

土耳其：特朗普大厦——花园和城市的微妙衔接

特朗普大厦位于伊斯坦布尔的商业和城市生活的CBD区，它与周围环境和城市景观相融合，停车场和地铁站相接驳。该项目是一个塑造城市形象和私人使用相结合的开放空间设计，概念设计包括露台和"塔"状建筑。

花园设计成带状，向四周散射状的为用户提供一个多功能的开放空间。虽然露台是作为一个散步和休闲的过渡区，但是还是有相当数量的植物；其他的露台被设计成日光浴区，水在炎热的天气创建了多个休息平台。所有的花园提供不同的功能，同时维持整个小生态圈的运行。

http://gardens.liwai.com/content-26036.htm
来源：园林景观网

Green Camouflage花园——别样的泰式风情

这个梦幻般的"Green Camouflage花园"景观项目位于泰国曼谷最繁华福地的Blocs 77公寓中。5244m^{2}的用地中容纳了467户住宅单元。花园前方是喧闹的街道，后方是宁静的运河和住区。绿色从水平蔓延到垂直。地面层宁静的水池开启了绿色空间的前奏，立面上种

了植物，空中花园也种满了植物。设计师巧具匠心，为居民和城市带来美丽环境。

http://gardens.liwai.com/content-25965.htm
来源：园林景观网

加拿大Cap-Rouge记忆墙

在2003年，魁北克市Cap-Rouge悬崖部分塌陷，"Cap-Rouge记忆墙"早期是作为临时项目用来保护该地区毗邻的de la Plage海滩Jacques Cartier海角，内容包括清除树木、改变自然排水模式和土壤压实。由于土壤已经遭到侵蚀，为了保护考古遗址和当地居民与游客的安全，将它转变成一个长期的项目来支撑悬崖，逐步演变成一个景观建筑项目，将圣劳伦斯河悬崖沿路区域的历史以一种当代艺术作品的形式表现出来。Cap-Rouge记忆墙不仅仅是一个历史的回忆，此外，通过附近的雕塑和上面的灯光可以感受不同的四季。

http://gardens.liwai.com/content-25987.htm
来源：园林景观网

澳大利亚 Cranbourne 皇家植物园：人、景观与自然

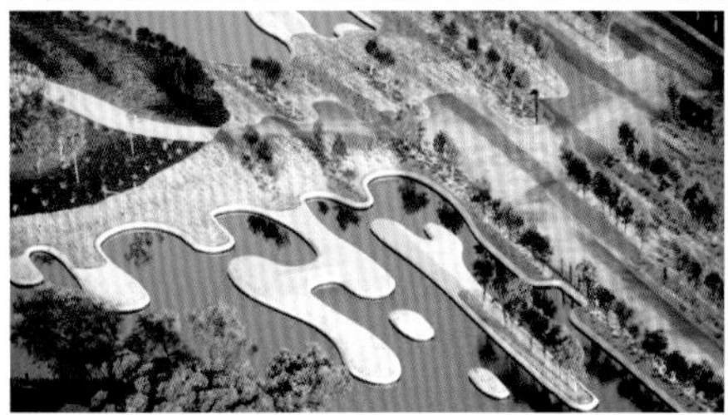

Cranbourne 皇家植物园位于墨尔本郊区东南部，通过“我们的景观和植物园林设计是探索和表达澳大利亚人之间关系不断发展的。”这一共同的主题，试图创建一个突出植物群特性的花园，突出自然景观之间的紧张关系，来激发游客进一步探索澳大利亚植物群与景观之间的关系。Cranbourne 皇家植物园成功地重新解释了“什么是澳大利亚景观？”。这也正体现了景观艺术的意义所在，景观设计不仅仅要满足美学的要求，更重要的是传达一种信念，从而使人与自然、人与景观和谐相处。

http://gardens.liwai.com/content-25982.htm
来源：园林景观网

芝加哥：把气候变化因素纳入城市规划

近些年来，气候变化使暴雨变得更密集，气温更极端。美国已有越来越多的城市不再等着联邦政府来处理气候变化问题，而是立足当地发现“无悔”方案，芝加哥就是其中之一。

自然资源保护理事会水资源分析师卡伦霍布斯说：“这些方案能帮居民省钱，节约许多事务开支，改善生活质量，而且还能带来额外的福利－减少辐射。”“风城”芝加哥正在实验以更环保的方式来进行城市规划，并把这作为“减缓气候变化影响计划”的一部分。

http://www.chla.com.cn/htm/2013/0411/164833.html
来源：人民网

意大利拉奎拉：古迹修复带动震后文化复兴

拉奎拉是位于意大利中部山区的美丽城市，城市周边是中世纪的城墙，城内不仅有迷宫般的狭窄街道，巴洛克或文艺复兴时期的建筑物和教堂，还有拉奎拉大学和众多文化机构，被视为阿布鲁佐大区的历史和文化中心。然而 4 年前的一场地震对这里的文化古迹造成了毁灭性打击。现在，一些修复项目已经陆续展开。值得注意的是，不少外国政府也伸出援手，共同拯救拉奎拉的文化遗产。由于拉奎拉及其周边地区严重损毁，修复重建工作将持续到 2021 年，共有 485 处需要修复，预计耗资 5.25 亿欧元。

http://www.chla.com.cn/htm/2013/0407/164282.html
来源：中国文化报

伦敦奥林匹克公园首次对公众开放

经过一年的改建之后，伦敦奥运会的主场馆奥林匹克公园将第一次向公众开放，得到票的幸运观众不仅能够进入奥林匹克公园，而且还能见证这个体育场馆群变成大型公共场地的过程。

经过改建之后的伦敦奥林匹克公园将于 7 月重新开放，不过在此之前，观众仍然能够提前进入，并且在大约一个小时的行程中看到奥运会体育场馆是如何变成一个能够容纳学校、商业等多种设施的多功能区的。组织者透露，目前已有 2.3 万张参观门票被售出。

http://www.chla.com.cn/htm/2013/0403/163894.html
来源：北方新报

圣彼得的眼泪

在巴西里约热内卢的一个文化中心，巴西艺术家维尼修斯·席尔瓦用了 6000 个装满水的灯泡去呈现下雨的形态，表现农夫们与降雨的关系。创意别出心裁，看似简单的设计却带给观看的人们与众不同的感觉。6000 个装满水的灯泡在视觉上给人以强大的冲击。图为巴西艺术家维尼修斯·席尔瓦的“圣彼得的眼泪”展览的一部分，一名男子在装着水的灯泡间穿行。

http://www.chla.com.cn/htm/2013/0325/162998.html
来源：新华社

小即是美——比利时小城德比

德比市位于比利时东部阿登高原地区，毗邻卢森堡，是目前世界上最小的城市之一。这座袖珍小城建于 1331 年，地处山坳，四面环山。它至今完整地保留着 14 世纪时的格局，小巷幽深，房屋古朴，一条条鹅卵石小道蜿蜒曲折。

比利时是一个精致到完美的国家，而德比小城则是这精致中的浓缩体现。天上是万里无云，地下是绿草茵茵，环顾四周，那浓浓的生活气息扑面而来，让你的心变得沉静而安宁。如此这般的欧洲小城，体现的是一种性格和人生观。

http://www.chla.com.cn/htm/2013/0325/162993.html
来源：新华社

苹果新总部替代设计方案：一座宜居城市

外观像一艘巨大飞碟的苹果新总部将隐藏在美国加州硅谷280号公路旁边的树丛中，建成后，它可容纳大约13,000名工作人员。新总部的主体为一个环形的4层建筑，它将是世界上最大的办公楼之一，占地面积

约175英亩（约合70万m^2），体积为五角大楼三分之二。

新细节的曝光增添了人们对苹果未来的无限遐想。新总部办公区域最多可容纳1.3万人。园区还有占地面积总计27870m^2的科研设施（半地下）和一个会议厅（全地下）。新总部被苹果称为是"与众不同、启发灵感的二十一世纪工作场所"。

http://www.chla.com.cn/htm/2013/0307/161115.html
来源：腾讯科技

另一种方式行走费城——树上漫步

这是一条长137m的道路网络，建造在15m高的树上，以一个巨大的"鸟巢"作为终点。该设计旨在引起游客们对森林和自然的关爱，传达这样一种信息：我们需要树木，树木也需要我们。

该设计中构造亭子使用的是环保装饰材料西部红雪松。这种材料是一种100%可再生材料，从种植到收获都十分环保。由于设计结构各部分都独立存在，因此即使有一棵树倒了弄坏了结构的某一部分，那么其他部分也是安全的。结构也可以进行重新组装和搬迁。

http://gardens.liwai.com/content-21110.htm
来源：园林景观网

葡萄牙创意迷宫

创意迷宫使用了1200HEMMA落地灯基地的1200 LEDARE灯泡，把贝伦文化中心一角变成了一个互动游乐场，供游人穿行照明和玩耍的迷宫。设计师巧妙地将立地灯组成的一个路径，每只灯泡以合适的高度存在，让进入的观众体验神奇的观感。该设计集合

了乐趣与创意于一体，让参与者体会到了乐趣，让观者享受到了创意的美感。

http://gardens.liwai.com/content-25637.htm
来源 园林景观网

越南石屋

这座环形的石屋位于去往河内下龙湾的路边，一座安静的住宅区内。上升的绿化屋顶和墙体由色彩柔和的深蓝色石头砌成，成为这个新住宅区一道漂亮的景观。所有房间都围绕着椭圆形的中心庭院设置，相互间成组成团的布置，内部交通空间沿着中庭设置，直到到达绿色屋顶，将室内所有功能房间串联在了一起。庭院和绿色屋顶形成了一个连续的花园，让室内外空间产生了丰富的联系。通过这个连续的花园，居民们体验着季节交替、四时变化，在自然中享受着丰富的生活。

http://gardens.liwai.com/content-25617.htm
来源：园林景观网

法国蒙彼利埃度假公寓设计竞赛揭晓

如果你喜欢高迪式建筑的丰富想象力与曲线美，那么你一定不能错过下面这则新闻。法国蒙彼利埃市最近宣布farshid moussavi

事务所赢得了南法一座假日公寓大厦的设计竞赛。基地位于一座与世隔绝的法国乡村中，建筑就建在一片度假别墅区里，这个竞赛希望为port-marianne区的jardins de la lironde滨海沿岸找到一个"疯狂"的建筑。来自伦敦建筑事务所farshid moussavi提交的方案是一座11层的大厦，每层平面都是一个优雅的曲线形，这些曲线形不断叠加，最终形成一个错综复杂的分层建筑形态。大厦共有36套公寓，其中还包括一座餐厅，它将成为当地一座显要的地标性建筑。竞赛评委特别赞赏这个方案对小度假别墅的重新诠释以及可持续的景观设计。这个方案预计于2014年开始动工建设。

http://www.szjs.com.cn
来源：中国建筑文化

北欧最大水族馆"蓝色星球"开放

水族馆也玩儿3D？这个理念已从理想走入现实。北欧最大水族馆"蓝色星球"3月22日在丹麦首都哥本哈根正式向公众开放。据悉，"蓝色星球"水族馆存水量为700万升，拥有450个物种，逾2万水生生物，其中包括锤头鲨、鹰鳐等珍贵水生物种。设计的灵感源自水流旋涡，无论是在建筑外观还是内部设计上处处都体现出水流旋涡这一理念。置身馆内，每个展厅似乎不是静止的，而是不断地被周围的"漩涡"卷起向前，充满水体流动的感觉。置身水族馆中，与各

种水生动物"同呼吸，共前进"，绝对非一般的体验。

http://www.szjs.com.cn
来源：中国建筑新闻网综合

东方园林股票代码：002310

城市景观艺术品
OrientLandscape
Urban landscape art

北京奥林匹克公园中心区景观
北京通州运河文化广场
首都机场T3航站楼景观
北京中央电视台新址景观
苏州金鸡湖国宾馆、凯宾斯基酒店景观
苏州金鸡湖高尔夫球场
上海佘山高尔夫球场
上海世博公园
海南神州半岛绿地公园
山西大同新城中央公园文瀛湖
湖南株洲新城中央公园神农城
辽宁鞍山新城景观万水河
辽宁本溪新城中央公园
河北衡水衡水湖及滏阳河景观
山东滨州生态景观系统及新城中心景观
浙江海宁生态景观系统
山东淄博淄河景观系统
河北张北风电基地及两河景观带
山东济宁微山湖及任城区中央景观
山东烟台夹河景观系统及特色公园

中国园林第一股
全球景观行业市值最大的公司
中国A股市场建筑板块、房地产板块前十强
城市景观生态系统运营商

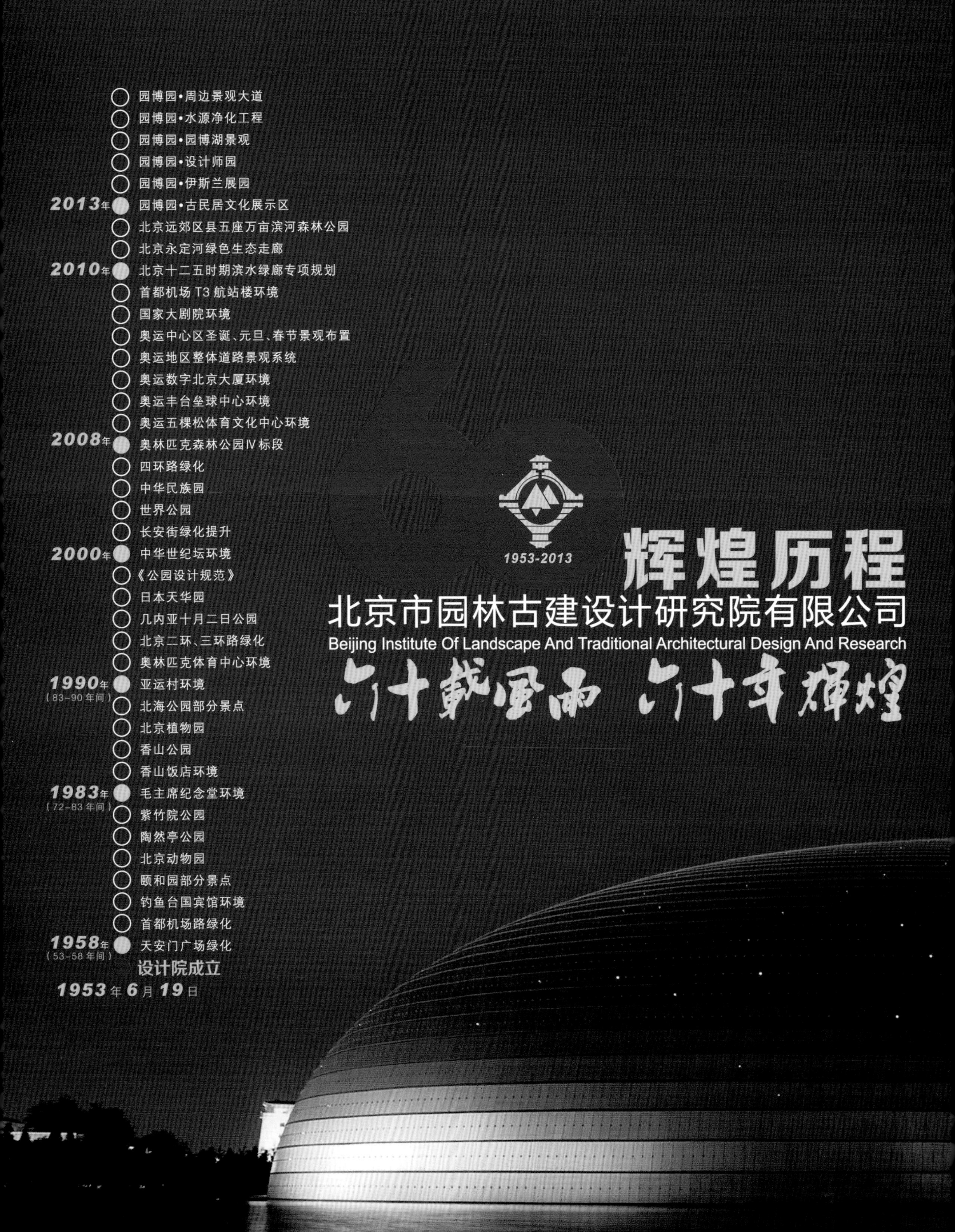

园博园·周边景观大道
园博园·水源净化工程
园博园·园博湖景观
园博园·设计师园
园博园·伊斯兰展园
2013年
园博园·古民居文化展示区
北京远郊区县五座万亩滨河森林公园
北京永定河绿色生态走廊
2010年
北京十二五时期滨水绿廊专项规划
首都机场T3航站楼环境
国家大剧院环境
奥运中心区圣诞、元旦、春节景观布置
奥运地区整体道路景观系统
奥运数字北京大厦环境
奥运丰台垒球中心环境
奥运五棵松体育文化中心环境
2008年
奥林匹克森林公园Ⅳ标段
四环路绿化
中华民族园
世界公园
长安街绿化提升
2000年
中华世纪坛环境
《公园设计规范》
日本天华园
几内亚十月二日公园
北京二环、三环路绿化
奥林匹克体育中心环境
1990年
(83-90年间)
亚运村环境
北海公园部分景点
北京植物园
香山公园
香山饭店环境
1983年
(72-83年间)
毛主席纪念堂环境
紫竹院公园
陶然亭公园
北京动物园
颐和园部分景点
钓鱼台国宾馆环境
首都机场路绿化
1958年
(53-58年间)
天安门广场绿化
设计院成立
1953年6月19日
1953-2013
辉煌历程
北京市园林古建设计研究院有限公司
Beijing Institute Of Landscape And Traditional Architectural Design And Research
六十载风雨 六十年辉煌

第九届中国（北京）国际园林博览会 开幕

THE 9TH CHINA(BEIJING)INTEMATIONAL GARDEN EXPO OPENING CEREMONY

摄影 / 崔勇

5 月 18 日上午，第九届中国（北京）国际园林博览会在北京市丰台区永定河畔隆重拉开帷幕。中共中央政治局委员、北京市委书记郭金龙，全国人大副委员长向巴平措，全国政协副主席林文漪，住房和城乡建设部部长姜伟新等领导同志出席了开幕式。开幕式由北京市常务副市长李士祥主持。

开幕式上，首先播放了由一张张笑脸组成的《美丽园博你我同行》的短片，表达了对建设者们在过去的两年中辛勤劳作的敬意；随后，一袭盛装的刘媛媛在鲜花盛开的舞台上演唱了园博会主题歌《天地一幅画》，现场志愿者及工作人员纷纷加入大合唱；伴随着跳跃的琴键，浪漫的音符仿佛飘散在空中，千余只色彩斑斓的蝴蝶从鲜花装饰的钢琴上翩翩起舞，《花间蝶舞》充分体现了园林文化、丰台花卉文化与音乐的有机融合，同时也庄严宣告第九届园博会正式开幕。

北京市市长、第九届园博会组委会主任王安顺在致辞中说，中国国际园林博览会是代表我国园林行业最高水平的国际盛会，在 16 年不平凡的发展历程中奏响了绿色交响、盛世园林的华美乐章，为推动形成新的生产方式、生活方式，促进城市转型发展，与自然和谐相融发挥了十分重要的作用。

住房和城乡建设部部长姜伟新在致辞中说，中国国际园林博览会举办 16 年来，已成为传播园林文化、扩大交流合作、展示园林绿化新成果的重要平台。本届园博会秉持尊重自然、保护环境的理念，采用生态修复技术，将北京市丰台区永定河西侧一处近百年的垃圾填埋场打造成精品园林，具有很好的示范性和推广价值。同时开幕的中国园林博物馆集展示、收藏、科研等功能于一身，是我国第一个以园林艺术为主题的公益性博物馆，是本届博览会的又一亮点。本届博览会国内展馆 94 个，国外展馆 34 个，共 128 个展馆，所展示的园林树木品种、类型为历届园博会之最。

据悉，展会期间，园博会将举行 3 类 21 项近千场次文化活动和演出。每天都有花车巡游，将有 24 辆花车为游客献上行进表演。开幕式后启动的城市周活动中，北京、南京、杭州等 19 个城市将以各具特色的文艺演出、展览展示，呈现不同城市的文化形象和风采。

2013

THE INTERNATIONAL FORUM OF LANDSCAPE ARCHITECTS OF THE 9TH GARDEN EXPO HELD SUCCESSFULLY IN BEIJING

第九届园博会风景园林师国际论坛在京成功举办

华中农业大学副校长高翅

朱育帆教授分享《流水印》设计理念

章俊华教授分享《明园》设计理念

Eva Castro 女士分享《凹陷花园》设计理念

Jake Walker 先生分享《皮特·沃克花园》设计理念

张新宇副院长分享《印象四合院》设计理念

Peter Latz 大师分享《彼得·拉茨花园》设计理念

5月19日，为展示第九届园博会建设成果，交流建园经验，加强国际间风景园林师的学术研讨与技术交流，第九届园博会组委会办公室、园冶杯风景园林国际竞赛组委会在丰台区大成路九号举办了风景园林师国际论坛。本论坛由中国风景园林网、世界园林杂志社、北京市园林学会、北京市园林施工企业协会及北京市公园绿地协会联合承办。国内多个省建设厅领导、十几个城市的园林局负责人，五十多所园林院校的师生，百余家优秀设计机构和园林施工企业的代表近千人参加了论坛。

上午的第九届园博会设计师创作交流论坛上，Peter Latz、Jake Walker 就大师园设计理念、创作思想以及施工过程进行了讲解，同时本届园博会设计师广场的建园设计师朱育帆、章俊华、Eva Castro、张新宇等也向与会嘉宾分享了各自展园的创造过程及设计理念。上午的会议由华中农业大学副校长高翅主持。高校长的主持风趣幽默，设计师的演讲精彩频现，现场观众反响热烈。

在下午的分论坛之一——风景园林师学术研讨会上，李建伟、朱建宁、白祖华、中里竜也等国内外知名专家和优秀设计机构的主创设计师先后上台演讲。研讨会由重庆市园林局总工石继渝女士主持。

北京林业大学园林学院朱建宁教授做了题为《历届园博会展园设计回顾》的精彩演讲，他介绍了自己参与的几届园博会的展园设计情况，并阐述了作为一个设计师在园博园这样的环境和要求下是怎样转变和成长的；EDSA 总裁兼首席设计师李建伟先生分享了自己的设计作品《五象湖公园》，他表达自己要做快乐景观以带给人们生活乐趣的设计理念；北京源树景观设计事务所总经理白祖华做了主题为《时空之美》的精彩发言，他通过述说自己与遗址公园结缘过程，以大明宫遗址公园的设计方案为例，分享了自己从项目中获得的启发和感悟；广州山水比德景观设计公司副总经理利征先生做了《传统与现代的景观对话》演讲，他所阐述的“现代体 传统心”新理念，引起了设计师们的关注；ATLAS（北京）首席景观设计师中里竜也以 2011 西安世园会及 2015 年武汉园博会作为实例，就博览会的场地设计特点做了解析；中国农业大学观赏园艺与园林系副主任孟祥彬做了题为《梦回水西，盛世钩沉——天津市水西庄公园规划设计方案》的发言，分享了自己对中国园林的内涵，对中国园林的历史、文化进行的探索和感悟；来自于台湾的涂智益博士结合自己多年来参加博览会的经验，给与会观众讲解了设计师的创作，如何与园博会组委会、客户需求相结合的心得体会；海韵天成总裁兼首席设计师顾志凌对体验式景观设计概念做了全面的分析，分享了自己对体验式景观设计的理解与启示；武汉园林设计院副院长、总工李芳从第九届园博会武汉园的设计谈到第十届武汉园博会的总体规划，她的发言使在座的观众对 2015 年的武汉园博会充满了期待。

重庆市园林局总工石继渝

EDSA 总裁兼首席设计师李建伟

北京林业大学园林学院教授朱建宁

ATLAS（北京）首席景观设计师中里竜也

北京源树景观设计事务所总经理白祖华

中国农业大学观赏园艺与园林系副主任孟祥彬

海韵天成总裁兼首席设计师顾志凌

台湾专家涂智益

广州山水比德景观设计有限公司副总经理利征

武汉市园林设计院副院长、总工李芳

参会嘉宾合影

曹南燕女士　李敏教授　马炳坚所长　景长顺会长　夏岩总裁　崔勇处长　马晓燕院长

瞿志副教授　唐学山教授　湛锦源先生　居阅时教授　刘桂林教授　王玉华巡视员　夏成钢副院长

在下午另一分会场——第二届国学造园主题论坛上，围绕真正的“传承”就是“创造”这一主题，与会专家、学者和企业家就文化与园林的关系展开讨论。论坛由南师大中北学院院长姚海明主持，夏岩文化艺术造园集团董事长夏岩做了《国学造园：真正的传承是创造》主题报告；华南农业大学城市规划与风景园林系主任李敏教授在发言中从三个层次论述了“国学造园”思想的精髓；北京园博会组委会协调处处长、首都经贸大学旅游发展研究中心研究员崔勇发表了题为《对国学造园和国学赏园的再认识，从第九届国博会谈起》的主题演讲；北京市古代建筑设计研究所所长马炳坚、北京农学院园林学院院长马晓燕、北京林业大学园林学院副教授瞿志等也发表了自己的真知灼见；最后，中国风景名胜区协会副会长、原住建部城建司副巡视员曹南燕女士从主题、收获、认识三个方面阐述了自己的看法。

当日晚上举办了对话大师活动，由日本千叶大学教授章俊华担任主持，本次活动邀请了欧洲景观设计大师 Peter Latz 和日本大师三谷徹，以及国内著名设计师朱育帆和李敏教授进行了对话与交流。现场气氛热烈，掌声不断，大师们智慧与思想的碰撞，给在座观众带来很大的启发。

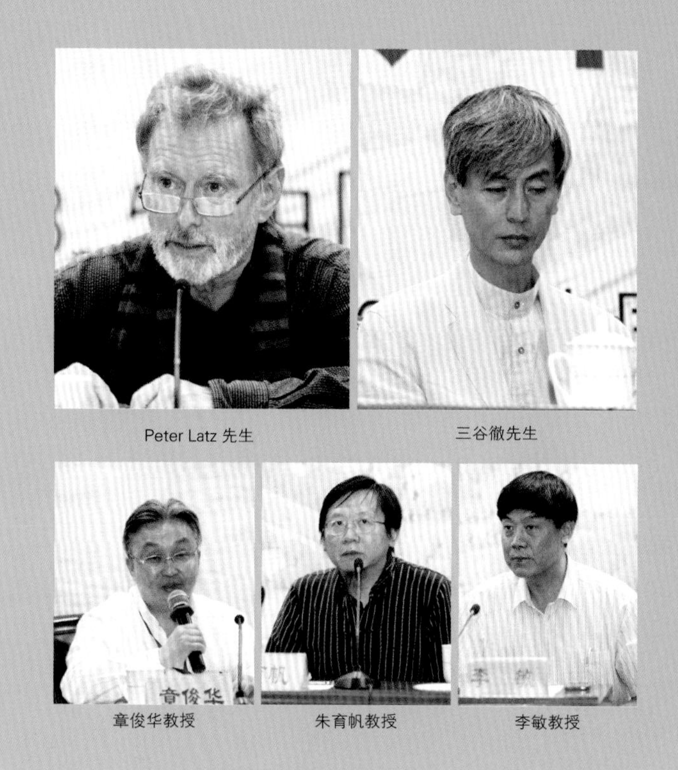

Peter Latz 先生　　三谷徹先生

章俊华教授　　朱育帆教授　　李敏教授

大师对话活动现场

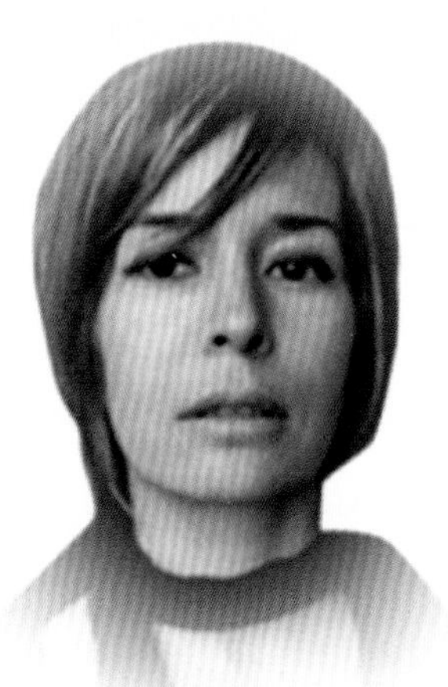

A DISTINGUISHED GATHERING OF DESIGN TALENTS OF WORLDWIDE LANDSCAPE ARCHITECTS

——THE MASTER GARDENS AND THE DESIGNERS' SQUARE IN THE 9TH BEJING GARDEN EXPO

全球设计师创意荟萃的舞台

——记第九届园博会的大师园和设计师广场

作者：王倚珺　张翀

大师报告会及对话现场

中国国际园林花卉博览会是由住建部和地方政府共同举办的国内最高水平的国际性园林盛会，到 2013 年已经成功举办了九届，第九届园博会特设大师园和设计师广场作为创意园区。2011 年 12 月，第九届园博会组委会办公室就设计师创意园区面向国内外公开招标，中国风景园林网在众多投标单位中脱颖而出，最终接受组委会办公室的委托，负责大师园和设计师广场项目的策划、组织、邀请和咨询工作。

在大师园建设方面，2012 年 1 月初，中国风景园林网面向国内外拟邀皮特・沃克等 5 位不同地域和风格的世界级大师来京建园，最终确定了极简主义代表人物、美国风景园林大师彼得・沃克，欧洲著名工业遗产景观大师彼得・拉茨，日本风景园林代表人物三谷徹参加大师园三个地块的创作，三个展园分别以大师的名字命名，这在全球首开先河。 从 2012 年 2 月开始，三位大师相继来京勘察现场，深入了解情况，并向组委会汇报了初步的概念设计方案。3 月底大师深化设计完成，通过组委会的审核确认。随后，东方园林、天开园林分别出资捐建了彼得・沃克花园和彼得・拉茨花园。

在设计师广场的建设方面，2012 年 2 ~ 4 月，园博会组委会通过中国风景园林网面向全球知名中青年设计师征集设计方案，经过资格审查，有 27 名国内外设计师

大师签名现场

大师勘察现场

入围参与方案评选，经过国际专家评审会的三轮论证，最后优选出 6 名设计师的作品实施建园，分别是朱育帆的《流水印》、章俊华的《明园》、戴安娜·巴莫瑞（美）《声波园》、伊娃·卡斯特罗（英）《凹陷花园》、克里斯托弗·康茨（美）《小径花园》、张新宇的《印象四合院》。2012 年 8 月，各展园陆续完成施工图设计，中国风景园林网组织多场专家评审会对每个展园的施工图进行审查，全力做好技术对接服务工作。

大师园和设计师广场的施工，自 2012 年“十一”节日期间陆续进场，2012 年底大部分展园基本完成了土建部分和部分乔木种植。由于北京气候条件限制冬季暂停施工，待 2013 年 2 月下旬再度开工时，距离开园只剩下两个多月的时间。各展园争分夺秒抢工期，很多设计师付出了大量的时间和精力在现场监督质量，和施工单位密切配合，确保了 4 月底如期竣工。

在设计创作和现场施工的技术指导期间，中国风景园林网在三位世界级大师和知名设计师来京之际组织了多场学术报告会和风景园林师沙龙，在行业内引起了很大的轰动，大大提高了本届园博会的国际影响力。

在技术对接和服务方面，中国风景园林网努力推进总体进度，把控各个节点。从项目启动之初，世界级风景园林大师的调研、筛选和邀请、园博会基础资料的整理和翻译、设计师勘察现场的接待、多场大师报告会的筹办，到中后期多次专家评审会的组织、技术对接会和协调会的举办。项目进行的整个过程中，协调主体层次多、涉及面广、利益交织复杂。需要协调解决的问题时效性强，有的宏观，有的具体而微，难度非常大。最终能够顺利完工，有赖于各方的通力合作和组委会办公室领导的大力支持。

5 月 18 日开园后，国内外游客和各路媒体纷至沓来，对设计师创意园区投注了极大的关注。毫无疑问，设计师创意园区成就了本届园博会最大的看点和亮点。

作为项目的策划组织方，在欣喜之余也看到很多缺憾和不足。大师和设计师在创意上虽然不受限制，但在资金、技术对接等方面受到很大的制约，尤其跨国沟通确实存在很大的困难，国外设计师通过 Email 和电话远程沟通很难把控现场，必然影响展园最终的品质和呈现的效果。在施工单位选择和配合方面，也有很多无奈，有的施工单位的技术水平确实很难满足设计师的要求。还有，展园后期的养护同样非常重要，如何保证长期的效果也是一个很大的挑战。□

大师及设计师勘察现场

印象
四合院

Oasis 无锡绿洲
景观规划设计院有限公司

无锡绿洲景观规划设计院有限公司成立于 2004 年，具备建设部颁发的风景园林设计专项甲级资质。我们的宗旨是在不同的专业领域中，力求景观设计的功能性、创新性、人文性以及环保性。绿洲坚持运用当代设计手法及语言，将自然、人性与艺术作为不懈探索的设计命题，以务实的态度和高度的热情参与实践。

我们始终注重博采众长，不断创新，并且通过我们与客户之间的合作，建造可持续发展的环境。在城市公园、绿地及水系、风景旅游区、住宅及商业区等领域的规划设计中，提供了独特的解决方案和优秀的服务品质，得到了广泛的认可。

Landscape Design
景观设计

Urban Planning
城市规划

Architecture
建筑设计

Environment
环境咨询

雾中叶脊 —— 彼得·拉茨的花园

THE SPINE IN THE MIST——A GARDEN BY PETER LATZ

彼得·拉茨

Peter Latz

项目位置：中国，北京，第九届北京园博会
项目面积：2,500m²
委托单位：第九届中国（北京）国际园林博览会组委会
设计单位：彼得·拉茨景观设计事务所
景观设计：彼得·拉茨
项目建筑师：苏菲·豪尔慈　拉茨景观设计事务所
施工单位：重庆天开园林股份有限公司
完成时间：2013 年 5 月

Location: The 9th China (Beijing) International Garden Expo, Beijing, China
Area: 2,500m²
Client: The 9th China (Beijing) International Garden Expo committee
Designer: Latz + Partner Kranzberg, Germany (www.latzundpartner.de)
Landscape Design: Prof. Peter Latz
Project architect: Sophie Holzer Latz + Partner,
Construction: Chongqing Tiankai Landscape Engineering Co.,Ltd
Completion: May, 2013

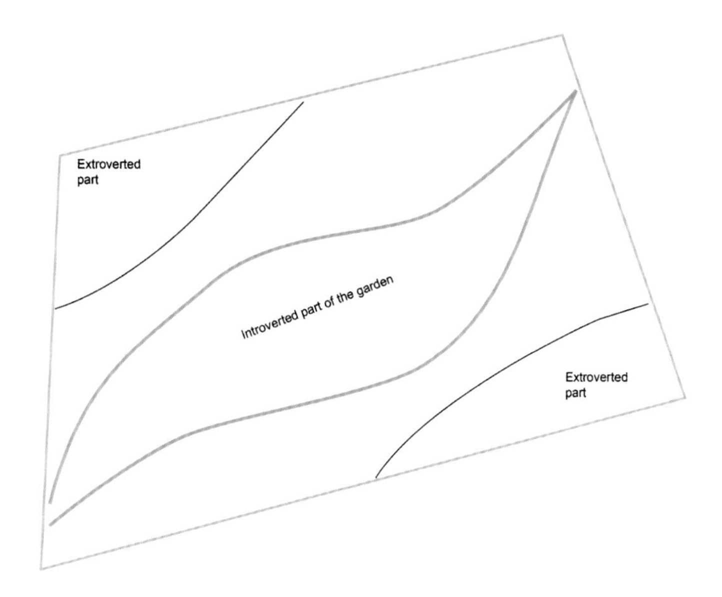

图 01 花园的内向空间和外向空间 (照片来源 : Latz + Partner)
Fig 01 Introverted and extroverted partial areas of the garden (Source: Latz + Partner)

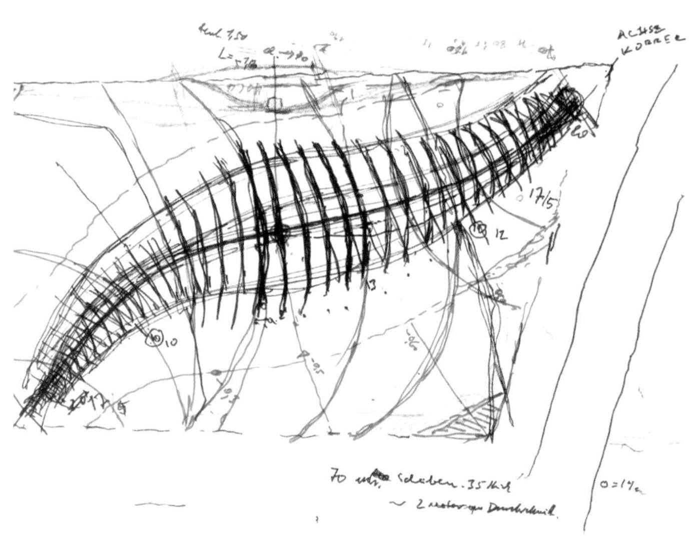

图 02 彼得 • 拉茨对大师园的初步研究 (照片来源 : Latz + Partner)
Fig 02 First studies of Peter Latz for the Mastergarden (Source: Latz+Partner)

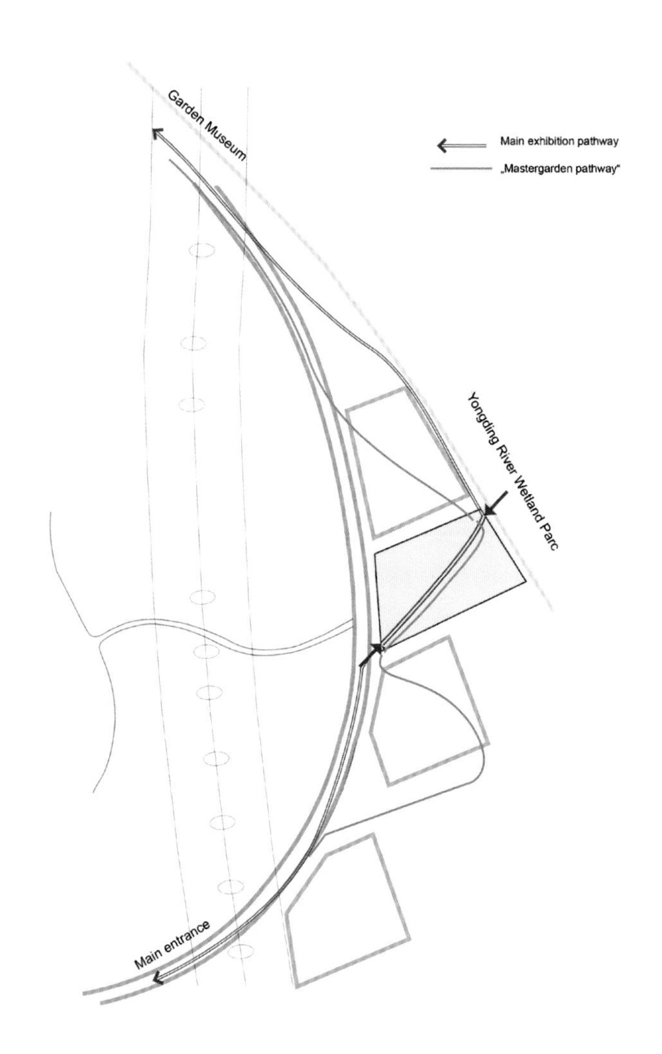

图 03 花园的入口和主路，以及与公园的道路系统的连接 (照片来源 : Latz + Partner)
Fig 03 Accesses and main pathway in the garden and connection to the road system of the park (Source: Latz + Partner)

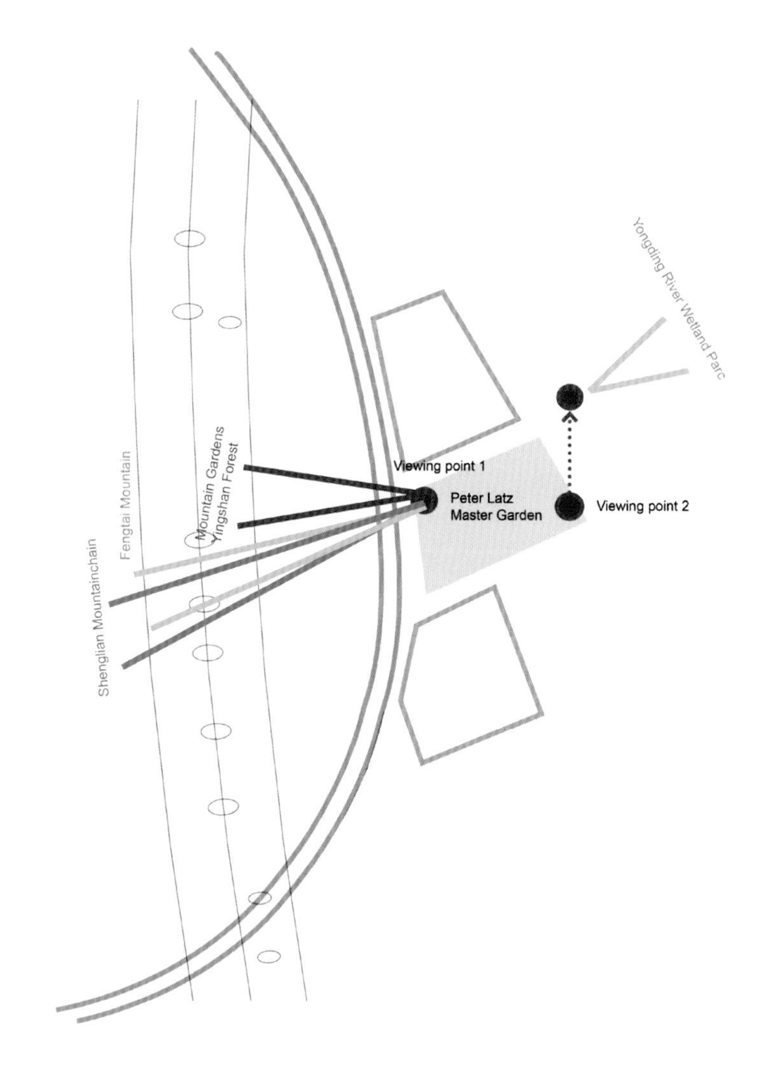

图 04 景点和视觉连接 (照片来源 : Latz + Partner)
Fig 04 Viewpoints and visual connections (Source: Latz + Partner)

理念与设计

在我们的思维里，花园主要意味着一种视觉感受。但事实上我们是用视觉 、听觉和触觉等全部的感官在体会一个花园。

在园林博览会或园艺展览上，人们感受一座花园的方式与在日常生活中是有所不同的。游客不是花园的主人，他们想欣赏一切，所以不会在同一个地方停留很久，通常仅仅花上几分钟，或者赶在闭馆前匆匆一瞥。他们在餐厅和咖啡厅反而会消磨更长的时间。同时，游客希望看到新颖、超乎设想的设计。就此而言，游客能了解并真正记住的只有极少的信息。

一般而言，花园由很多信息层次构成。在这里它们被减少到几个能被即刻理解的层面。

一般而言，花园被设想为内向性构筑。但地势较高的花园本身在位置上就展示了一种外向性。

花园通常被用作独自休闲或朋友聚会的场所；人们进入花园，在其中停留，他们可以休息片刻或在其中工作。但大部分的游客对于花园的这项功能都持有相反的态度。

因此，我找到一个很易理解的解决方案，在可以容纳大量游客的基础上，使其兼具了花园的内向性和外向性的双重特点。最后，“脊状结构”看起来比我（彼得•拉茨）另一种集中式的几何形状的设计更有效果（图01–02）。

设计理念是将园博会花园解读为一个打开封闭花园的通道。因花园设有两个出入口，可以使游人免于走回头路。对于那些乐于在园中多做停留的人，还会发现许多通幽小径，也可在长凳上小憩（图 03–04）。

Idea and design

Thinking of gardens mostly means a visual perception for us. But we experience a garden with all our senses: Seeing, hearing and feeling.

In a garden-exposition or a horticultural show, gardens are experienced in a different way than in everyday life. Visitors are not owners. Visitors want to see everything, so they can not stay in the same place for very long, usually some minutes only, near closing time only seconds. Just in restaurants and cafes one stays for a longer time. At the same time they want to see all that's new and unexpected. In this way they can grasp and really take in or keep in mind only little information.

- In general gardens are built with a lot of information layers. Here they are reduced to a few layers which can be understood immediately.

- In general the garden is imagined to be mainly introverted. But the position of the garden on a high plateau suggests just as well an extroverted situation.

Usually a garden is rather used by oneself or together with friends; one goes into the garden, stays there, has a rest or works there. But a large number of visiting people can conflict with the idea of the garden.

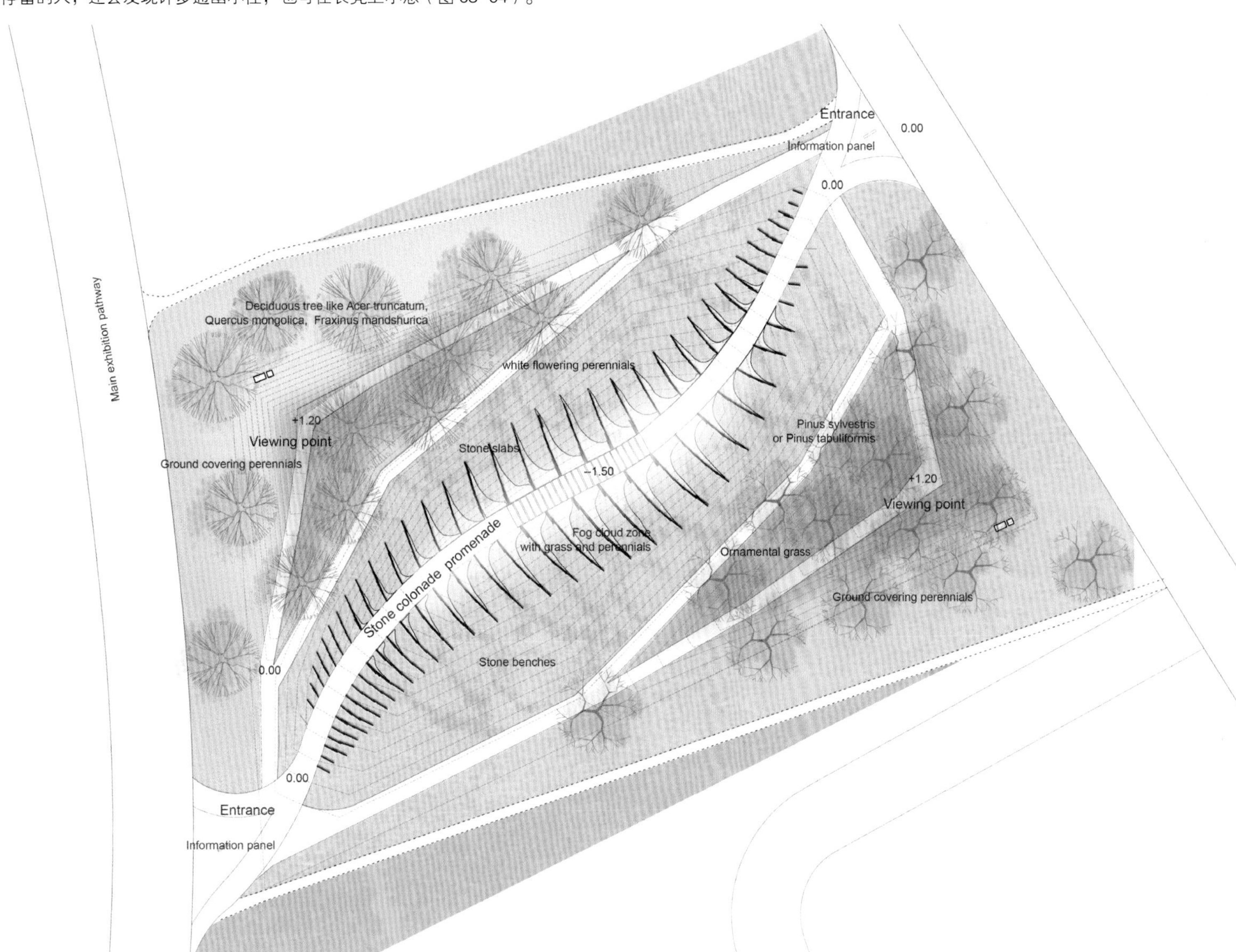

图 05 路径系统和不同园林要素的规格概览（照片来源：Latz + Partner）
Fig 05 Overview with path system and specification of the different garden elements (Source: Latz + Partner)

图 06 实景图 1
Fig.06 The built garden1

正是地形解决了不同游人通行、逗留以及渴望私密空间之间的矛盾（图 05–08）。

主路顺着轻缓的斜坡下抵花园中心，后又缓缓上升至第二入口。一个由垂直安放的天然石板搭建的脊状结构走廊，像一个峡谷一样伴随着逐渐下降的通道。它的颜色和纹理塑造了花园的特色。在入口处，柱廊的高度及膝。在低洼的中心部分，石板的高度可达 3m。从那里通道地势再次升高，而石板也沿着一个动态的曲线，再次降低到及膝高度（图 09–10）。

每走一步，周围被隐藏的景物都变得越来越清晰。视线最终落在花园内向核心中的“薄雾之枕”上， 它位于花园通道中间部分的石板之间（图 11）。

通过对于花园东面和西面地形的抬升，人们站在观景台上，便可以看到位于高架铁路下方山谷中的花园。山顶的针叶林和落叶乔木林风景如画，是花园重要的特色观景点，并且这些林带将在未来不断地增加山体的高度（图 12–13）。

在园博会期间，新种植的植物显然不能达到设计者所预期的效果，这是对于设计的另一个至关重要的影响。因此，石头形成的空间和水雾的凉爽清新取代了未来才成型的树荫（图 14–15）。

在最初迟疑地踏入水雾笼罩中之后，你会感到幸福，以及从炎热和尘埃中解脱出来的怡然。

雾是由特殊的高压喷嘴喷射出的水珠形成。这些喷嘴设计极为精妙，使无数水滴浮荡在空气中。白色的水雾在阳光下折射出彩虹一样的色彩。这将成为视觉与触觉的双重体验。水分能立即降低温度，令呼吸也顿感轻松。水雾是花园的特色，并且它是独立于植物的灌溉系统之外。

材料、科技、技术

喷雾体系

高压喷嘴被安装在柱廊前面和地面上。多种程序调节着喷雾的频

Therefore I looked for a well understandable solution which offers both an introverted and an extroverted character with a capacity for many visitors. Finally “the spine” seemed to be more efficient than my alternative centralized geometries(Fig.01-02).

The idea is to understand the exposition garden as a passage, to open the enclosed garden. One is not forced to turn back, as the garden offers two entrances and exits. Who wants to stay there longer, finds several smaller paths and benches(Fig.03-04).

It is the topography which resolves the contradiction between passing through and lingering and the desire for the introverted(Fig.05-08).

The main passage way slopes gently down to the centre of the garden and up again to the second entrance. A gallery with the form of a spine made of vertically mounted natural stone slabs accompanies the lowered passage way like a canyon. Its colour and texture characterize the garden. At the entrance, the colonnade starts knee high. In the low lying centre, the slabs reach a height of up to 3 metres. From there, the passage way rises again, whereas the slabs reduce their height in a dynamic curve to knee level again(Fig.09-10).

With every step the surroundings get hidden more and more visually. The perspective ends with a “pillow of mist” in the introverted core of the garden which emerges between the slabs in the middle of the passage (Fig.11).

By raising the ground to the east and to the west, lookout points enable visual contact to the gardens in the valley below the elevated railway. Picturesque groups of conifers on the one hilltop and of deciduous trees on the other are important features which will multiply the height of the hills over the years(Fig.12-13).

图 07 路径
Fig 07 The path

图 08 石板基础
Fig 08 Base of the slate

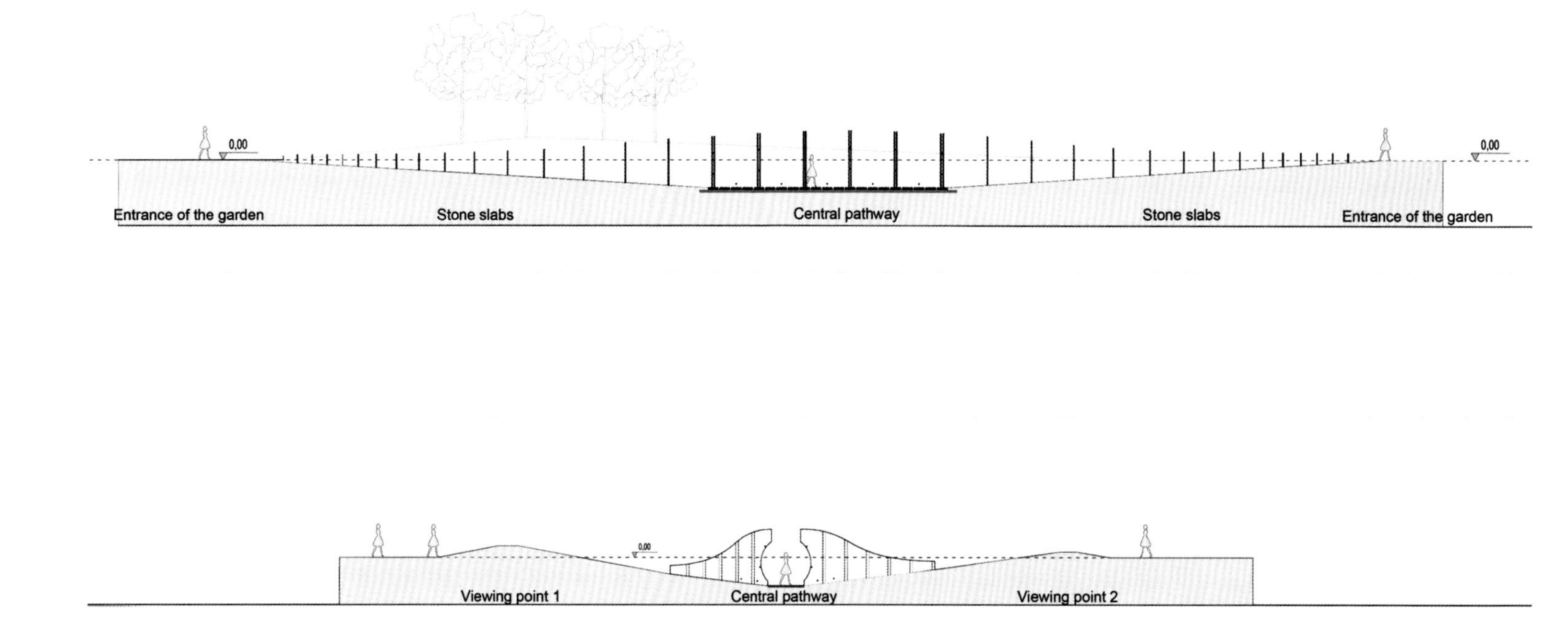

图 09 主路的立面和剖面图（照片来源：Latz + Partner）
Fig 09 Longitudinal and cross section of the main pathway (Source: Latz + Partner)

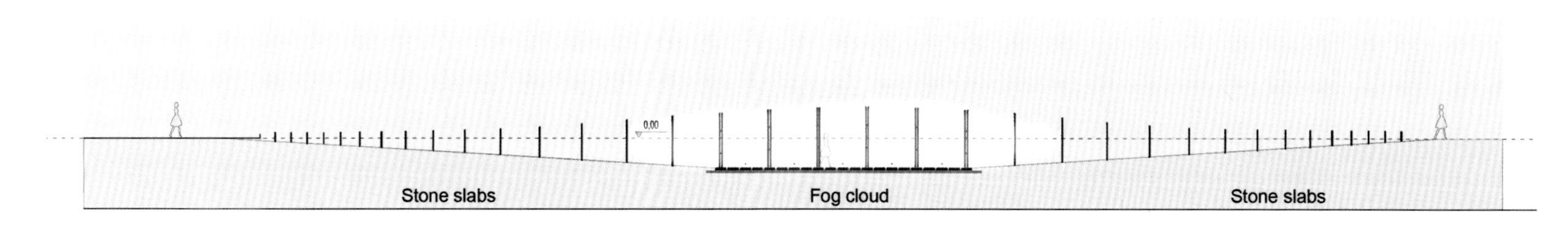

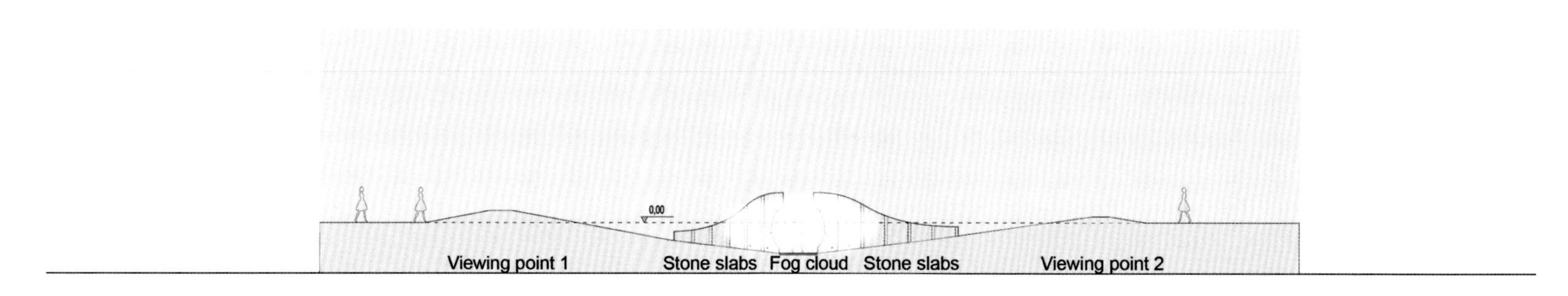

图 10 花园中心雾云图案的立面和剖面图（照片来源：Latz + Partner）
Fig 10 Longitudinal and cross section with pictorial representation of the fog cloud in the centre of the garden (Source: Latz + Partner)

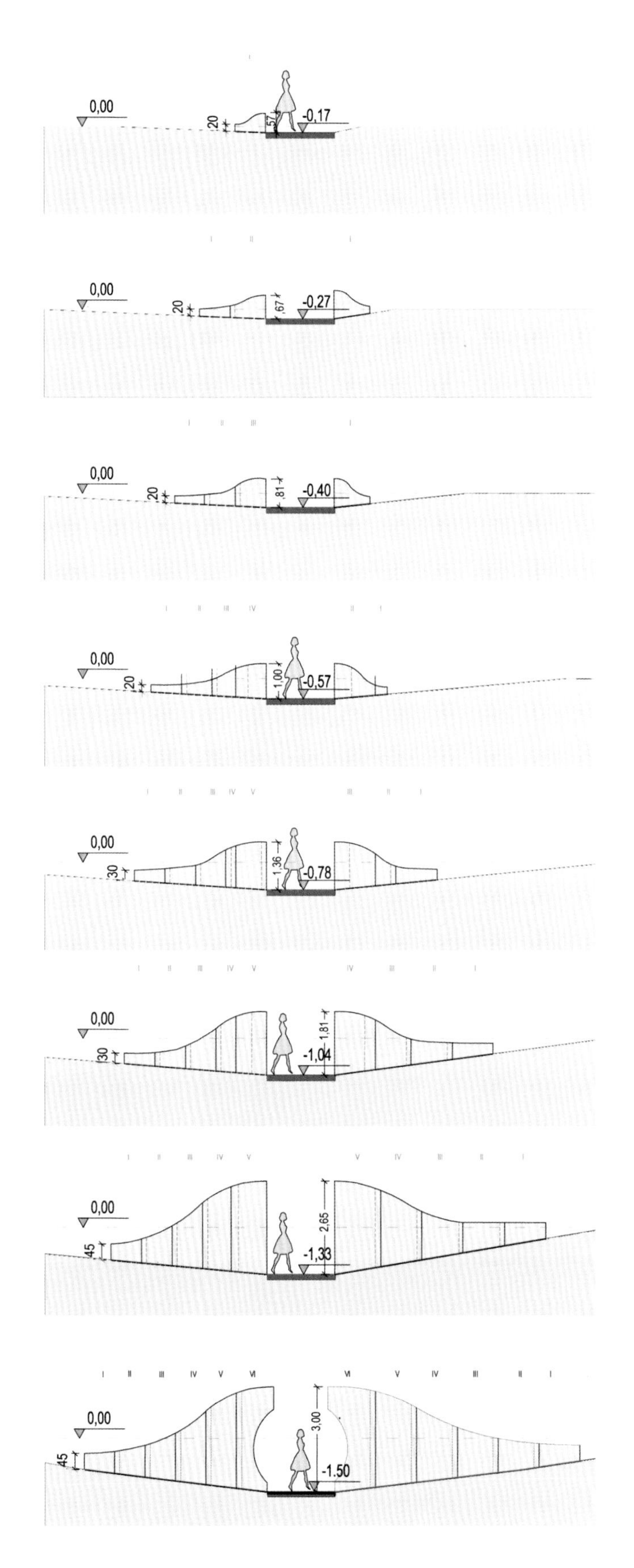

The fact that new plantings don't reach the wanted impression during the exhibition has another essential impact on the design. Therefore stone is forming the space and the coolness and freshness of mist replaces the future shade of trees(Fig.14-15).

Having entered the shrouds of mist hesitantly at first, one feels happy and liberated from heat and dust,

The mist is created by water ejected with high pressure from special nozzles. These nozzles are extremely fine, so that millions of drops remain in the air. The white mist gets coloured by the sunshine like a rainbow. It will become a visual and tactile experience. The moisture lowers the temperature immediately and has a relieving impact on the respiration system. The mist represents the characteristic feature of the garden and is independent from the irrigation system for the planting.

Materials, technologies, skills

Mist system

The high pressure nozzles are mounted at the front of the colonnade and on the ground. Various programs regulate the frequency of the mist events, the quantity of the water as well as a special illumination for staging the veils of mist at performances in the evening(Fig.16-17).

The technology was tested successfully in two projects: with a spiral – like figure at the Garden Festival 1998 in Chaumont-sur-Loire, France and with a turbine – like figure at the Federal Garden Show 2005 in Munich, Germany(Fig.18-20).

The garden is situated on a layer which covers a flattened landfill and contains a high percentage of silt. Vegetation did not exist.

Existing situation, cut and fill, topography

The garden is situated on a layer which covers a flattened landfill and contains a high percentage of silt. Vegetation did not exist.

A depression 1,50m deep is excavated in the centre. The cut forms the Eastern and Western mounds with a height of 1,20m. The fill remains 5 cm below the final level which gets set up by mineral surface material(Fig.21).

Spine and passage ways

The material for the stone colonnade is granite called "Green Symphony" with a fine sawn surface. The slabs are fixed in the ground with a reinforced concrete foundation. The first row shows a composite construction which allows for higher stability and protected installation of the mist system(Fig.22-23).

The surface of the central "stone promenade" is built with bead-blasted asphalt and granite aggregates, changing to granite stepping stones in the low lying centre(Fig.24-26). The

图 11 不同层面的石头柱廊的剖面图 (照片来源 : Latz + Partne)
Fig 11 Cross sections at different levels of the stone colonnade (Source: Latz + Partne)

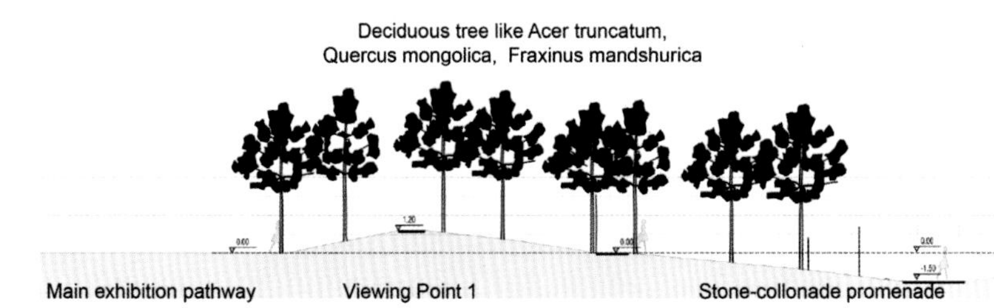

图 12 带有等高线的主路剖面图 (照片来源 : Latz + Partner)
Fig 12 Cross section of the main pathway with elevations (Source: Latz + Partner)

图 13 实景图 3
Fig 13 The built garden3

图 14 新种植的植物
Fig 14 New plants

图 15 雾中叶脊
Fig 15 The spine in the mist

图 16 用电脑制作的彼得·拉茨大师园最终形态动画展示 (照片来源 : Latz + Partner)
Fig 16 Computer animated illustration of the final state of the Peter Latz Mastergarden (Source: Latz + Partner)

图 17 水雾效果
Fig 17 Water mist

率、喷水量，控制着在夜晚演出中也能凸显轻纱般雾韵的特殊照明设施（图 16–17）。

这个技术目前已在两个项目中测试成功：1998 年法国肖蒙花园节的螺旋形状（喷嘴）；2005 年德国慕尼黑联邦园艺展上的涡轮形状（喷嘴）（图 18–20）。

现状、挖方和填埋、地形

花园的地层是填平的垃圾填埋场并且充满淤泥，不存在植物。

在中心挖掘一个深 1.50m 凹陷。挖掘出的土将东、西的地形垫高 1.2m。填埋后保留 5cm 的高度铺设含有矿物质的覆料（图 21）。

脊状结构和通道

石头柱廊的材料是被称为“绿色交响曲”的具有细锯齿纹理的花岗岩。这些石板被钢筋混凝土基础固定在地面里。 第一行的复合结构确保水雾系统的安全性和更高的稳定性（图 22–23）。

中央的“石头长廊” 表面以喷丸沥青和花岗岩骨料建成，在低洼的中心改为花岗岩的垫脚石（图 24–26）。其他路径都是开敞的碎石表面，最后一层则由高品质的花岗岩碎石组成（图 27–28）。

植被区域

新的植被系统包括两个层次：第一层由山上的树木构成，第二层由草和多年生草本植物组成。

树木的修剪使游客的视线不被遮挡。在西边的山坡上，可发现欧洲赤松和油松；在东边的山坡上有元宝枫、蒙古栎等阔叶树木（图 29–31）。

鉴于现有的小气候条件，在第二层种植能够忍受中心低洼地带潮湿的植物，如苔草和花蔺；耐旱植物则种植在山丘的斜坡上。植物按照石板的方向种植成行。为了简化维护，我们一致使用行间 60cm 的

other paths are equipped with an open gravel surface, the last layer consisting of high quality granite chippings(Fig.27-28).

Planted areas

The new vegetation consists of two layers: the first one is formed by the trees on the hill, the second one by grasses and perennial herbs.

The trees are pruned in a way that the visitors can look through. On the western hill we find Pinus silvestris and Pinus tabuliformis, on the eastern one broad-leafs like Acer truncatum and Quercus mongolica(Fig.29-31).

Due to the existing microclimatic conditions, the second layer shows plants which tolerate the humidity in the moist low-lying centre, like Carex aurea and Butomus umbellatus, and drought-resistant plants on the slopes of the hills. Planted in lines they follow the direction of the slabs. In order to simplify the maintenance, we use consistent planting grids and methods with spaces 60 cm wide between the lines. Ornamental grasses go along with the Pinus species, foliage plants like Geranium sanguineum with the deciduous trees. On the central line, a selection of geophytes like Allium “Mont Blanc” is blooming already in spring(Fig.32-35).

Mineral mulch

Stone chips called “mineral mulch” cover all the surfaces. Their typical scrunching sound accompanies each step and contributes to the acoustic impression of the garden.

In the year of the exposition, the mineral mulch will dominate

图 18 法国肖蒙花园展上的雾云
（照片来源：Joelle Caroline Mayer & Gilles Le Scanff）
Fig 18 Fog cloud event at the garden exhibition in Chaumont-sur-Loire, France (Source: Joelle Caroline Mayer & Gilles Le Scanff)

图 19 1988 年法国肖蒙花园节“雾之花园”中的雾云
（照片来源：Keiichi Tahara）
Fig 19 Fog cloud event in the "FoG Garden" at the Garden Festival 1988 in Chaumont-sur-Loire, France (Source: Keiichi Tahara)

图 20 2005 年德国慕尼黑联邦园艺展“天气变幻”花园中的石板长凳（照片来源：Latz + Partner）
Fig20 Benches made of stone slabs in the exhibition garden "Change of Weather" at the Federal Garden Show 2005 in Munich, Germany (Source: Latz + Partner)

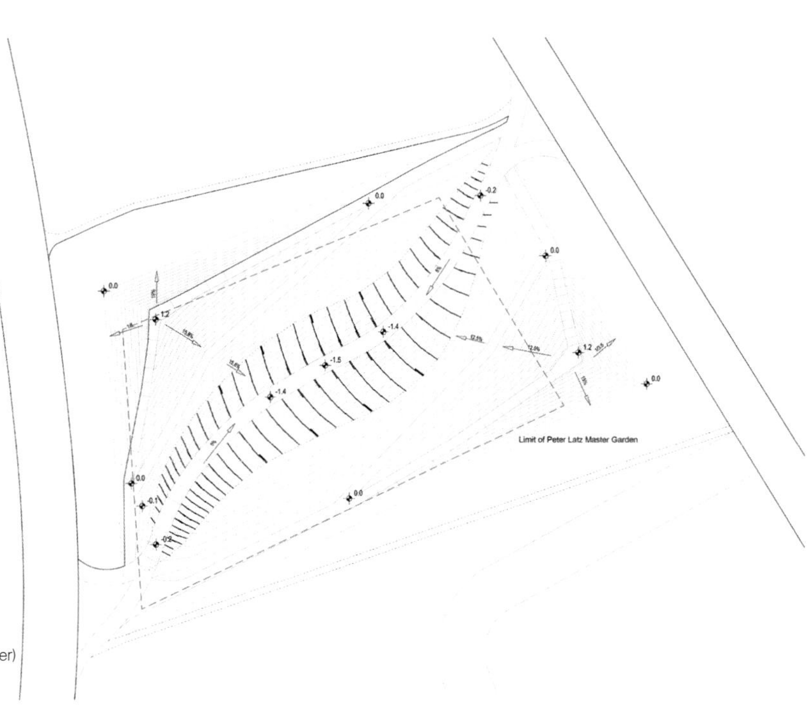

图 21 带有等高线和斜面的地形平面图（照片来源：Latz + Partner）
Fig21 Topographic groundplan with contour lines and inclines (Source: Latz + Partner)

图 22 正在建设中的天然石板路 (照片来源 : Wang Xiaoyang)
Fig22 The natural stone slabs under construction (Source: Wang Xiaoyang)

图 23 正在建设中的石板柱廊 (照片来源 : Wang Xiaoyang)
Fig23 The stone slab colonnade under construction (Source: Wang Xiaoyang)

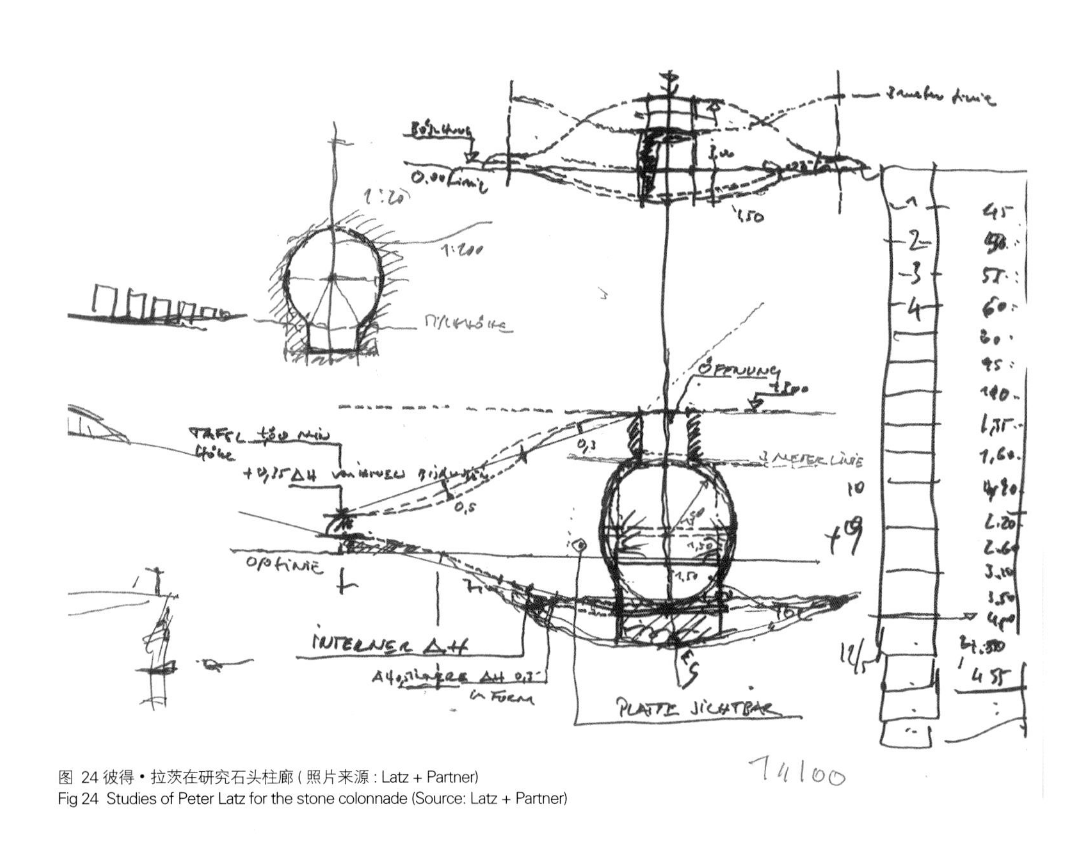

图 24 彼得 • 拉茨在研究石头柱廊 (照片来源 : Latz + Partner)
Fig 24 Studies of Peter Latz for the stone colonnade (Source: Latz + Partner)

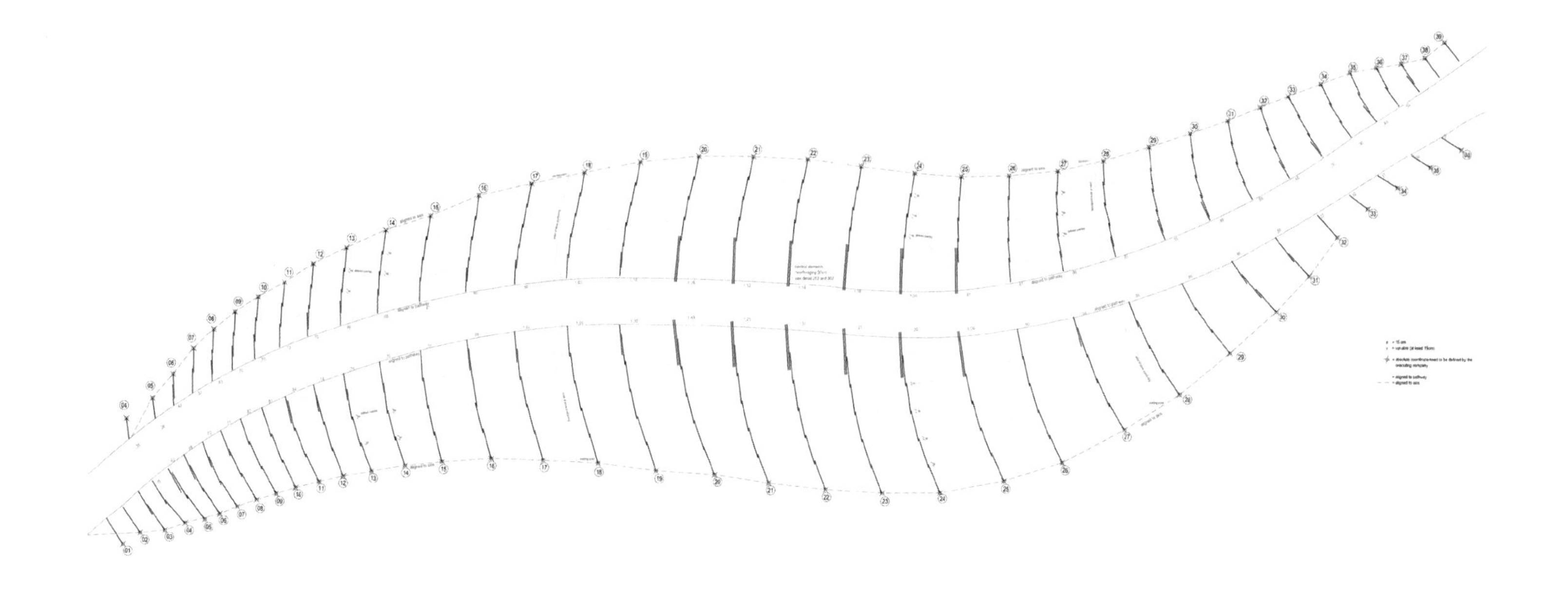

图 25 石板的定位，与路径和中轴校齐，以及单块石板的交叠 (照片来源 : Latz + Partner)
Fig25 Positioning of the stone slabs, alignment to pathway or axis and overlapping of the single slabs (Source: Latz + Partner)

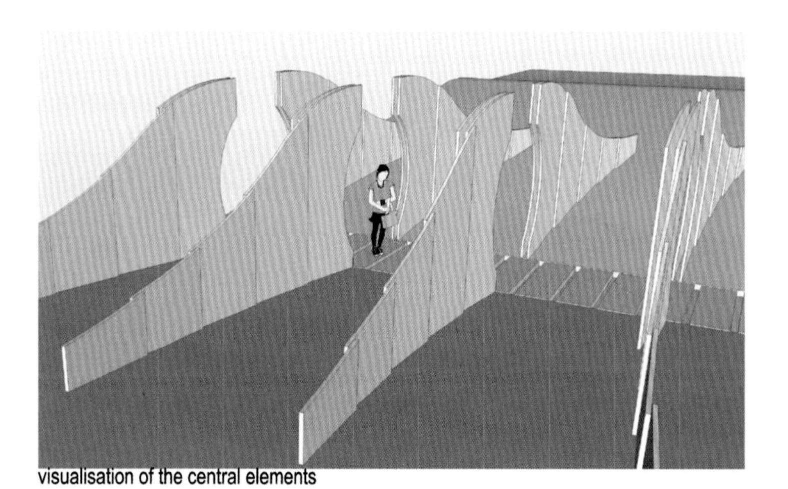

1,10 1,20 1,30 2,00 2,00 1,50 1,40 1,30 1,20

I II III IV V IV III II I

20cm overhanging stone

15cm overlapping

15cm overlapping

3,00

,45

2,00

0,06m x 4m (3m apparent) x 2,00m

0,06m x 4m (3m apparent) x 2,00m

section 1:50

图 26 具象化的中心元素细节图 (照片来源 : Latz + Partner)
Fig 26 Detail drawing of the central element with visualization (Source: Latz + Partner)

图 27–28 在北京进行第一次展示的大师园设计模型 (照片来源 : Latz + Partner)
Fig 27–28 Working model of the Mastergarden for the first presentation in Beijing (Source: Latz + Partner)

图 29–31 在坡顶上栽植针叶树和落叶树（图片来源：拉茨景观设计事务所）
Fig 29–31 Conifers and decidous trees characterize the hill tops (Source: Latz + Partner)

图 32–37 开篇印象（图片来源：拉茨景观设计事务所

fig 32–37 Impressions of the opening (Source: Latz+Partner)

种植网格和种植方法。观赏草和松类植物、观叶植物如红花老鹳草，以及落叶乔木相映成趣。在中央线上，选择种植在春天开花的勃朗峰大花葱（图 32–35）。

矿物覆料

被称为“矿物覆料”的石屑覆盖花园表面。它们嘎吱作响的声音伴随着游客的每一个脚步，为花园带来声之印象。

在园博会举行的这一年，矿物覆料将裸露在刚种植完植物的地面上。数年后，规划方案中的绿色植物将逐渐覆盖地面。

基础技术设施

除了水雾系统，滴灌也将保证植物的健康生长。控制、定时以及配电系统被安装在高压设备安装盒内。

我希望你能享受这座花园，和水雾扑面而来、将你包裹在白色轻纱中的那种惊喜。它令你全身心放松，你会渴望希望重复这个令人赏心悦目的体验。□

within the freshly planted surfaces. Over the years, the green colour shown in the plans will prevail more and more.

Technical infrastructure

In addition to the mist system, drip irrigation guarantees the successful growth of the plants. Controlling and timing as well as the electrical distribution are installed in the boxes for the high pressure installation.

I wish you to enjoy the garden and the sudden surprise when the mist reaches for you and smothers you in its white veils. It will release you - and you will be desirous of repeating this exhilarant experience.■

作者简介：

彼得·拉茨生活和工作在德国慕尼黑附近的克兰兹贝格。他毕业于慕尼黑工大景观设计专业，在 1968 年他在亚琛工大完成了城市规划专业的研究生学习和实践，并在这一年与妻子 Anneliese 成立了自己的景观设计和规划事务所（自 1988 年更名为拉兹 + 合作伙伴事务所，网址 www.latzundpartner.de）。在2011年3月，他们将事务所交给同为建筑师和景观设计师的儿子蒂尔曼管理。

自他创办事务所和从事教学工作之始，彼得·拉茨就一直重点关注着自 20 世纪 80 年代中期以来尤其集中于后工业时代的生态城市重建。其获奖项目杜伊斯堡景观公园把前蒂森钢铁厂脱胎换骨成宜人的公园和这个城市美丽的一隅，从而为他博得了世界性的声誉（该作品曾多次获得国际大奖，如巴塞罗那 2000 年度罗莎·芭芭拉园林设计奖，巴黎 2001 年城市建设学术奖章，EDRA2005 年广场规划奖，2009 年绿色优秀设计奖）。目前在特拉维夫的两个项目是展示他有条不紊的设计方法和对当代环境新型设计的能力的代表之作：“阿里埃勒·沙龙公园”，既是一个巨大的洪水保留盆地，也是具有全国意义的景观公园；另一个项目是“Hiriya 垃圾场修复”（获 2010 年绿色优秀设计奖）。

在 2010 年，彼得· 拉茨作为“绿色设计的领袖、先锋和创新者”，被授予绿色优秀设计人物奖。

彼得·拉茨的学术生涯始于在荷兰马斯特里赫特的 van Bouwkunst 学院担任讲师，在 1973 年他被任命为卡塞尔大学的全职教授。1983 年至 2008 年，他担任慕尼黑工大景观建筑和规划专业的教授，然后以荣誉教授的身份离开了这个他从事教学和研究的地方。彼得·拉茨多年来也担任美国宾夕法尼亚大学兼职教授、哈佛大学客座教授。

Biography:

Peter Latz lives and works in Kranzberg near Munich, Germany. He graduated from the Technical University of Munich as a landscape architect and completed his studies in 1968 after postgraduate research and studio work in urban planning at the RWTH Aachen. In this year, he established the studio for landscape architecture and planning together with his wife Anneliese (since 1988 Latz + Partner www.latzundpartner.de). They assigned the office to their son Tilman, architect and landscape architect, in March 2011.

Since the beginning of his office work and teaching, a main concern to Peter Latz has been ecological urban renewal, which concentrates since the mid eighties on post industrial sites. With the award-winning project "Landscape Park Duisburg Nord", the metamorphosis of the former Thyssen ironworks into a people's park and vivid part of the city, he has gained world-wide reputation (a. o. First European Prize for Landscape Architecture Rosa Barba Barcelona 2000, Grande Médaille d'Urbanisme of the Académie d'Architecture Paris 2001, EDRA Place Planning Award 2005, Green Good Design 2009 Award). Two current projects in Tel Aviv are representative for his methodical approach and his ability to develop new expressive forms of contemporary environmental design: The "Ariel Sharon Park", becoming at the same time a gigantic flood retention basin and a landscape park of nationwide significance, and the "Hiriya Landfill Rehabilitation" (Green Good Design 2010 Award).

In 2010, Peter Latz received the Green Good Design People Award for being "a leader, pioneer, and innovator in Green Design".

Peter Latz started his academic career as a lecturer at the Academie van Bouwkunst in Maastricht (Netherlands) and was appointed full professor at the University Kassel in 1973. 1983 until 2008 he held the chair of landscape architecture and planning at the Technical University Munich and left the place of his teaching and research activities as an Emeritus of Excellence. Peter Latz was an adjunct professor at the University of Pennsylvania for many years and was a guest professor at Harvard University.

有限 / 无限
FINITE / INFINITE

雅各布・施瓦茨・沃克 Jacob Schwartz Walker
蒋侃迅 Kanxun Jiang

图 01 花园全景
Fig 01 Overall View the Garden

项目位置：中国，北京，第九届北京园博会
项目面积：3,534 m²
委托单位：第九届中国（北京）园林博览会组委会
设计单位：彼得・沃克合伙人公司与百安木设计公司
景观设计：彼得・沃克合伙人公司与百安木设计公司
完工时间：2013 年 5 月

Location: The 9th China (Beijing) International Garden Expo, Beijing, China
Area: 3,534 m²
Client: The 9th China (Beijing) International Garden Expo committee
Designer: PWP+BAM
Landscape Design: PWP+BAM
Completion: May, 2013

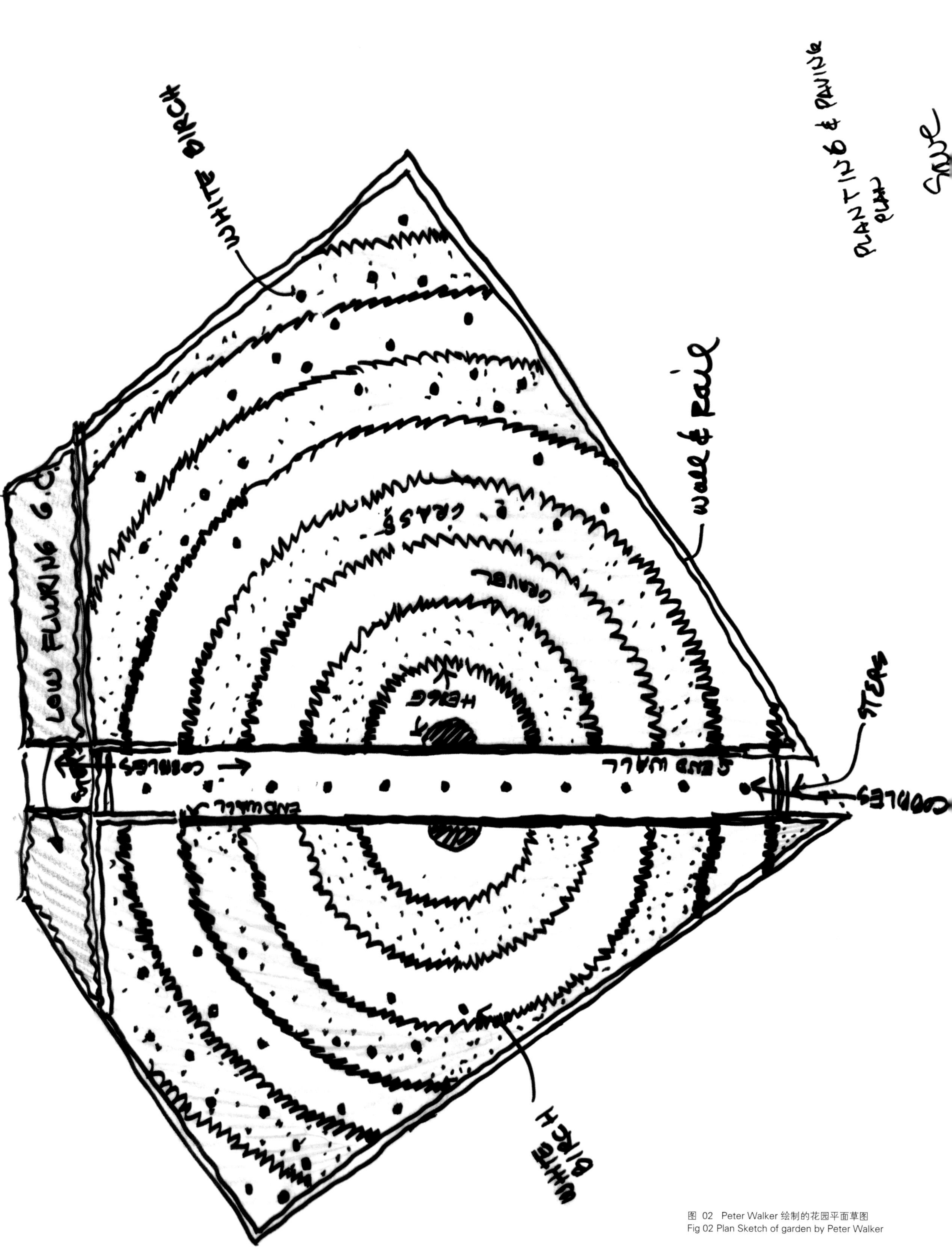

图 02 Peter Walker 绘制的花园平面草图
Fig 02 Plan Sketch of garden by Peter Walker

图 03–04 早期带有镜面的工作模型
Fig 03–04 Early physical model with mirrors of garden

花园描述

以一个物件，在同一空间中创造出两种截然不同的环境，这就是花园所呈现的效果。

花园被简单的环形绿篱所覆盖（图 01–02），一条笔直的步道从绿篱中央穿过将花园一分为二，步道两侧都被一堵双面镜墙所界定，而步道中央是一排 5m 间隔的悬铃木。

如此一来，对于镜墙内部的人，由于倒影反射倒影的“理发店效应”（图 03–04），狭窄的步道被变化成为一个“无限的景观”：单排的悬铃木被不断复制成为一片无边的树林（图 05–06）。

而与此对照，由镜墙外侧的人看来，被墙体切成两半的绿篱又在镜面中与各自的倒影合并为完整的圆环，由此，一个清晰、完整的景象在镜墙外侧被创造出来。

缘起

倒影是园林景观中的常用手段，长久以来它都以水作为介质，存在与池塘、小溪、喷泉等形式中（图 07）。玻璃与水具有一些共同的属性——透明与反射，它们总是一直并存着，在应用上水多用于园林，而玻璃多用于建筑。

玛莎·施瓦茨在 2011 年西安园艺博览会的大师园中曾探寻过园林与建筑中的反射对比的应用（图 08），在此过程中 BAM 同样扮演着重要的角色。这个花园由一系列玻璃、单向镜子以及中式砖墙构成，既是一个迷宫，同样也是一个趣味十足的房子，在这里玻璃与镜子这种建筑语言被用来营造反射与透明的景观效果，而正是这一次尝试促使了沃克在今天的花园里进行了更进一步的探索。

在“有限 / 无限”花园中半圆环被还原的现象让沃克想起了他早期的景观尝试，那时的他正致力于寻找园林景观中的极简手法。例如 1983 年，他在波士顿马尔堡街的屋顶花园中，就曾用小镜子来反射蓝天，镜子就是当时他在试验性花园中所采用的普通材料中的一种（图 09）。相比之下“有限 / 无限”花园的概念更宏大、更完整，也更加清晰。马尔堡街屋顶花园的小镜子们仅仅像补丁一样反射出天空的颜色；而今天的花园则反射出人们所感知的自我。正如艺术家丹·格雷汉姆的

Garden description

The garden is defined by one device that creates two contrasting conditions. A simple circular parterre garden (Fig.01-02) is divided down the center by a pathway lined with two walls, each covered with mirrors on both sides. Centered in the central pathway is a line of sycamores on a five-meter grid. For the viewer inside the two walls the pathway and its trees are transformed into an “infinite landscape” by the “barbershop effect” (Fig.03-04), in which reflections are reflected on top of reflections. The one row of sycamores becomes an orchard that multiplies into infinity (Fig.05-06). By contrast, the circular parterre is divided in half by the walls, until the viewer looks into the mirrors on the outside of the wall. . . and sees the parterre completed in its reflection. The mirrors on the outside of the walls thus create a defined, whole, and complete object.

Origins

Reflection is a phenomenon often used in landscape architecture, usually in some form of water—a pond, a river, a fountain (Fig.07). Glass shares many qualities with water, transparency, reflectivity and opacity—all apparent all the time at the same time. Typically water is associated with the landscape and glass with architecture.

Martha Schwartz's garden for the 2011 Xian Horticultural Expo, for which BAM also played a key role, explored these reflective contrasts in landscape and architecture (Fig.08). The garden was a kind of maze—or fun house—composed of traditional Chinese wall typologies intertwined with glass and one-way mirrors. The adoption of the architectural method for achieving reflectivity and transparency in what was essentially a landscape condition prompted a further exploration in the Walker garden.

图 05 镜墙外侧完整的“有限景象”
Fig 05 “Finite Object” seen from the outside of the Walls

图 06 镜墙之间创造的“无限景象”
Fig 06 “Infinite Landscape” created between the walls

图 08 西安园艺博览会大师园，Martha Schwartz
Fig 08 Martha Schwartz , Xian Horticultural Expo Garden

图 07 南海岸广场项目的镜面池，PWP
Fig07 PWP South Coast Plaza Reflecting Fountain

图 09 马尔堡街屋顶花园中设计的反射镜，Peter Walker 与 Martha Schwartz
Fig09 Peter Walker & Martha Schwartz, Reflecting Mirrors at Marlborough St. Roof Garden

图 10 纽约中央公园的爱丽丝奇境记雕像，PWP
Fig 10 PWP, Alice in Wonderland, Central Park NYC

图 11-12 Peter Walker 在 BAM 北京的设计工作营（Peter 与 Daniel Anthony Gass、张媛讨论设计）
Fig 11-12 Peter Walker workshop at BAM in Beijing (Daniel Anthony Gass and Yuan Zhang pictured working with Peter)

名言所表达的那样，如果没有观众艺术将不复存在，而让观众能够从作品中观察自己对于艺术十分重要，因此格雷汉姆在众多作品中使用了玻璃和镜子，使人们在观赏的过程中不但能够看到自己也能看到他人，从而创造出艺术品本身。沃克也从路易斯·卡罗尔 1871 年的文学作品《镜子王国里的爱丽丝》中获取过灵感，书中描写了爱丽丝穿过客厅的镜子来到镜子国里所遭遇的奇妙经历（图 10）。源于以上众多启发，沃克在“有限 / 无限”花园中进行了一种关于光学原理在园林中新应用的尝试。

过程

项目初始，沃克与 BAM 的讨论时表达出他对玛莎大师园中镜子、玻璃与光学反射的兴趣，以及将这种尝试更进一步发展的想法。当沃克来到北京查看场地后（图 11-12），BAM 为他准备了镜子、镜面纸、模型树等材料用于随后的设计研讨，由此产生了“有限 / 无限”的基本要素。

BAM 的主要任务是将沃克的草图转化为一个能够被建造的花园，并负责监督整个施工过程（图 13-16）。“质量就是一切”、“上帝存在与细节”、“三思而后行”等等这些名言在国内园林景观的施工过程中常常不被理解，更不用说如何落实。而在这个花园里对于失误的容忍度又极度微小，因为所有的东西都会在镜面墙中反射出来，施工中一旦出现错误——哪怕是一个非常小的失误——都将被无限倒映，成为巨大的缺陷。

这个花园的效果将在很大程度上取决于细节的工艺与质量，成败一线之间，要么获得巨大的成功，要么沦为糟糕的失败品。正因如此，BAM 在每一个环节中毫不松懈的把关至关重要。

在设计深化阶段，BAM 发展出非常精致的细节，这些细部处理不仅需要制图过程中的精确与细致，也需要在工厂制作与现场安装阶段付出相同的努力。而且质量控制的重要性不仅仅存在于硬景细节里（图 17），同样也存在于植物材料的选择和运用之中（图 18）。

在北京，人们对于杨树多少有所偏见，很多景观设计师都将其视为不上档次的乔木，并不适合应用在展示型的花园里。尽管杨树在北京已经有很长的栽植历史，但大多数都应用于道路及防护绿化等公共项目，加之业主与设计师都缺乏“用平凡材料创造高价值景观”意识，导致在花园树种选择时遇到了一些困难。根据设计花园里需要竖直向上的柱状树形乔木，新疆杨是首选树种，但难以置信的是在北京甚至河北的所有苗圃里都未能找到理想的植株。我们最终在山西大同西郊找到了所需的杨树，那里的土地或许足够贫瘠而保持着对这些廉价速生树种的需要，但这个过程着实给 BAM 带来了不小的麻烦。庆幸的是，

The completion of the circle in Finite/Infinite also recalls Walker's early landscape experimentations, which were aimed at finding a minimalist expression for the landscape. For example, in the 1983 roof garden on Marlborough Street in Boston, the sky was reflected in small mirrors, typical of the modest materials used in these early experimental gardens (Fig.09). Finite/Infinite is at once more grand, more global, and more cerebral. The little mirrors of the Marlborough Street garden reflected little patches of sky; the mirrors of the Expo garden reflect the perceiving self of the viewer. Artist Dan Graham has been famously quoted for stating the importance of the viewer who sees himself in the art work, for, of course, without the viewer the art would not exist. Not coincidentally much of Graham's work uses glass and mirrors to allow the viewer to see not only himself but others in the action of viewing—and thus creating—the work of art. Walker also found literary inspiration in Lewis Carroll's Through the Looking Glass and What Alice Found There (1871), in which Alice tumbles through the parlor mirror to discover a new Alice in a thoroughly atypical garden of talking flowers (Fig.10). Inspired by these various sources, Walker explores new visions of garden reflectivity in Finite/Infinite.

Process

In discussions with BAM, Walker expressed his interest in furthering Martha Schwartz's exploration of reflection via mirrors and glass. Thus, when he came to Beijing to see the site (Fig.11-12), BAM presented him with a kit of elements: Mirrors, reflective paper, model trees, all were used in the work shop where Walker outlined the fundamentals of the design.

BAM's principle tasks were to turn Walker's preliminary drawings and sketches into a constructible garden and then oversee its construction (Fig.13-16). Quality is everything. God is in the details. Measure twice, cut once. These terms are neither easily understood nor achieved when working in China. Unfortunately the design for this garden had very little tolerance for mistakes because everything in it was reflected. One mistake was not one mistake, but infinite mistakes. Thus there could be no mistakes.

图 13 各部分组件的电脑模型
Fig13 Digital model of detail assembly

图 14 Jacob Schwartz Walker 现场查看样板段的施工情况
Fig14 1:1 mock-up inspection at site by Jacob Schwartz Walker

图 15 镜墙单元块交接处的施工细节
Fig15 Mirror wall joint detail with reflection of Jiang Kanxun

图 16 铺设中的花岗岩石钉和砂石垫层，以及暴露出来的镜墙底部不锈钢排水槽
Fig16 Gravel & sand setting bed for granite cobbles with exposed mirror wall drainage detail during construction

图 17 花园边界的细节（木坐凳与不锈钢栏杆）
Fig17 Details of the Garden Edge (Wood Bench & Stainless Steel Railing)

图 18 中心步道上悬铃木的树影
Fig18 Shadow of the Tree Casted on the Center Street

图 19–20 黄昏的中心步道
Fig 19–20 The Center Street in the Twilight

20

图 21 树干上的白漆条纹与镜子的单元网格相吻合
Fig 21 White Paint Stripes on the Trunk Match to the Mirror Grid

我们选择的悬铃木要相对名贵一些，具有更高的种植价值，因此更容易获得。

从各个细部，到植物选择，再到现场配合与监督，BAM 在业主与施工方眼中似乎始终都是一个麻烦的对象，因为我们拒绝任何形式的妥协，始终坚持着维护设计的完美。如果为了达到完美而需要我们自己爬上树顶去剪枝，我们也会毫不犹豫。实际上我们就是这么做的。

今天的中国到处是精彩的建筑作品，但园林景观的发展却相对滞后。在本项目中面对异常严格的 BAM，东方园林的施工团队非常专业地完成了施工任务，将精美的设计变为了现实。这次实践成就了一个设计者、建设者和管理者紧密协作的成功案例，向世界展示了在中国同样可以实现高质量的园林景观（图 19-23）。□

The garden could either be an abysmal failure or a great success, and it all was dependent upon the quality of the details—the craftsmanship—and, hence, on BAM's unrelenting oversight of every step of the process.

The details BAM developed are highly refined and required precision and care in the drawing, manufacturing, and installation. And not only are the hard details important (Fig.17). The quality of the planting and type of planting are key (Fig.18).

In Beijing there is a prejudice against poplars. Many landscape architects in China consider them low-quality trees, far too common for a show garden. Although the tree is found and was historically used in Beijing, this prejudice against using the common to create great value—and the tree's frequent use in public projects—made it difficult to find the right trees for the garden. The pyramid white poplar or Xingjiang Poplar, which was specified, could not be found anywhere around Beijing or Hebei Province. This presented BAM with quite a problem because the garden required a columnar tree. Eventually we were able to find the right trees just west of Datong in Shanxi Provence, a place poor enough to appreciate the useful cultivation of cheap fast-growth trees. Happily the sycamore is considered a much more worthwhile tree. They fetch higher prices and are thus thought worth cultivating—and thus were easily acquired.

From the detailing, to the choice of plants, to the on-site supervision BAM was frequently a thorn in the side of both the client and the contractor because we chose to fight for the perfection of the design. We found it necessary to resist any compromises, especially ones based on offers of future work, and if that demand for perfection meant we had to go to the site and trim the trees ourselves, then we would. And we did.

China abounds with architecturally impressive structures, but the landscape unfortunately lags far behind. Although BAM had to be tough on the contractors, the Oriental Land team executed the design beautifully and professionally, and now all of us—designers, managers, and contractors alike—have a beacon that reveals to the world what landscape can achieve in China (Fig.19-23).■

图 22 镜中倒影创造的树林与光之天蓬
Fig 22 The Orchard and Light Canopy Created by the Reflection

图 23 一堵墙划分出两种截然不同的风景
Fig 23 Two Entirely Different Conditions divided by the Wall

作者简介：

雅各布·施瓦茨·沃克 / 男 / 风景园林师 / 百安木设计有限公司 / 中国北京

蒋侃迅 / 男 / 风景园林师 / 百安木设计咨询（北京）有限公司 / 中国北京

Biography:

Jacob Schwartz Walker / male / Landscape Architect / Ballistic architecture machine / Beijing, China

Kanxun Jiang / male / Landscape Architect / Ballistic architecture machine / Beijing, China

“初源之庭”——倾斜土地，创造空间

A BEGGINING OF THE GARDEN—TILTING THE EARTH AND MARKING THE PLACE

三谷徹

Mitani Toru

项目位置：中国，北京，第九届北京园博会
项目面积：2,395.98m²
委托单位：第九届中国（北京）国际园林博览会组委会
设计单位：日本千叶大学，Sukiyaki 设计事务所
景观设计：三谷徹
初步设计协助：Heartland Engineering 有限公司
设计开发与工程文件：北京源树景观规划设计事务所
完成时间：2013 年 5 月

Location: The 9th China (Beijing) International Garden Expo, Beijing, China
Area: 2,395.98m²
Client: The 9th China (Beijing) International Garden Expo committee
Designer: Chiba University, Sukiyaki Design Architects
Landscape Design: Mitani Toru
preliminary design assistance: Heartland Engineering
Design development and Construction document: Beijing Ltd
Completion: May, 2013

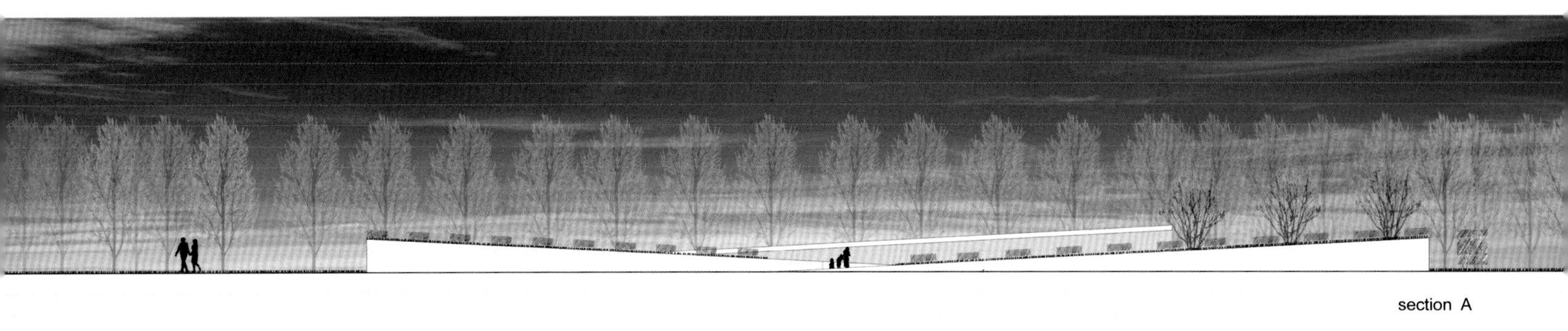

section A

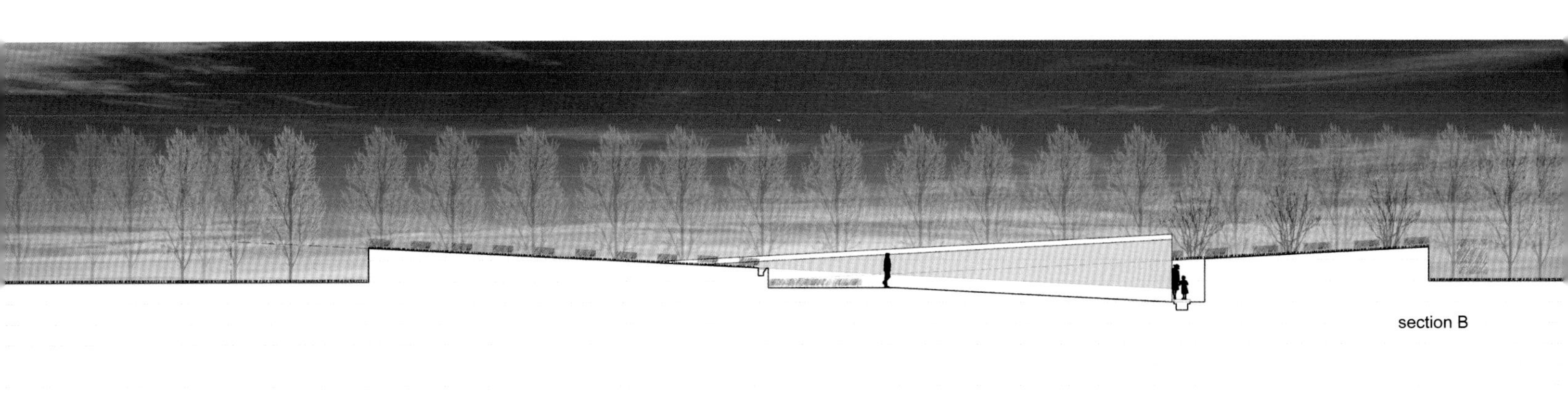

section B

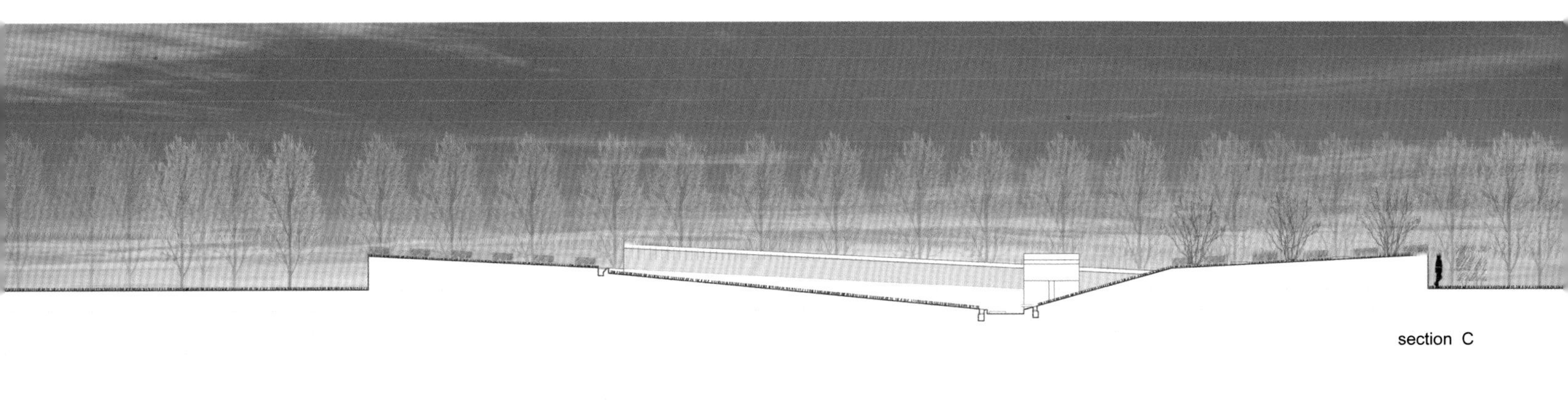

section C

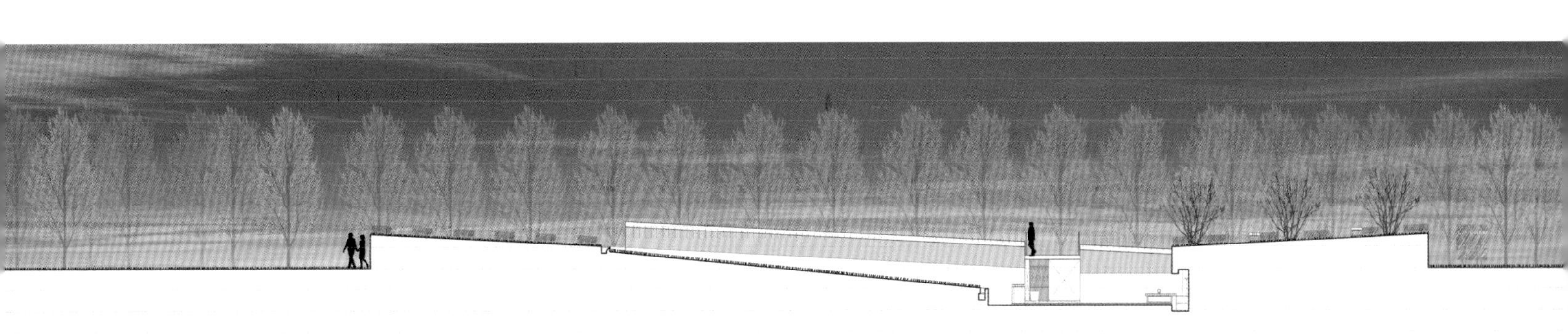

图 01 剖面图
Fig 01 Section

section D

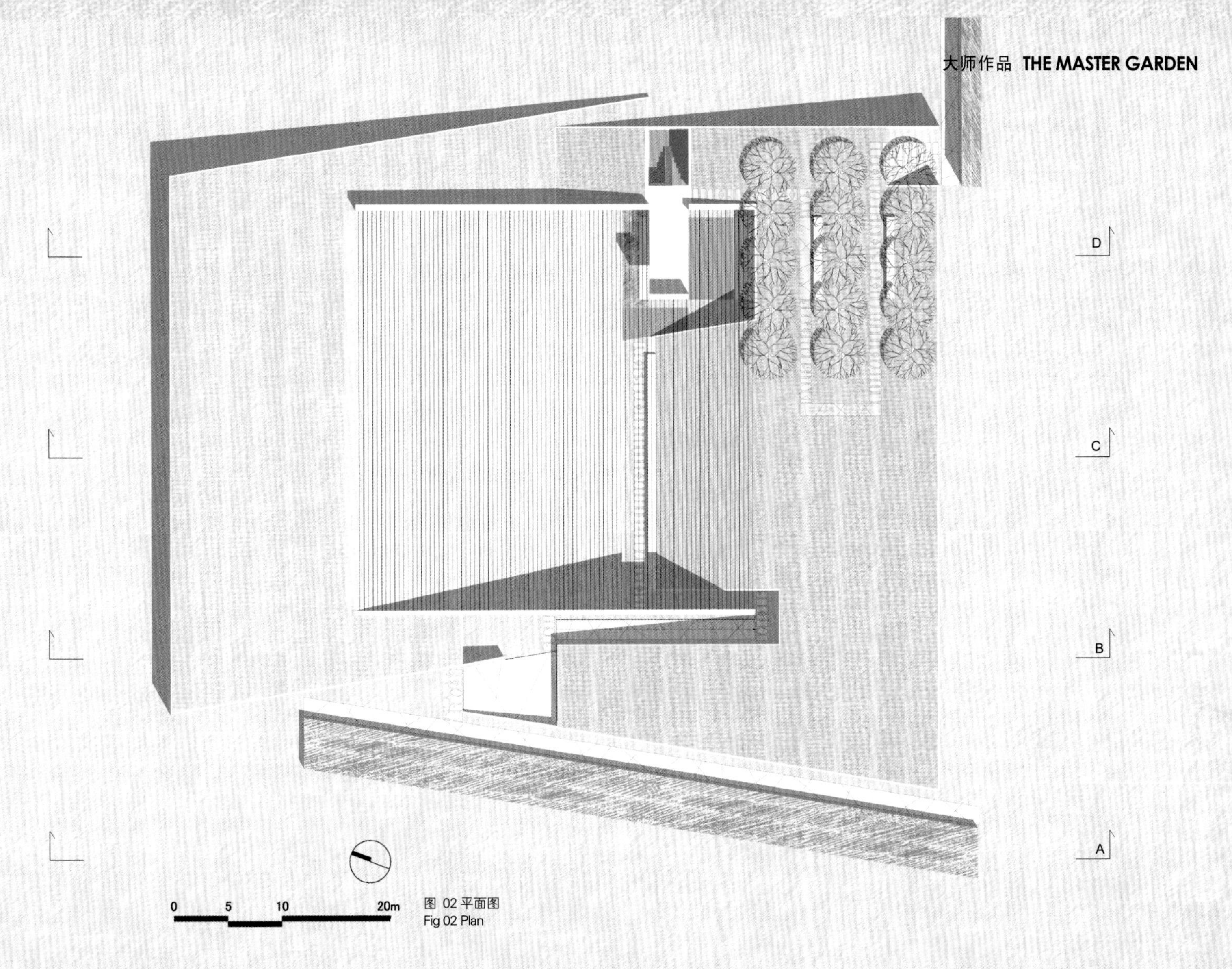

图 02 平面图
Fig 02 Plan

当我被邀请为北京园博会设计一个现代花园时，我从碗架中拿出了一个大盘子。这个简单的，有着优雅曲线的圆盘象征着一个等待着盛放菜肴的空间。我发现这个形式很好地将原料、大地以及与艺术擦出的火花结合在一起。因此，这次花园设计的主旨就是创造一个像空盘一样的空间，等待着人们的到来。

花园的本质是我们脚下的土地，这个花园是一个超越了文化隔阂的对大地的赞颂。在地球上，一些区域幸运地拥有着潮湿温暖的气候，而另一些地方却在忍受着寒冷和干旱，相比之下，植物或水已经成为次要的因素。每一个园林都需要建造在土壤之上，因此，地势而非气候是在任何地方、任何国度建园最初考虑的问题。

我们都曾见过世界上大地艺术的遗址。虽然由于地理位置和当地文化的不同它们的形式不一样，但它们处理场地的手法却极其相似：它们或者抬高土地，或者凹陷土地。英国古代的怪圈遗址就是这种手法最好的例子。同样，在它的南方，在古希腊，祈祷的地点也仅仅简单的被一块在山体中切割出的凹陷场地标示出来。很多园林考古学家都热切地关注着古代园林的地形处理，这是因为这些地形的处理手法反映着古代人们对宇宙的认识。地形处理同样为游览园林带来了乐趣。就像在意大利的山坡上爬上爬下，在英国的冰川中漫步或是在法国的巴洛克式平台上放松。

当我第一次走进北京的设计场地时，我发觉自己是站在一块平台的中央。这里没有树，没有路，除了脚下的碎石，其他地方都是空荡荡的。在我看来，这表现了空间的缺失。如果是一个建筑师，他或许会在这里设计柱廊来明确空间，但是我作为一个景观师却不会这么做。我决定通过改变地形来创造空间。在现有技术的基础上，我重点设计了两个倾斜的平台。

设计的方法是两个地形平台，它们各自以百分之五的角度轻微的倾斜成面对面的形态。平台的最高点为 1.6m，相当于普通人的视线高

When I was requested to design a garden in Beijing as one of the contemporary landscape architects, I pulled out a large dish plate from the kitchen shelf. The simple round shape with an elegant curve represents a space to wait for the cuisine. I found this form to be an ideal match for the material; the earth, and the result of one touch on the art; fire. Therefore, the objective of the design was to create a space like a dish, inviting for people.

A garden is, at first, a matter of the mother earth. It is a form of celebration of the earth in any culture throughout the world. As some regions are blessed with warm and wet climate while others with cool and arid conditions, planting or water feature is secondary elements. Since every garden is to be designed upon the earth, the garden is a part of the earth. Therefore, regardless of the climate, deciding on the earth form is the very beginning of garden design in any country and any era.

We often come across ancient ruins of the earthwork in the world. Though the appearance is different, depending on the location and local culture, the way the earth is handled is simply the same: it's either cut or filled up. The earthworks – ancient circles in England are the most representative example of this technique. While further down south, in Greece, the praying locations are marked by just a concaved landform carved in the hillside. Most garden archeologists today focus intensively on the landform of ancient gardens. The reason behind this is that the landform in the garden reflects people's attitude towards the universe at that time. Another delight provided by the landform in the garden is experiencing it; climbing up and down the terraces

度，而最低点则为地面下 1.5m。平台前方低处的挡土墙被漆成白色，象征着它背后地面的高度。它也同样寓意着船的吃水线，为人们创造了一种花园是漂浮在地面上的感觉。在平台的对角线上是一条供人们漫步的路径和停留休息的场所。沿着小路游览，人们会发现他们慢慢走入了地下并且以一种全新的角度观赏园林——一种新的，不同于在两个倾斜平台之间的视角（图 01–02）。

园林是一处封闭的场所，但它同时也面向天空。

园林是它们所处环境中的一部分，但同时它们也创造了一种与周围环境不同的感受。这是被进园的路径创造的。走进园林的过程就是走入另一个世界的过程。同样，在走出园林后也会发现周围的环境发生了变化。

在东方传统园林中，茶室总是被安排在园林的最深处。游人被一条精心设计的曲折路径引入。在传统设计手法中，对路径的设计意味着对人们入园体验的设计。在游览苏州园林时，园路的石板上会雕刻精美的图案。日本的桂离宫就好像镶满宝石的匣子，里面安置着各种被精心设计，挑选的踏步石。在欧洲也有很多被艺术品装饰的园路，比如在意大利冈贝里亚庄园规则式绿篱中的马赛克铺装。即使在奥比昂公墓这样的现代杰作中，也精心的设计着水平的流动空间，让我们在墙，草坪和水景之中游览。通过设计一个转向另一个的直角，设计师 Carlo Scarpa 创造了一种进入更深层空间的体验。

我的设计目的是让游览者体验在两个倾斜挡土墙之间的强烈的视觉效果。在进入园林，转过几个转角之后，人们会被踏脚石和 'Nobedan' 小路引入地面之下的空间。在花园最低最封闭的地方，一个小广场在土地中被切割出来，作为一处供沉思，私密聊天或者休息的场所。我希望这里能安静的让人们可以联想到他们背后水滴滴落的声音。

园林最终的目的是带引人们走进天堂。

花园的形式创造出一种通往天空的意向，虽然它一直是大地的一部分。前面提到的英国怪圈运用的是最原始的抬高或降低地面的手法。一些观点认为这样做表现了人们对于宇宙天象的敬畏。在之后的几个世纪中，园林作品表现着一种真挚。巴洛克园林中的倒影池起到镜子的作用，将无限天空映射在其中。然而在中世纪教堂的庭院中，则是将天空浓缩在了抽象的画作中。我想起了在布罗德保护区的一个由几位美国风景园林大师建造的园林，当我站在树林边上方形的绿篱中时，感到设计中体现着的纯粹的天空之美。

这个由盘子带来灵感的花园同样向天空开放着，它好像在张开双臂向天空祈祷。一块薄薄的反射板在倾斜的平台上延展，就位于休息区的旁边。这块反射着太阳光的板子由精致的不锈钢制成，承受着 500kg 的拉力。反射板的两边同样以不同的方向倾斜，创造了一种壳的形态，就像是在花园中的船帆。在踏上探索之旅后，人们可能会觉得他们漂浮在土地上的另一个地方。这样，在有着对大地的归属感的同时，他们也能沉浸在天空带来的永恒的宁静中（图 03–09）。

花园是为了这次园博会建造的，但它也可以作为一个未来的遗址。在时间的流逝中，设计种植的植物会枯死，水景会被人们忘却，大量的沙尘会覆盖在场地上。但或许在未来的某一天，人们会挖掘出它，发现它在大地上书写的形式。这又是一个对地形处理的作品，进一步地说，这种几何的形式会成为当代人思想的验证。最后，它代表着我最深切的愿望，那就是让园林传达我们对于未来永恒和宁静的祈望。□

on the hillsides in Italy, rambling around the glacial topography in England, or just loosing yourself in the middle of the flatness of the French baroque.

When I first visited the design site in Beijing, I found myself standing in the middle of a flat landfill. No trees, no path, but vast plain with gravel scattered around. In my perception, this meant an absence of a place. An architect would build a column to mark the place, but I, as a landscape architect, decided to move the earth. At the same time, due to the contemporary approach and techniques, I put the focus on the earth surface by tilting it carefully.

The design solution for the site consists of two earth plates, slightly tilted by 5 % in order to face each other. The highest point is GL+16m; almost the same as eyelevel height; while the lowest point is GL-15m. The frontal, lower part of the retaining walls is painted white, indicating the height of the soil behind. This is metaphor for the Plimsoll-line marked on sailing ships, which creates the sense of a garden floating on the earth. Between the two angles of the earthen surface, a path for ramble and a place for rest are positioned. Walking down the path, the visitors will find themselves under ground level and witness the landscape from a completely different perspective, beyond the new horizon of two tilted earth plates(Fig.01-02).

A garden is an enclosed space, but it is also open to the air.

Gardens are part of their immediate environment, yet they provide a unique sense of being somewhere else. This is achieved through approach design. The path leading into a garden is a way to another world. Similarly, when leaving a garden space, the world outside appears different.

In traditional Asian gardens, the tea pavilion is located in the furthermost part of the garden. This pavilion is accessed through carefully designed path, with many turns and curves. In the traditional manner, designing the approach means designing the behavior of the feet. While visiting the gardens in Suzhou, an impressive palette of icons on the footpath can be seen. Villa Katsura in Japan is like a jewelry box filled with many types of experimental, but precious and beautiful stepping stones. Europe also has wonderful examples garden paths adorned with artwork; such are the mosaic tiles among the topiaries in Villa Gamberaia, Italy. Even in recent times the Garden of Brion Cemetery, a modern masterpiece, has a careful configuration of the horizontal circulation taking us along the walls, the grass area and the water. By setting right-angled turns one after another, the designer, Carlo Scarpa, created the sense of entering into deeper and deeper space.

The intention with my design is to invite visitors by revealing the strong perspectives generated by inclined retaining walls. Once inside, and after a few turns, the stepping stones and the 'Nobedan' path lead the visitor bellow, under ground level. At the lowest and enclosed part of the garden, a small square is cut out in the earth, intended as a room for solitary meditation, chatting with friends or even as a resting place. I hope this space will create the quietness for imagining the sound of invisible water dropping in the background.

A garden is, finally, an offering to the heaven.

Its form has a predisposed orientation towards the sky, all the while it is part of the earth. The circles in England, mentioned above, were made by means of the most primitive technique of cutting and filling up the earth. This was done, in some opinions, for the people to express their faith and respect to the astronomical phenomena. Gardens in later eras show the same sincerity. Baroque reflection pools act as mirrors to absorb the infinity of the sky, while the courtyards in the medieval abbeys frame up the sky into one abstract painting. I recall a beautiful experience while standing in the square hedge among the woodland of Bloedel Reserve: the garden designed by few master landscape architects in the United States. All that is expressed through that design is the beauty of the sky.

图 03 花园实景图 1
Fig 03 The built garden1

图 04 花园实景图 2
Fig 04 The built garden2

图 05 小径
Fig 05 Trail

图 06 花园实景图 3
Fig 06 The built garden3

图 07 花园实景图 4
Fig 07 The built garden4

图 08 石凳
Fig 08 Stone bench

图 09 花园实景图 5
Fig 09 The built gardens

spreads its arms as if it's waiting for a blessing from the heaven. A thin layer reflecting sunshine is spread above the tilted landform, right above the resting space. This layer of light is composed of ninety stainless steel strings stretched by 500kg of tension. The two perimeters of this layer are also tilted in the opposite direction, thus creating a HP shell form that appears as a sail in the garden. When stepping onto the observatory, the visitors may feel as floating on another surface on the earth. In this way, while embracing the sense of belonging to the earth, they would contemplate the peaceful eternity of the sky over the world (Fig.03-09).

This garden is designed for the international exposition today, but also as an ancient ruin in the future. With time, the planted greenery will die and disappear, the water feature will be forgotten and tons of sand and dust will cover the place. But maybe, someday, future generations will excavate it and find the form inscribed on the earth. Again it is a matter of the landform; precisely this geometric form will be a proof for the existence of intellectual minds today. Finally, it is my deepest wish that this garden conveys our message of eternity and peace of the world in the future. ■

作者简介：

三谷徹（Mitani Toru）：日本景观设计大师，哈佛大学（GSD, Harvard University）硕士，东京大学（University of Tokyo）工学博士。早年曾就职于皮特·沃克与玛莎·施瓦茨合作景观事务所（Office of Petere Walker & Martha Schwartz），现为studio on site事务所合作伙伴，并任日本千叶大学（Chiba University）园艺学研究科教授。

Biography:

Mitani Toru: Japanese landscape designer，Master of Harvard University, Doctor of Engineering of University of Tokyo, the early years worked at Office of Petere Walker & Martha Schwartz). Currently the partner of Studio on site firm, and served as professor on Horticulture studies of Chiba University.

骆思同（编译）
Translated and edited by Sitong Luo

全球化时代的两重方式

A WAY OF DUALITY IN THE TIME OF GLOBALIZATION

三谷徹 Mitani Toru

三谷徹：日本现代风景园林师。通过与坂茂、槇文彦等诸多日本著名建筑设计师的合作，推出了一系列的现代风景园林设计作品。作为“studio on site”的合伙人参与设计活动的同时，以教授的身份在日本千叶大学从事现代风景园林的设计教育工作。设计作品多次荣获日本造园学会奖，日本土木学会奖以及好设计奖。在翻译地景艺术以及近现代风景园林史的同时，通过《风景之旅》、《场所设计》等著作表达自己的思想。此外还发表了关于日本庭园空间特征的系列论文。

Mitani Toru: A contemporary landscape architect.He puts forward a series of contemporary landscape works by the cooperation with Shigeru Ban, Fumihiko Maki and many other famous Japanese architects. He is involved in the design activities as a copartner of the "studio on site", and at the same time engaged in the education work of contemporary landscape design as a professor in Chiba University. His design works have gained award of Japanese Society of Landscape Architecture, award of Japan Society of Civil Engineers and Good Design Award. He has translated Earthwork Art and History of Contemporary Landscape. In the meantime, his expression of ideas are shown to the world *through Pilgrimage in the 20th Century American Landscape, Designing Places: Dialogue with Fumihiko Maki* and some other writings. In addition, he has published a number of papers on the characteristics of the Japanese garden space.

图01 与事务所成员及合作合伴照片，（右下图，从左）长谷川浩己、铃木裕治、户田知佐、三谷徹
Fig 01 With members and partners in 'studio on site' , (lower right,from left) Hiroki Hasegawa、YujiSuzuki、Chisa Toda、Toru Mitani

王小璘（以下简称"王"」)：三谷先生、各位与会嘉宾和贵宾下午好。很荣幸担任这个座谈的主持人，一起主持的还有章俊华教授。今天座谈会的方式是先请台上的嘉宾提问请教三谷先生，过程中如果各位贵宾有问题请教，我们随时欢迎。今天全场用中文发言，有位非常杰出的翻译先生将为我们做现场日语翻译。

回顾刚才三谷先生介绍的一些作品，我们可以发现跟过去两场看到的西方园林设计师风格不太一样，有着相当程度的东方文化底蕴和内涵，尤其细节部分十分细腻，不仅以简单的设计手法呈现出深刻的氛围，并且和城市有相当程度的链结，这给了我们启发，就是园林师不仅做一个单项的园林设计，还需要考虑和整个城市与环境的结合。接下来就请在座的嘉宾提问。

朱玲(以下简称"朱")：各位先生、各位女士下午好，非常高兴今天能够在这里聆听三谷先生的报告，并且能够在这里一起互动。今天先生讲了几个他的作品，我也有很大的启发，很重要一点我也想跟三谷先生有一些探讨。首先就是三谷先生说到了景观和城市的关系，然后建筑和地景一体化设计等一些方面，我们知道大师本科、硕士、博士都是以建筑专业为背景，我本人也是原来学建筑的，现在一直在从事景观设计，那么很想跟三谷先生来探讨一下，这可能是一个问题的两个方面。一个方面就是在刚才看到大师作品里面很精美的线条，尺度，还有比例，都是非常严谨、精美，包括景观和空间之间互动，像刚才说到两个设计，一个是出云大社（英文名：Izumo-taisha 日本

Xiaolin Wang (Name as 'Wang' below) : Mr.Mitani, Ladies and gentleman, Good afternoon. It is a great honour to be the host of this conference, and Professor Junhua Zhang will host together with me. Firstly, we will invite the honoured guests on the stage to ask questions. At the same time,we welcome every distinguished guest here to ask questions at any time if you want to seek advice from Mr.Mitani. We will speak in Chinese today, andourexcellent Japanese translator will be the site interpreter for us.

Looking back on these projects that have been introduced by Mr.Mitani, we can find that the style is a little bit different from that of western landscape architects'. They have considerable Eastern cultural information andintensions, especially some of the fine details. They do not only present profound atmospheres by simple design methods, but also connected with the city fairly well. They demonstrate that projects can both be viewed as independent expressions as well as integral parts of their larger surroundings.

Now we will proceed to questions from our honoured guests. Please.

Ling Zhu (Name as 'Zhu' below) : Good afternoon, ladies and gentleman. I'm really happy to be here to learn about the work in Mr.Mitani'report and to have the opportunity to interact with

图 02-03 岛根县立古代出云历史博物馆
Fig.02-03 Shimane Museum of Ancient Izumo

设计版权：studio on site
Project Design Credit：studio on site

岛根县出云市的神社）的入口大厅完美地诠释一个通透性的空间关系，还有本田研究所中景观和城市紧密结合的设计理念，包括互相碰撞产生鲜明空间对比这样一些理念。那么我想问大师，就是说在这样一些设计中，是不是这样的结果是跟您这样一个建筑师的背景有关系？这是问题的一个方面，就是建筑和景观在设计当中是怎样去结合的？第二个就是可能我们现在面临更多的一种情况，就是在景观设计的时候已经完成了规划设计，完成了建筑设计，那么在这种情况下我们景观设计如何跟建筑来形成一个完整的景观整体？

三谷徹（以下简称“三谷”）：我是建筑背景，所以对景观与建筑两者都非常感兴趣。朱老师的提问如果用一句话来回答的话，那就是建筑是“存在的事物”，而景观是“消失的事物”。建筑的形态是显而易见的。景观由于其广阔性加之没有界限，所以不易于被察觉。就一棵树下的美丽树荫来说，大家都不知道在这种情况之下的设计范围到底在哪里吧。此外，究竟是谁设计的、亦或是小鸟播下的种子生长而成的，没有答案。这样的消失方法十分有趣。

如果先回答第一个问题的话，我经常下意识地认为景观应该包含建筑，也就是说景观的尺度应当远远大于建筑。所以我不会将建筑的比例用于景观。我仅想从非常大的尺度与非常小的细节这两点来处理景观。

接下来就第二个问题，也就是景观与建筑的结合来谈谈我的看法。

all of you today. Mr.Mitani introduced several of his own projects today, and these projects inspired me a lot. I really want to discuss with him about some important things. Firstly, it is the relationship between landscape and cities as Mr.Mitani mentioned just now as well as the integrated design of architecture and landscape, etc. We know that Mr.Mitani studied architecture in the undergraduate, postgraduate and PhD stages. I also learned architecture but work in the landscape field now. So I want to discuss with Mr.Mitani two aspects of one question. On one hand, we can see the exquisite lines, scale, and proportions in Mr.Mitani'sprojects as just showed, including the interaction between landscape and spaces. Just like these two projects mentioned before, one is the entrance lobby of Izumo-Oyashiro, which perfectly shows the permeability of the relationship between spaces, the other is the integration of landscape and city design concepts in the HONDA research laboratory, including some concepts such as crash spaces to create obvious contrast of spaces. I would like to ask if is there any relationship between this prospect and your background in architecture? How do you combine architecture and landscape in your design? The second situation, which probably happened more in our careers, is when the city planning and architectural design are already complete when we start the landscape design. In this case, how can we create an integrated landscape in unity with buildings?

Toru Mitani (Name as 'Mitani' below) : I have an architecture background, and I am extremely interested in both landscape architecture and architecture. If I were to use one sentence to describe teacher Zhu's question, it is that architecture is something existing and the landscape is something disappearing. The shape of buildings is obvious, but not of the landscape. Because of the boundlessness of landscape, it is not easy to define a clear edge. Take the beautiful shade of one tree for example, it seems no one can define a clear edge to it. Besides, there is no answer as to who designed this shadowor was responsible for a tree that has grown after being seeded by birds. This phenomenon of disappearing is really fascinating.

To the first question, I usually think landscape contains architecture, by that I means, the scale of landscape is much bigger than architecture. That is why I would not use the scale of architecture in landscape. I just want to deal with landscape in two aspects, both huge scale and tiny detail.

To the second question, let me share my opinions concerning the combination of landscape and architecture. To think of how to make the inner and outer spaces of buildings continuous is the most essential thing. It means considering how to design transitional areas. I hold the opinion that much attention is also paid to the relationship with outdoor spaces in Chinese traditional architecture. Take screen for example, the spaces are separate but continuity is maintained. In Japan, the semi-outdoor spaces created by eaves and Verandas are also representative of this to some extent. As far as I amconcerned, the method to connect the exterior and interior has been established in each respective culture, over time.

Zhu: I definitely agree with Mr.Mitani in the interaction of architecture and landscape. In fact, it may be easier if we combine architectural design and landscape design together.

我觉得最重要的是如何使得建筑的内外部空间具有连续性，也就是如何设计中间领域。我想中国传统建筑也应该是很注重与外部空间的关系。就像是用屏风等在分割空间的同时又具有一定的延续性。在日本，屋檐与走廊所组成的半外部空间具有一定的代表性。我想，诸如此类的内外部连接方法在各国的文化及时代样式背景中得以确立。

朱：我非常同意刚才三谷先生讲的建设和景观互动的观点，实际上如果建筑和景观一体化设计可能这个问题相对容易解决一些，我们遇到的困难就是因为建筑已经形成，它的封闭性已经形成，它自身独特性已经完成，那么我们景观再去做这样一个互动可能有一定的难度。

三谷：这是一个具有挑战性的课题。现代建筑有一个很大的特征、同时也是一个课题，那就是玻璃幕墙所形成的“透明性”。在视觉上形成了最大限度上的连续性，但是完全遮断了人与空气。建筑与大地的结合部分最为重要，我每次都就这个结合部分与建筑家做缜密的研讨。但是我认为现代景观当中还没有找到与玻璃幕墙的理想相处方法。

沈守云（以下简称“沈”）：很高兴听三谷先生精彩的演讲，刚才在演讲里面听到他讲了一句话，就是感到很快乐。他说实际做这个设计就像是在大地上涂鸦，那么涂鸦以后感觉人留下一些美好的东西，他在这里面感觉到很大的快乐。那么我们知道，咱们任何一个项目，任何一个事情，任何一个工程，它的地域性是不一样的，没有一个是一样的，从欧洲的到美洲的，从日本的到咱们中国的都不一样。那么我就是想我们一直碰到这样的话题，对于园林怎么显现地域性的东西？地域性包括地貌、文化、宗教等很多，都是地域方面所包含的。我想跟三谷先生一起探讨地域性的观点你有哪些新的理解，一些感受，另外一个我们大家都知道这次第九届园博会您有 2,600m^2 的园，我们都想知道三谷先生将怎么涂鸦，让中国人都感到非常的快乐。

三谷：关于园博会的庭园还是一个秘密。此外，请大家为我在日本的项目中在地上作画的事保密。

地域性是一个非常重要且有趣的话题。对我来说最让我能感受到地域性的就是植物。我想没有比植物更能够反映地域气候风土的素材了吧。我从建筑转到景观学的时候，发现了一件有趣的事。那就是哈佛的著名景观学教授去了西海岸就成了一个业余爱好者，也就是说植物完全不同。所以我在做设计的时候会非常重视当地的植物专家的意见。此外，我所关注的另外的一个地域性就是在雨水的处理以及在排水设计上随着地域与文化的不同有着很大的差异。打个比方中国与日本的降雨方式有所不同。日本是长时间的缓慢降雨，而北京则听说是下雨的次数少但是雨量比较大。所以说由于排水设施的不同导致了地面的表情也会产生差异。排水对于景观来说是一种构造物，地面的氛围也会自然地产生差异。

沈：谢谢。在中国，地域性的概念不一样，中国从南到北差异性很大，我觉得我们这一块以后要加紧这个研究。三谷先生昨天把他的作品模型从日本带来了，就是为了分析这个地区，为了表达这个地域的概念，就是为了在中国留下美好的东西，谢谢。

包志毅（以下简称"包"）：今天非常高兴，听到三谷先生的报告，我的背景和三谷先生还不太一样，所以我觉得听到报告有很多启发，当然也有很多碰撞。因为我以前主要从事园林植物的研究，然后大概在 1999 年开始就尝试怎样应用植物进行规划设计，所以可能考虑自然的，考虑生命的因素更加多了，有这样一个机会我很高兴，要向三谷先生请教几个问题，可能问题会比较多一点。我想请教第一个问题，就是我看三谷先生设计，普遍的来讲植物材料种类用地比较少，也就是多样性是比较少的。第二个我觉得总体用的数量，特别是树木这一块，用量也比较少，那么我很想向三谷先生讨教，出于什么样的考虑运用这种植物景观的模式，这是第一个问题，先这个吧。

The problem we face now is that because architectural design creates a feeling of enclosure, the unique characteristics are all contained within. Thus it is probably a little bit difficult for landscape architects to interact with them.

Mitani: It is a challenging issue. Contemporary architecture has a notable characteristic, which is also a big issue, being the transparency created by glass walls. They create a maximum visual continuity, but also totally separate humans from the outside air. The design areas that connect the architecture with the land are the most vital parts. I discuss this very carefully with architects every time. But I do not think contemporary landscape design finds an ideal approach to get along with glass walls.

Shouyun Shen (Name as 'Shen' below) : I am really pleased to listen to Mr.Mitani's wonderful speech. I noticed one sentence he mentioned just now, it is that he feels happy. He said that designing is, in fact, like scrawling in land. He feels extremely happy when such scrawling can leave something glorious in the world. As we know, any project, or any program, has its own regionalism. The projects from Europe to America, or from Japan to our China, are all different from each other. So I was wondering how we can show regionalism in our landscape project?Regionalism contains topography, culture, etc. They are all included in regionalism. I want to discuss with Mr.Mitani about what his new understanding and feeling of regionalism is. Another thing is that you have a garden of 2600 square meters at the 9th Landscape Architecture Expo. We all want to know how you are going to scrawl on it, and make Chinese people feel happy.

Mitani: The garden in the Landscape Architecture Expo is still a secret. And please keep the secrets that I painted in my project in Japan!

Regionalism is an important and interesting topic. In terms of myself, plants are the most obvious symbol of regionalism. I think nothing can be better than plants as the material to reflect on regional climate and culture. When I changed my major from architecture to landscape architecture, I found an interesting thing. That is, if a famous landscape professor goes to the western coast, he becomes an amateur,because of the east coast plants there are totally different. That is why I pay a lot attention to opinions from local plant experts when I am working on my design projects. Aside from this, another aspect of regionalism I am concerned with is the infiltration of rainwater and the design of the drainage system. They are equally as different from each other as the varied regions and cultures. Take the characteristics of rainfall in China and Japan for example, they are different. In Japan it rains slowly over a long period of time. However, in Beijing it rains less frequently but with a higher capacity. The different drainage infrastructure results in varied appearances of land surface. As drainage is a kind of landscape in frastructure, the sense of land surface will be unique as well.

Shen: Thank you. The concept of regionalism is different in China, as there are huge differences from the south to the north. I think we need to intensify our research on this phenomenon. Mr.Mitani brought his model from Japan to analyse this region, show the concept of regionalism, and leave something wonderful in China. Thank you.

6

图 04-06 奥多摩森林浴林道
Fig 04-06 Therapy trail in Okutama
设计版权：studio on site
Project Design Credit：studio on site

7

三谷：我经常被问到关于如何运用植物之类的问题。确实如您所说景观与建筑最大的不同就是在于景观使用植物形成空间。此外，植物也非常重要。植物聚集在一起会形成柔软的、具有幻想性的空间。就像是云一样的得以显现的空间感觉。这是建筑绝对模仿不了的。

关于如何进行植栽设计这一个问题，我的方法就是与其注重多样性还不如彻底追求一种植物素材的可能性。这可能与我对于景观中的现代主义感兴趣有关。现代主义将环境根据其形成要素将其分开，试图看透各个要素的本质。科学与艺术也是一样。我经常所做的就是想尝试使用一种树木来如何创作空间。

即便是只有一种树木，无论是 A 植物还是 B 植物，也会形成不同的颜色、不同的树枝密度、不同的树荫，我希望人们能够彻底地感受到这些差异。我就是在做这样的实验。

在做景观的时候，最困惑的就是觉得只要有大量的植物在，大家就会高兴这样的观点。我一直希望简单明了地表现通过植栽是在尝试怎样的空间设计。植物的种类以及数量也尽可能少地使用，试图发现其中的可能性。我现在就像是在为钢琴或者是小提琴的独奏作曲，当然 20 年以后我想会尝试为所有的乐器谱写交响乐。我现在还在学习之中。

包：这个问题我想进一步往下探讨，因为讲的都是围绕特殊的场景，比如说火葬场，比如说美术馆，是一种围绕建筑为核心做的景观，所以我比较理解用比较简洁的手法来表示场地的特殊性，但是假如说它是公园的话，我不知道因为今天没有介绍这种类型的项目，我不知道三谷先生是用什么样的一种办法来做。因为今天我们前面看到几个项目都是围绕特殊的产品，围绕建筑来做的，没有看到就是像我们在城市中公园的形态，这是不一样的，所以我想问三谷先生的意思就是

Zhiyi Bao (Name as 'Bao' below) : I'm very happy to listen to Mr.Mitani's report today. My background is not the same as Mr.Mitani, so I get a lot of inspirations and of course at the same time see many contrasts with my own work. I began mainly worked on the research of garden plants, and then I started try to plan and design with plants in about 1999. So I probably think more about natural and biotic elements. I am really happy to have a chance like this to ask questions to Mr.Mitani, maybe I will pose a couple more questions. The first question is, in Mr.Mitani's design projects, the number of species of plants is generally quite low, and therefore have a low level of diversity. And I also think the total number of plants, especially trees, is relatively low. I really want to consult Mr.Mitani as to the strategy behind this mode this planting. That is the first question.

Mitani: I have been asked questions like this about how to use plants quite often. The biggest difference between landscape and architectural design is indeed that landscape uses plants to create different spaces, just like you said. Besides, plants in themselves are also very important. The grouping of plants can create soft and imaginative spaces, like the sense of a cloud. It is this characteristic that architecture cannot imitate at all.

In terms of how to design a planting plan, my method is to pursue the different possibilities of one species thoroughly rather than focus on the variety of them. It is probably related to my interest in modernist landscapes. Modernism detached the design

说假如说有机会做一个城市公园的项目，它的景观想怎么做呢？如果做过它是怎么考虑的？

三谷：实际上刚才讲到的风之丘火葬场也是公园。美术馆周边的景观也是公园。都是在城市规划范畴内的城市公园。

包：我讲的是没有主体建筑的景观项目。

三谷：是的，我想我也会用同样的方法。在城市公园中，有各种各样的人有着各种各样的使用方法，如果都去逐一解决的话就会成为一个各种功能的结合体。与其这样，我想还不如首先赋予他们一个空间，让他们自由使用。

最重要的应该是让人们共有来自空间体验的记忆。也就是说公园必须有一个能够令人感动的主要空间。老人的使用、带孩子的家庭的使用等等，可以在周边用一些小的心思予以附加。我想在这种情况下的共有空间就成为了没有明确的实际意义，或者是没有实际功能的空间设计。

庄子的教诲中，有“无用之用”一词。日本人在小学的时候就开始学习中国的传统思想。“无用之用”一词，对于景观来说也是非常重要的教诲。

章俊华（以下简称“章”）：我补充一点，日本人情况可能跟中国人也一样，就是做景观有本身背景，有一开始景观出身背景的设计师，还有从雕塑转到这个行业里的，所以各个背景不一样，每个人的风格也不一样，三谷先生从建筑背景转到园林背景来说算是很成功的。我想提一个问题，这个问题就是每人做设计的时候一开始的出发点，或者是最重要的东西，可能刚才说的每个背景都不太一样，但是我想三

from the environment using the elements that form it. Science and art are the same. What I sometimes do is try to create spaces with only one species of tree.

Even with only one species of tree, no matter if it is plant A or plant B, they will result in varied colour, density of branches, and shadow. I hope people can truly feel these differences. That is the experiment I am doing.

When I am working on landscape, the most confusing thing is that if there are a large number of plants there, people will feel happy. I was always trying to make it obvious that designing with plants is designing with space. I try to find the possibilities behind plants within a minimum number and quantity of species. What I am doing now is like writing solo piano or violin. I think I will definitely want to write symphonies for all instruments 20 years from now. I am still in the learning stage now.

Bao: I want to discuss about this problem further, because it is mainly focused on special areas like crematorium and galleries. This kind of landscape put architecture at the core, so I understand the approach of using simple ways to present the special characteristics of it. However, if it is a park, I do not know how Mr.Mitani would deal with it, because he has not introduced any projects like that today. It is because these several project introduced today are all in special areas and designed around the architecture, the form of them is different from what we can see in a park, so I want to ask that if Mr.Mitani had a chance to design a city park, how would he want to deal with it? And if you

图 07–08 中津市风之丘殡仪馆
Fig 07–08 Kazenooka Nakatu
设计版权：studio on site
Project Design Credit：studio on site

9

谷先生在做设计的时候，您认为您最最重要的出发点，或者最重要从头到尾贯穿下来的，你觉得最重要的核心是什么？能不能简单的给予介绍。

三谷：章老师比较了解我，所以说章老师的提问是最难回答的。但是，作为设计核心，一直有所考虑的有一点，那就是，与其说是为了现在的人们创造空间，还不如说是为了百年以后的人们留下信息。

当然，首先应该满足现在的人们的要求，制作相应的活动项目。但是，50 年后这些项目是否有必要，谁都不知道。如果说我的作品在 50 年后没有上述的活动项目以及没有目的的情况下不是一个好的作品的话，就成为了一堆大型垃圾。

50 年后的人们如果能够在那里发现一些有趣的形状，然后又将那些形状另作解释、或者是另作他用，那我就会感到十分欣慰。

章：谢谢，实际上我补充一下三谷先生的作品，我认为他的特点就是从空间来做，他怎么从空间考虑做这个东西，这个我个人认为是最大的特点。

王：三谷教授是在日本接受大学教育，之后到美国深造，并且和美国和很多公司合作过。请教三谷先生您如何面对东西文化的差异，并将他们融合及转换成就您的作品。

三谷：我不希望被认为是我是日本人就应该做日本式的作品。但是，与之前提到的地域性的问题相同，在一个地方，如果不学习当地

have already done some projects like that, how did you think about it at that time?

Mitani: The ‘Crematorium of kazeno-oka’ mentioned just now is actually a park. The areas near the galleries are also parks. They are all city parks in terms of the city planning.

Bao: What I mean is a landscape project without main buildings.

Mitani: Yes, I think I would use the same approach. Different people use city parks in different ways. If we deal with this problem one by one, the park will become a combination of different functions. Instead of this, we had better give people a space to use however they want.

The most important thing is letting people have the experience of sharing the same spaces. It means even a park must have a main space that touch peoples spirits. It provides usage for the elderly and young families with children, etc. We can consider about some details to support the main space. I think in this circumstance, shared spaces become spaces with no specific meaning or specific function.

There is a saying by Zhuangzi: “ useless things can be useful”. (We Japanese learned the traditional thinking of China in

图 10 大阪西梅田广场
Fig 10 West Umeda Plaza, Osaka
设计版权：studio on site
Project Design credit: studio on site 10

传统的排水方法以及屋檐的制作方法、树木的种植方法等，设计庭园以及景观是不现实的。就算像是我虽然在日本从事工作，如果去日本的古庭园的时候，也会对以前的人们所做的努力而感到惊讶。设计手法适应日本的风土，通过简单的技术完成了许多设计上高难度的细节。日本庭园中，对于飞石的意匠有着各种各样的研究。日本是一个多雨的国家，由于地面潮湿，就有必要保证人们在穿着和服走路的时候不易被淋湿、而且易于行走。这在技术上就发展成为了飞石，而这些飞石有时候又形成了小河的形态。我对于这些努力非常感动，并且自发地运用这些技术。这些是不是日本式的我也不太清楚，就好像是地域风土所要求的技术，从而使得设计文化得以传承。我不知道是否是回答了您的问题。

王：是的，谢谢。接下来请在座的来宾们提问。

何友锋：由于气候变迁，全世界的动植物受到很大的伤害。根据专家和 IPCC 国际组织的研究认为，地球温室效应的主要原因之一是地球释放太多 CO_2，因此地球上的人们纷纷大量种植能够吸收 CO_2 的乔木。根据统计，乔木一棵 40 年每平方米可吸收 CO_2 1,200kg，草地大概 20 kg，您的设计大部分是用草地，是否表示您的设计对降低 CO_2 的贡献比较少？另外，地球气候变迁，环境异常，造成世界各地重大自然灾害，请教以极简的景观设计手法，如何有效应对日益严重的气候变迁？

三谷：显而易见环境问题对于景观师来说是一个非常值得重视的

elementary school.)This saying: " useless things can be useful" is vital instruction for landscape design.

Junhua Zhang (Name as 'Zhang' below) : One more thing I want to mention here, perhaps the situation in Japan is the same with China. The situation is some landscape architects have landscape background, while some other transferred from other majors like sculpture. Different backgrounds result in varied styles. Mr.Mitani is really successful in terms of transferring his architectural background to landscape architecture. What I want to ask is that everyone has a starting point or something he focuses on most when he designs. Everyone has a different background as I mentioned before,so I want to know what is the most essential starting point for Mr.Mitani, or what is the most vital thing from start to the end? Could you please talk about that?

Mitani: Teacher Zhang knows me very well, so the question he asks is the most difficult to answer. But at the core of design, we always consider about one thing; that we need to leave some messages to our descendants rather than simply creating spaces for contemporary people.

Of course, we need to create suitable activities for contemporary people to satisfy them. But no one knows whether it will be necessary 50 years later. If my design work could not be

图 11-12 东京电机大学
Fig 11-12 Tokyo Deki University
设计版权：studio on site
Project Design Credit：studio on site

问题。但是这个问题的探讨必须建立在理清尺度的前提下。大尺度的地球环境问题对于科学家以及城市规划来说是非常重要的。

今天的讲演主要是围绕小尺度的设计。在这样的小尺度空间设计中，我与其说是在改善物理环境，不如说时常想到的是如何使得人们的意识向环境方面转移。特别觉得在日常生活中应当常有作为孩童们生长场所的、能够接触自然的、普通的庭园。

朱：其实我觉得问题的关键是像刚才何教授提的问题我们应该怎样去看？中午跟三谷先生探讨的主要话题是生态方向，我非常同意三谷先生的一个观点，就是我们在不同的环节去解决不同的问题。在小尺度的环境里面可能我们更多的是需要解决人们的参与性，我们提供给周边使用的人们什么样的愉悦的空间环境，这是这个尺度要解决的基本问题，或者是首要的问题，当然在这个过程当中我认为还是可以运用一些生态理念。比如说刚才三谷先生做的一些东西也是运用了生态的理念，实际上是很好的一种生态观念，只不过他是从形态方面而没有从这个角度去诠释这个问题，实际上他用到的手段已经跟我们现在生态理念的基本概念相吻合的。种草皮的问题我也确实有不同意见，因为草皮确实它在生态效益方面作用相对来说要小一些，并且它的维护费用，或者维护的环节要很多，增加很多的负担，尤其是缺少淡水的城市。虽然草皮是我们景观园林非常重要一点，因为它的平坦性是一般的植物达不到的效果，它对建筑的衬托作用是一般大型植物达不到的，所以我们在景观当中会大量运用这个草坪，但是在当今的环境情况下要慎用。在生态上面要有一个平衡，刚才三谷先生说的角度，大量尺度还是我们需要更多地去探讨的相关问题。

章：我补充一下，日本的环境本身绿地覆盖率就比较高，所以在城市里新增绿地的话还是希望要开放的绿地，要开放怎么做呢？可能三谷先生的手法正好比较适合日本城市发展的一个潮流吧，谢谢。

useful in 50 years, they will become a mass of rubbish.

If people could find some interesting forms, give new explanations, or take some other aspects of my design works, I would feel very delighted about it.

Zhang: Thanks. Let me add something more about Mr.Mitani's projects. I think his design strategy is to starting from spaces and how he thinks about a project begins from the perspective of space.

Wang: Mr.Mitani received his undergraduate study in Japan and went to American for further study, and then cooperated with many companies. I want to ask Mr.Mitani how can you combine eastern and western culture and transfer them into your work when you face the gap between them?

Mitani: I do not want to think that I need to design Japanese style projects because I am a Japanese designer. However, if we do not learn local drainage practices, plants or things like that, it is not realistic for us to design garden and landscape of one place, just like the former question about regionalism addressed just now. Even when I work in Japan, I am surprised by predecessors' efforts when I go to classic gardens. Their design approaches are suitable for Japanese local climate and context, and finish highly complex design details through simple techniques. In Japanese gardens, they there is a wide array of research on the functions of the Tobiishi. Japan is a rainy country, because of the wet ground cover, it was necessary to ensure it was convenient for people

2007.07.24 12:10:40

who dressed in kimonos could walk and not get wet in the rain. In terms of techniques, they developed into Tobiishi, and sometimes these flats took the shape of streams. I am really moved by these efforts and use these techniques spontaneously. I am not sure if it is Japanese style or not, it is like the requirements of a local physical condition, continue into the required techniques of the culture of design. I do not know if I answered your question or not.

Wang: Yes, you have. Thank you. Now would our honoured seated guests please ask questions.

Youfeng He: Animals and plants all over the world have suffered due to climate change. The research of specialists and IPCC international organisation indicate one of the main reasons is the huge emissions of carbon hydrate from the earth. That is why people plant a large number of big trees that can absorb carbon hydrate.

According to the statistics, every tree can absorb 1200 kilograms of carbon hydrate per square meter in 40 years, and grasses can absorb about 20 kilograms in the same situation. The majority of what you use is grass, do you think your designs contribute little to reduce the quantity of carbon hydrate? Also, climate change and unusual environmental conditions result in serious disasters all around the world. Could you please share some extremely simple design approaches to deal with the increasingly severe climate change?

Mitani: It is obvious that environmental problems are issues worth a lot of attention for landscape architects. But the discussion about this problem must be based on a clear scale. Large-scale environmental problems are most important for scientists and city planners.

Today's speech is mainly focused on small-scale design. I improve the physical environment, at the same time, I usually think about how to remove people's consciousness of an environment in a small-scale designed space like this. Especially the spaces for children to grow up in, so they can get in touch with natural and ordinary gardens.

Zhu: In reality, I think the core of this problem is how we think about it. What I discussed with Mr.Mitaniwas mainly about ecology. I seriously agree with his opinion that we need to solve different problems at different scales. Probably we need to work more on the participation of people in small-scale projects. How we provide pleasing environments for people to use is the fundamental and primary problem of projects on this small scale. Of course we can use some ecological concepts in the processes. For example, what Mr.Mitani introduced to us just now also uses the concept of ecology, and is actually a good ecological concept. He just did not express this consideration in the morphology, but the methods he used are actually the same with fundamental ecological concepts. I also hold a different attitude on the use of grasses. It is because grasses are of little ecological benefit. They also have a cost in terms of maintenance or management which is often higher than with other alternatives in cities which are short of water. Even though grasses are very important in landscape design, because of the outstanding sense

图 13-14 东京银座斯沃琪大楼
Fig 13-14 Nicolas G. Hayek Center
设计版权：studio on site
Project Design Credit：studio on site

14

王：从不同角度看待不同的问题，这是非常精彩的对话！

提问者一：我从三谷先生身上看到一个很朴素、干事很踏实的日本人，也看到一个传统的日本设计师，请讲一下您的创业史或者是您自己的人生哲学，谢谢。

三谷：作为景观师，畅想自己的未来是一件非常重要的事。我的答案就是不要一次性的有太多的欲望。比方说我的实际情况中，大学会让我搞科研、写书等等，会有各种各样的事。人生是有限的。所以我想慢慢地、好好地做我的设计。

沈：我讲一句，今天中午跟三谷先生去吃饭，电话打了十几次还没来，他说就在调试演讲的屏幕，饭吃了几口，他又跑回去调文件，不吃了，我想这就是敬业。

章：我补充一下，刚才说的一句话，他专注的做一件事情，他的答案是这个，但是他成功我觉得是必然的，为什么必然的呢？看他的简历了，因为他受过的教育可能是最好的教育，这是第一个。他最初是与槙文彦一起工作的，就是当时他在东大上学的时候是一个学生跟老师的关系。你要知道槙文彦是一个什么概念，他曾获得了普利兹克奖，就相当于建筑行业的诺贝尔奖。所以要成功一定要跟大师一起做，这就是成功的秘密。

王：除了三谷先生的专一之外，就是他很坚持，中午我们找他吃饭他就是不来，因为他要把下午的报告的每一张图片都要弄到最好，所以，他坚持的精神也是成功的因素之一了！

of smoothness and flatness they create. That is why grass is so often used in landscape design. But we need to be more careful to use it in situations at present. We pursue ecological balance. What Mr.Mitani mentioned about scale just now, we still need further discussion about some problems related with it.

Zhang: Let me add a little bit more in support. The greening rate of Japan is relatively high, so if we want to create new open green spaces in cities, what we can do? Perhaps the method Mr.Mitani uses corresponds with the trend of urban development in Japan. Thanks.

Wang: Considering different problems from different perspectives. What a brilliant dialogue it is!

Questioner 1: I can see an unadorned, traditional Japanese designer in Mr.Mitani. Could you please talk about your career history or your philosophy of life? Thanks.

Mitani: My answer is that one should not have too much desire at once. Take myself for example, the university let me do different things like research, writing books, etc. But life is limited. So I just want to work with my design slowly and carefully.

Shen: We had lunch together with Mr.Mitani and we called him ten times before he came for lunch. He told us he was adjusting

图 15-16 福井县立图书馆
Fig 15-16 Fukui Prefectural Library
设计版权：studio on site
Project Design Credit：studio on site

提问者二：请问什么原因步行空间项目做了 14 年，从开始的概念到后来的实施有没有一些变化？谢谢。

三谷：那个空间看上去是一个整体的公园，但是实际上是由十几个民间地块所组成的。并非公共空间，而是私人性庭园的一个组合。这些企业的意见统一（基本规划）用了 10 年，设计与施工用了 3、4 年。我自始至终主张的就是应当用树木将那个空间堆满。

在这样大型的工程项目会议上，发生了非常令人困惑的事。所有企业的社长们对于一体化设计的想法都觉得非常优秀，表示赞同。但是会议结束以后，会把我叫到其他的办公室商议是否可以把自己大楼前的设计得特别一点。也就是说形成了总论赞成各论反对的局面，大家都想在自家门前做不一样的东西。这也是现代城市问题的缩影。

提问者三：请问您看北京公共环境树有哪些优点，以及北京哪些地方您感觉不太好？

三谷：有一点十分感动，那就是连接机场与北京的绿带以及环状绿带等等，大型的绿色基础设施正在与城市建设同步进行。这对于现代城市的营建来说是非常重要的。那种大规模尺度的景观在日本是难以实现的，非常羡慕。反观在小尺度及其质量上就感到有些遗憾，就好像是设计方与施工方在做不同的工作一样，设计意图与施工精度不一致。将设计师的设计图纸百分百加以实施的施工方不是好的施工方。最重要的是能够与设计师进行沟通，将重要的部分予以好好施工。看了北京的工程项目，有时候觉得是否施工方没有理解设计方的意图。

the screen for his speech. After a short time for lunch, he came back for adjustment again. I think it demonstrates his dedication to his work.

Zhang: I think his success has been inevitability. Why? After looking through his curriculum vitae, firstly, the education he received is the best. As well, he worked together with Zhen Wenyan initially. It was the relationship between a teacher and a student, when he studied at the University of Tokyo. Zhen Wenyan won the Pritzker prize, which is essentially the Nobel Prize in architecture. So the secret of success is working with great designers.

Wang: Along with Mr.Mitani's focus, it is also because of his perseverance. We called him to have lunch together, but he didn't come. That is because he was busy with arranging the images for his speech in the afternoon. So, his spirit of persistence is also one of the reasonsfor his success.

Questioner 2: What is the reason of the project Walking Spaces took 14 years to complete? Was there any change from the initial concept to the subsequent implementation? Thank you.

Mitani: That space looks like a whole park, but actually it was

章：我补充一下，他认为可能是设计和施工的接口有问题，但是这个问题出在什么地方他搞不清楚，所做的东西不是像图纸反应的那么好。我自己是设计师，我以前也觉得只要是做不好就是施工的问题。但是最近我做新疆的项目，做完以后有一点改变，我觉得再差的施工队，再恶劣的气候，只要你设计到位，只要你现场跟到位都能够做出好东西。

提问者四：请问三谷先生在比较密集的城市建筑区来做景观规划的时候，您主要考虑哪些方面？这种类型的景观设计需要哪些特点呢？

三谷：不是树木，我采取了将地面与建筑立面的微小倾斜予以强烈对比的手法。所以，我没有在建筑的周边配置多余的植栽。此外，我有时会在地面中刻意地放入完全水平的物体。我没有在幻灯片中介绍的诸如"三原文化中心"以及"福井图书馆"项目中，广场的正当中嵌置了供游人歇息的、完全水平的水镜以及平台。由此能够发现那样的微小倾斜就会成为一种错位。与先前的尺度问题相关联，景观中非常有趣的事情就是这样的数厘米的错位以及高差，对于大尺度的风景有着非常重大的影响。这可能与水平眺望这一种景观中人们特有的视线有关。

三谷：六本木（Roppongi）中城开发与品川（Shinagawa）中央公园的类型是不同的。品川是大家一致同意建造一个大型空间而来的。而六本木则是一个将剩余的土地逐一缝合加以连接的项目。具有历史性的城市原本就是一个建筑密集的场所，如何将建筑之间的空隙加以连接，使之成为一幅新的图案这也是景观师的工作。

章：补充一下，实际上你说的问题每一次讲演会每个人都会问。这应该怎么做？实际上这个问题三谷先生没有一个标准答案，同样是城市但内容都不一样，地块不一样，而且投资者也不一样，都不一样。

提问者五：请问当您和甲方的景观设计有冲突的时候，您如何坚持自己的景观理念？

三谷：我也想知道问题的答案。如果说是有一个方法的话，可能如下所述。满足甲方的梦想以及想法是当然的。但是，可以将自己的主题隐藏在其后进行设计。比方说，我的主题是针对百年以后的地上巨画等等，也就是说作品中包含了双重的价值。

与被要求庞大功能以及程序的建筑不同，景观所被要求的不是一个城市性层面上的功能。比方说安乐或消磨时间等，是一种人们心理状态的创造。所以说，只要不偏离这一点，那就与甲方一般没有冲突，同时也可以赋予自身的价值。

王：年轻人怎么样说服甲方就是你自己要有一些墨水，才能让他接受，这是起码的条件。

章：这是比较大的一个课题吧，包括我，包括王老师一样一定要满足甲方的要求，你在满足的情况下去表现你的设计，这是最好的方法。

提问者六：请问您是怎么把西方的理念引用到亚洲来的，怎么让中方人接受从西方吸收的这些思想和做法？

三谷：这是一个难以回答的问题。但是与之前的问题也有个别相同之处。即便是其他的形式以及样式，在一个场所扎根的时候，素材是这个场所特有的物体。在此也体现了景观意味的双重性。形式通常反映理念。素材通常反映场所。

如果想要做好的话，我想可能对于历史的学习是最好的方法。深刻理解、大量了解各个国家、各个时代的庭园设计师究竟是想要实现什么。传统样式不仅是形状的问题，也是空间以及意味的问题。

made up of ten land parcels. It is not a public space, but a combination of private gardens. It took ten years to integrate different opinions from different companies, and it took 3 to 4 years to design and build. I always held the opinion that we needed to use trees to fill that space.

Some confusing things happen in the meetings of such large-scale projects. All the managers of these companies spoke highly of the idea of integrated design. However, when we were alone in another meeting room they discussed with me about whether the design of the spaces in front of their buildings could be unique. It demonstrates the situation where peopleas a group generally agree with an idea, but they are against it individually as they want different things in front of their building. It is a modern urban problem in miniature.

Questioner 3: What are the well-designed aspects of the tree plantings in Beijing? And which aspects are not that good?

Mitani: One moving thing I found is the green belt connecting the airport and Beijing's city centre, I also noticed some ring-like green spaces. The construction of large green spaces is keeping pace with urban development. It is vital for the construction of a modern city. It is unrealistic to do landscape projects in such a huge scale like this in Japan. That's really admirable. When we see these and then look back at our small-scale projects, it feels a little regretful. It seemed like the designers and contracters are working on different things, the intention of designers and the precision of contracters went in different directions. The contracters that construct exactly as the designers' drawings depictare not the best. The most essential thing is that they can communicate well with designers and work carefully with the most important areas. After having a look at some projects in Beijing, I think probably some engineers are not getting the real intention of designers.

Zhang: Perhaps he sees a problem in the connection between design and construction, but he cannot figure out where these problems arise. Some works are not as good as the drawings. I am also a designer, I used to think that it was the problem of construction when some work did not come out as well as I expected. However, my mind gradually changed when I finished the project in Xinjiang. I think no matter how disappointing the workers are or how terrible the weather is, good work can emerge when you design very well and control the construction process correctly.

Questioner 4: What do you mainly consider when you work on landscape planning in projects with high building density? What are the characteristics of this type of landscape design?

Mitani: I use the method of enforcing a strong connection between the ground and the building elevations, instead of trees. That is why I do not use that many trees around buildings. Also, I will put some totally horizontal elementson the ground for some purposes. I placed a totally horizontal platform in the middle of a square for people to a rest in some projects like Mihara Cultural Centre and the Fukui Library, which I have not introduce in this powerpoint presentation. From this situation, we can find that a very minor tilt can result in malposition. Related with the scale issue we discussed about before, it is very interesting that several

图 17-18 国际佛教大学
Fig.17-18 International College for Postgraduate Buddhist Studies
设计版权：studio on site
Project Design Credit：studio on site

图 19–20 町田市新市府楼
Fig 19–20 New government office in Matida City

设计版权：studio on site
Project Design Credit：studio on site

20

王： 我们知道，要作为一个成功的园林师不是一天可以造就的，他需要很多的条件，包括经验，学识等，而且每天要学习，要专注一个原则，今天从大师的谈话中相信各位都能得到很多的启示。由于时间的关系今天的论坛就到此结束，谢谢各位的参与。□

centimetres of malposition in altitude matters a lot in landscape. It is perhaps related to the unique horizontal sight-lines of people related to the landscape.

The project types of RoppongiMidtown and Shinagawa central Park are different. All parties agreed with the construction of a large space in PINCHUAN. However, Liubenmu is a project wherewe sutured various pieces of open land together. Historical cities are areas with high-density buildings. What landscape architects need to do is connect the gaps between buildings and create a new image of them.

Zhang: Actually your question has been asked in every previous symposium. What do we need to do? In fact, there is no standard answer to this question. Although they are all cities, their content, site, and sponsor are all different.

Questioner 5: When you hold contrasting design opinions with developers, how can do you push for your landscape concept?

Mitani: I am always looking for this answer. Perhaps it will be like this, if there is a method. Of course we need to satisfy developers' requirements and opinions. Nevertheless, we can hide our own theme in our design. For example, I focus on the large-scale image that will be imprinted in the land a hundred years from now. It means there is double value in my work.

As opposed to the strong focus on function and process in architecture, the main requirement for landscape is not utilitarian function in the urban sphere.It is the creation of mental feelings, like peace and happiness, or simply to provide a place for killing time. So, if we keep working like this, we won't be at odds with developers and we will give value to ourselves at the same time.

Wang: Young designers should have basic qualifications, so that they can be accepted by developers. It is a fundamental requirement.

Zhang: It is an extensive topic. We all need to meet the developers' requirements, including teacher Wang and myself. The best way is to present asatisfactory design to the developers.

Questioner 6: How can we introduce western concepts to Asia? And how can wepersuabe people to accept these ideas and practices that areabsorbed from the western world?

Mitani: It is a difficult question to answer. But it is partially similar to the former questions. Even though the shapes and forms of a site can be varied, the materials only belong to one site, once they are installed. They also show the dual nature of landscape. Usually, the shapes and forms in a design reflect the concept, and the materials reflect the site.

I think learning history is the best way to deal with it. So that you can deeply understand what garden designers really aimed for in different countries and times. Traditional is not only found in forms, but also in the meaning of the space.

Wang: As we know, it is impossible to suddenly become a successful landscape architect. It requires a lot, including experience, knowledge, learning every day, and keeping focused. I think the people here can all gain inspirations from this speech. And now it has come to an end. Thank you for coming.■

大师报告会对话嘉宾简介：
Dialogues with the guests at the presentation:

嘉宾主持：王小璘 教授／《世界园林》总编
Guest: Xiaolin Wang Professor / Editor–in–Chief of Worldscape

嘉 宾：朱 玲／沈阳建筑大学城市规划与建筑学院书记副院长
Guest: Ling Zhu / Secretary，Associate Professor，ShenYang Jianzhu University Urban Planning and Architecture.College

嘉 宾：沈守云／中南林业科技大学风景园林学院院长
Guest: Shouyun Shen / Professor，Central South University of Forestry and Technology Landscape Architecture College.

嘉 宾：包志毅／浙江农林大学园林学院院长
Guest: Zhiyi Bao / Professor, Zhejiang Agriculture and Forestry University.

嘉 宾：章俊华／日本千叶大学教授
Guest: Junhua Zhang / Professor of Chiba University

大师报告会及嘉宾对话时间：2012 年 4 月 27 日
地点：北京新大都饭店国际会议中心
Date: 27 April, 2012
Place:International Conference Hal of Beijing Capital Xindadu Hotel

张安／清华大学建筑学院景观学系 博士后（日语翻译）
Japanese translated by An Zhang
罗洁梅（中译）
Translated by Jiemei Luo
查尔斯・沙（校订）
English reviewed by Charles Sands

项目设计版权：studio on site
Project Design Credit：studio on site
项目照片拍摄：吉田诚
Photographor：Makoto Yoshida

南京万荣立体绿化工程有限公司

NANJING WANROOF CO., LTD

屋顶绿化／墙面绿化／坡面绿化／水面绿化

绿空间 wanroof

全国统一客服电话：400-025-9800
网址：www.wanroof.com　　www.wanroof.cn
地址：南京市察哈尔路108号　传真：025-5808752

企业简介

公司成立于2007年，是专业从事特殊空间（屋面、墙面、水面、室内等）绿化的设计、施工及相关技术、工艺的研发和推广的高新技术企业，是全国立体绿化工程联盟成员单位、国际立体绿化促进组织常务理事单位。

公司主营轻质屋面绿化系统，墙面绿化系统，及其相关植物、基质、资材、灌溉等相关配套产品的销售与技术咨询服务，在南京建有立体研发中心和生产示范基地500余亩，建有现代化温室、自动浇灌设施和各种试验设备。年可供屋面绿化专用草卷10万㎡，墙面绿化模块2万㎡。

公司秉承“以客户为中心， 以质量求生存 ，以创新求发展”的服务理念，自成立以来，已为南京乃至江苏的政府和企业提供近百次服务，受到客户的广泛赞誉。公司荣获2011年度江苏省特殊空间“扬子杯”优质工程奖；2011年度南京市园林绿化工程“金陵杯”；2009年获 “世界屋顶绿化优质工程奖”等。

轻质屋面绿化系统--绿色屋面

我公司主营的轻质屋面绿化分为生态型、景观型及组合型三种形式。生态型屋面绿化平均荷载≤80kg/㎡，适合荷载小的轻钢屋面和老建筑屋面绿化项目；景观型屋面绿化平均荷载≤120kg/㎡，适合有休闲游憩需要的普通上人屋面绿化项目；组合型介于生态型和景观型之间。

南京市容局
景观型轻质屋面绿化

南京紫东创意园
组合型轻质屋面绿化

南京南大太阳塔
生态型轻质屋面绿化

墙面绿化系统--生态绿墙

我公司的墙面绿化系统分为简易型和精细型两种模式，简易型墙面绿化系统的纵向平均荷载为40-60kg/㎡，适合简单覆绿工程；精细型墙面绿化系统的纵向平均荷载为40-120kg/㎡，适合墙面绿化美化工程。

南京玄武大道环陵路匝道口
简易型墙面绿化

南京万荣立体绿化研发中心
精细型墙面绿化

南京朗诗钟山绿郡
室内墙面绿化

凹陷花园
SUNKEN GARDEN

伊娃・卡斯特罗　王　川

Eva Castro　Chuan Wang

项目位置：中国，北京，第九届北京园博会
项目面积：1,000m^2
委托单位：第九届中国（北京）国际园林博览会组委会
设计单位：普玛建筑设计事务所 | 大地实验室
景观设计：伊娃・卡斯特罗、土川、阿尔弗雷德・拉米雷汝
完成时间：2013 年 5 月

Location: The 9th China (Beijing) International Garden Expo, Beijing, China
Area:1,000m^2
Client: The 9th China (Beijing) International Garden Expo committee
Designer: Beijing Plasma Studio | Ground Lab
Landscape Design: Eva Castro, Chuan Wang, Alfredo Ramirez
Completion: May, 2013

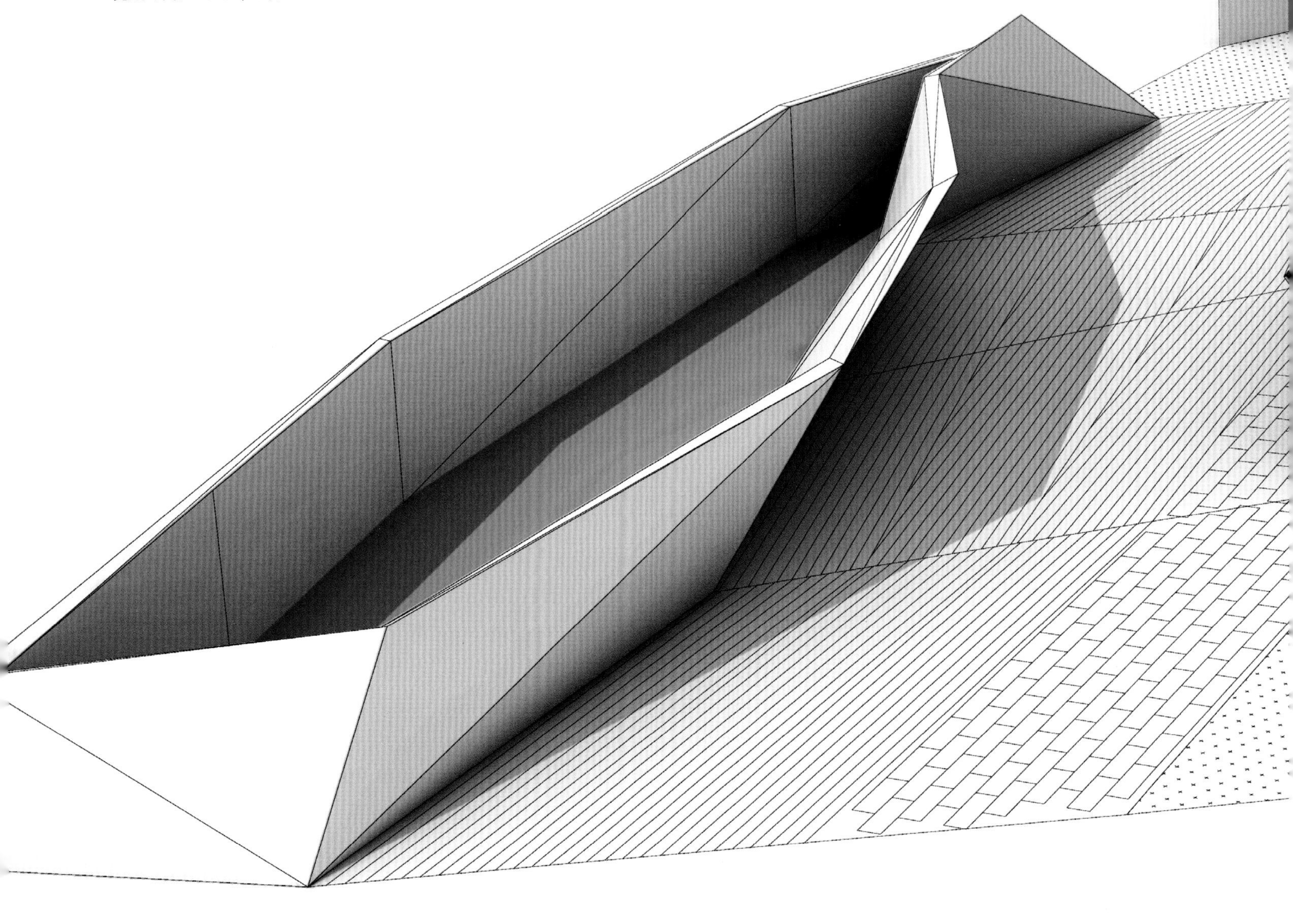

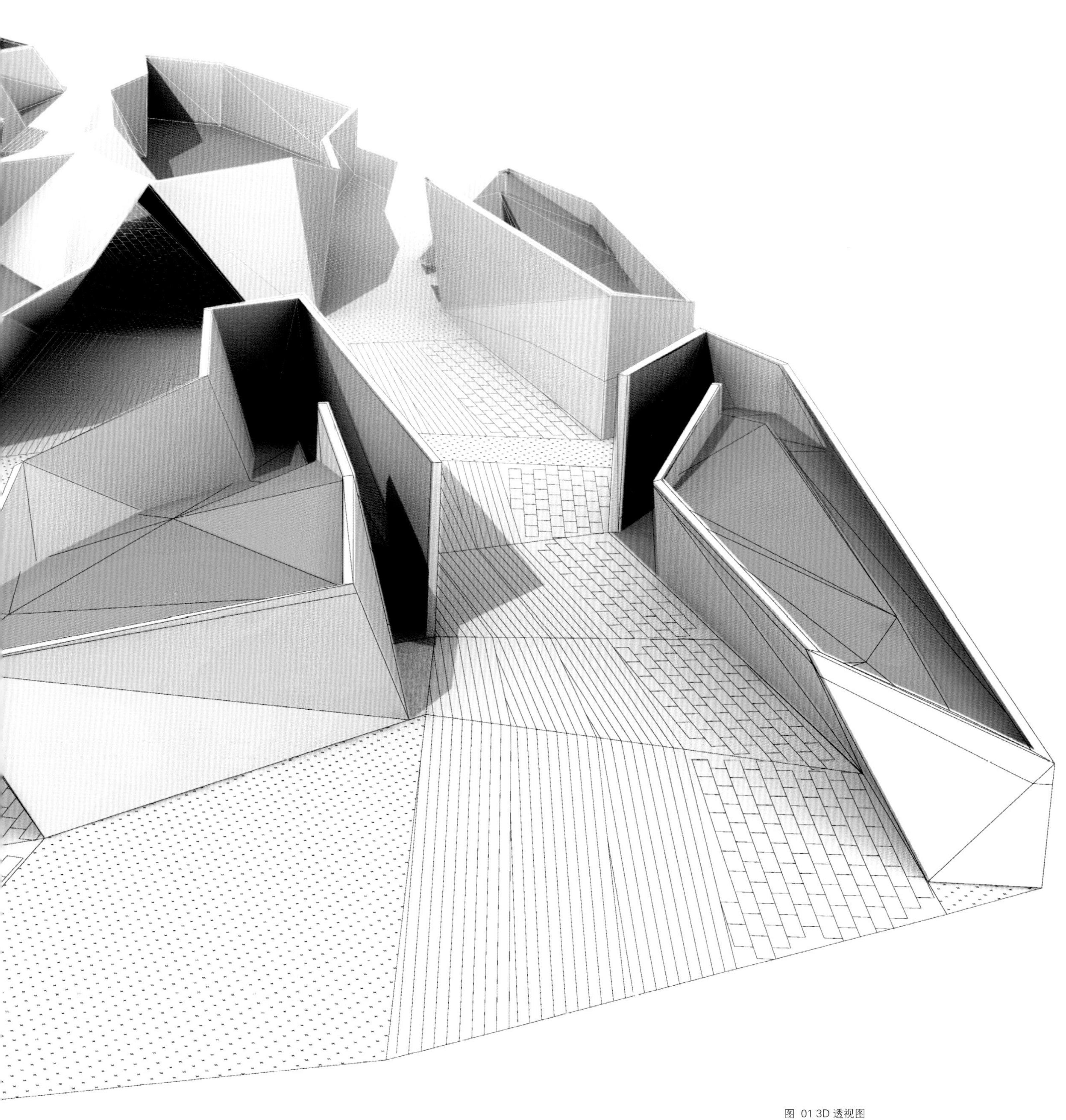

图 01 3D 透视图
Fig 01 3D perspective drawing

注：本文图片由普玛建筑设计事务所提供
Photography：Plasma studio offfered

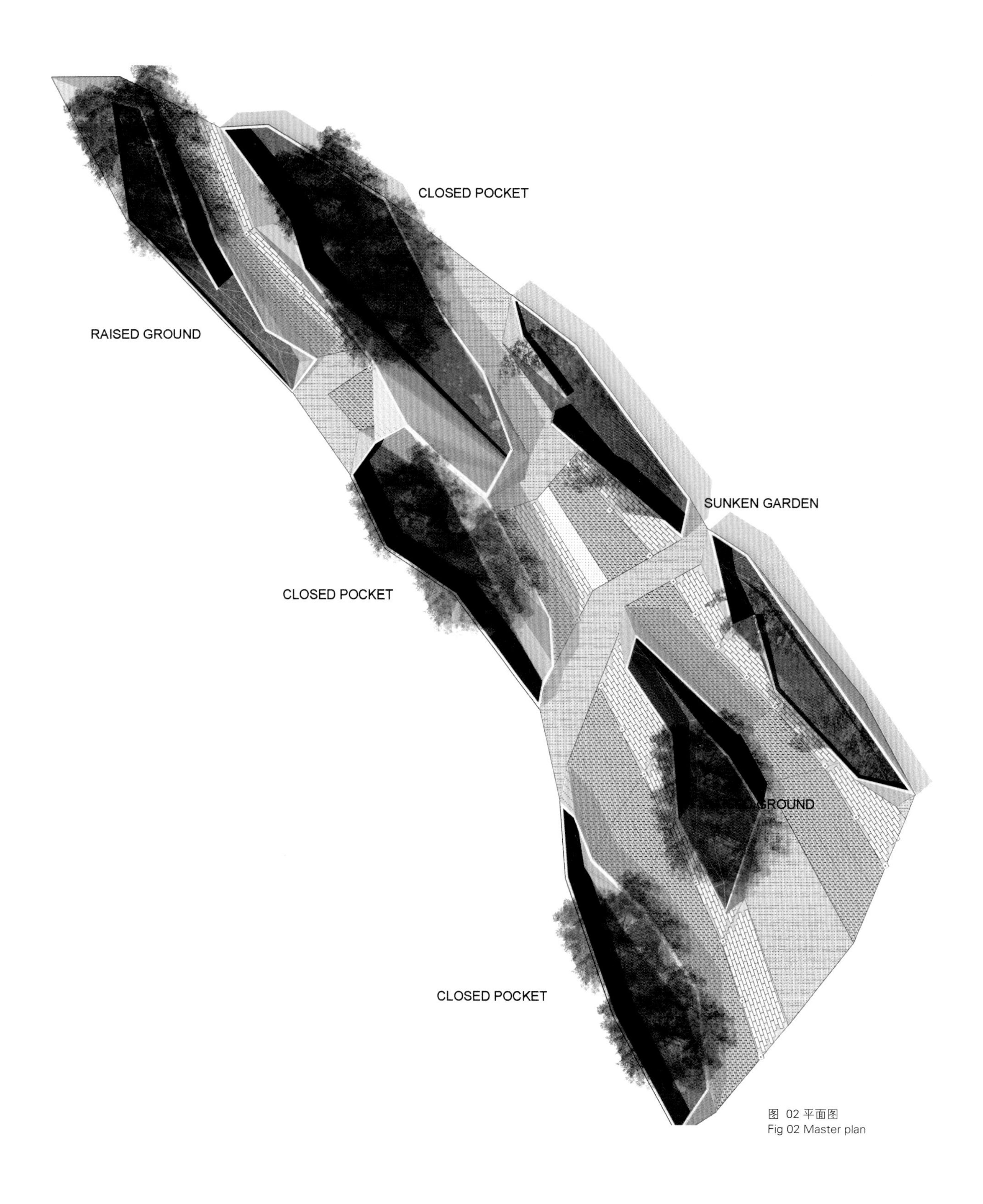

图 02 平面图
Fig 02 Master plan

本次设计的理念是凹陷的地表。这个设计理念来自于希望创造一种与环境亲密并激烈相融的空间感觉。凹陷地表从传统的中国苏州园林（假山的堆叠）以及西方的石窟中吸取灵感。虽然在空间和时间上它们有所不同，但这种空中花园以及下沉式庭院的概念却是异曲同工。本次设计敏锐地扭曲了地表空间，从而形成了强烈的三维感，使得人们感觉自己被水泥、土壤和植被包裹着，并可以自由探索一系列“口袋景观”，而正是这些“口袋景观”为游客提供了放松嬉戏的机会。

最初的设计想法是峡谷：压缩景观到一个走廊空间。天然峡谷都会有强烈的封闭空间品质，继承了地质形成的层状特征，显示出地层裂缝和基岩断裂的信息。我们希望利用这种信息厚度来探讨景观建筑元素如何创造一段由材料、光影以及封闭外墙决定的空间序列体验。

The concept of working with a sunken ground comes from a mission of fabricating an experience both intimate and intense combined with a strong feeling of harmony with the environment and intimate contemplation. Sunken ground looks at concepts present in Suzhou gardens tradition, such as the rock, the outcrop and the occidental equivalent of the grotto to then travel in space and time to the image of hanging gardens and further on to the concept of a sunken courtyard. The result is a combination both of intense distortions of the ground and an acute sense of three dimensionally, where one feels itself draped by concrete, steel and vegetation, free to explore a thickened version of

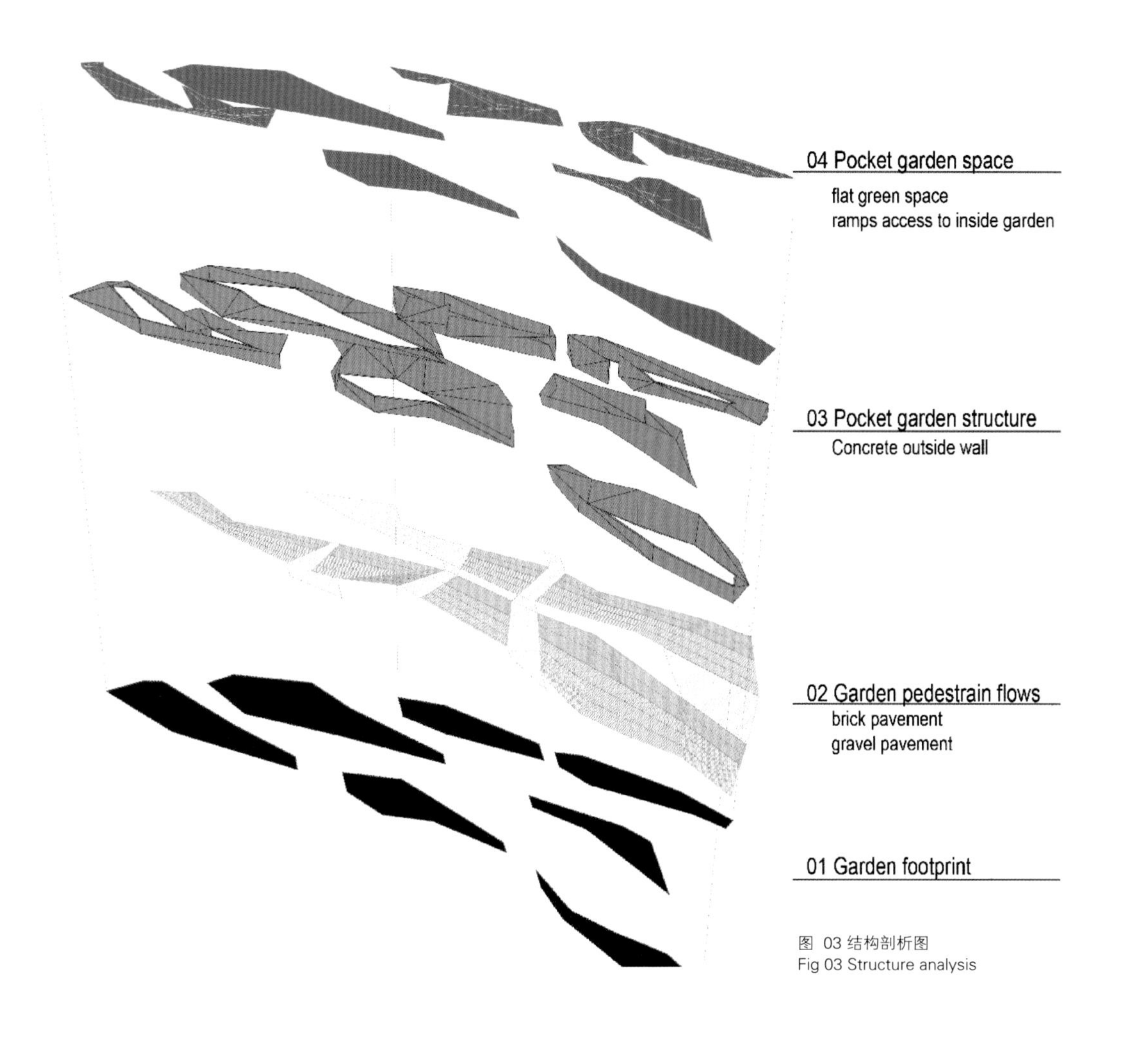

图 03 结构剖析图
Fig 03 Structure analysis

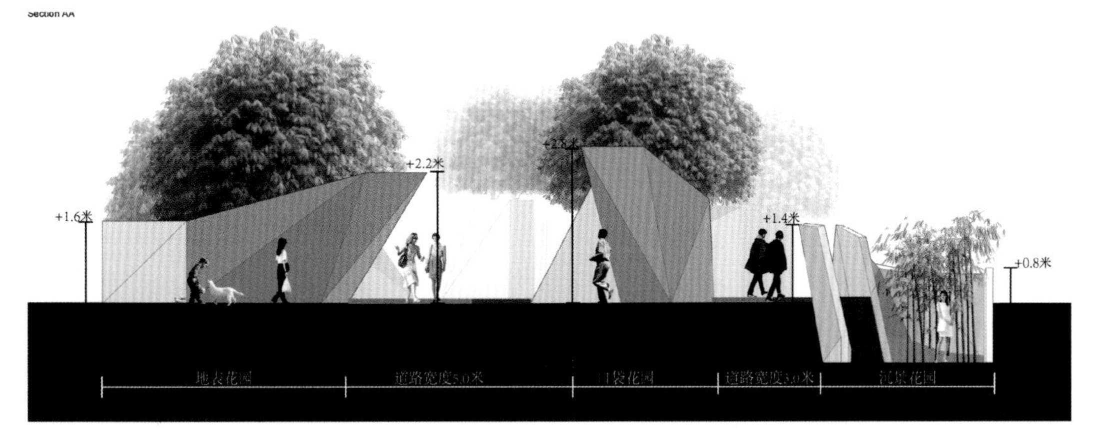

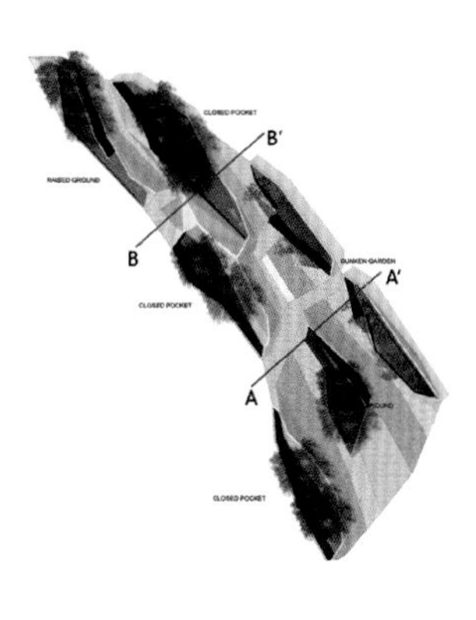

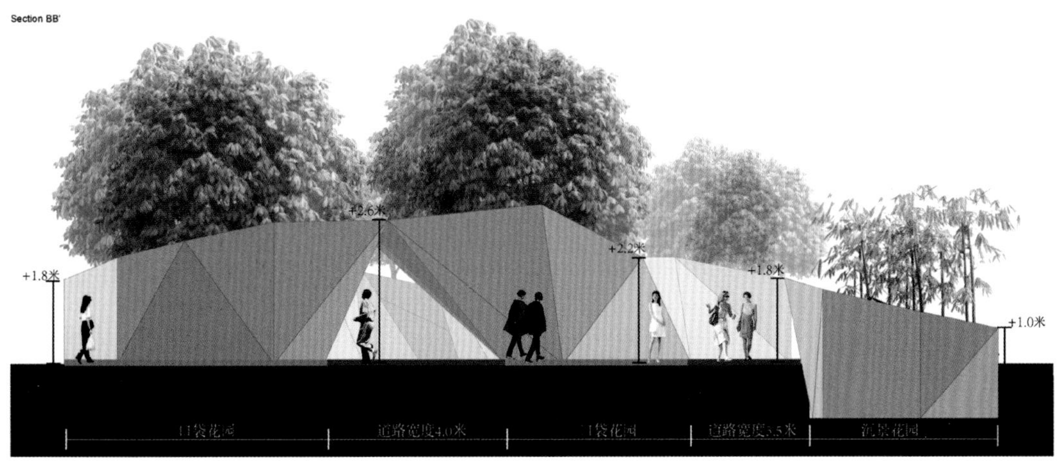

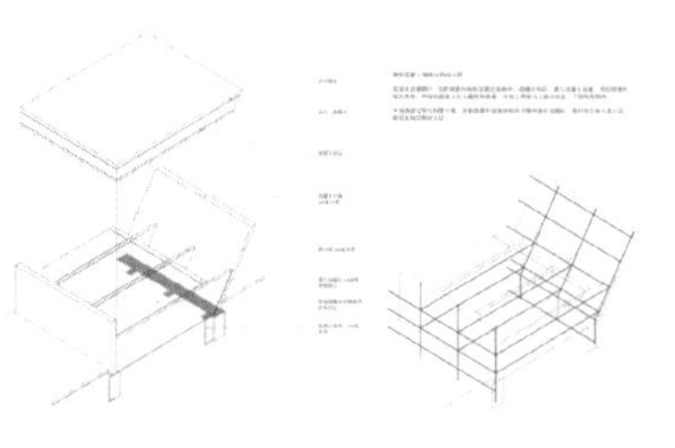

图 04 剖面图
Fig 04 Section View

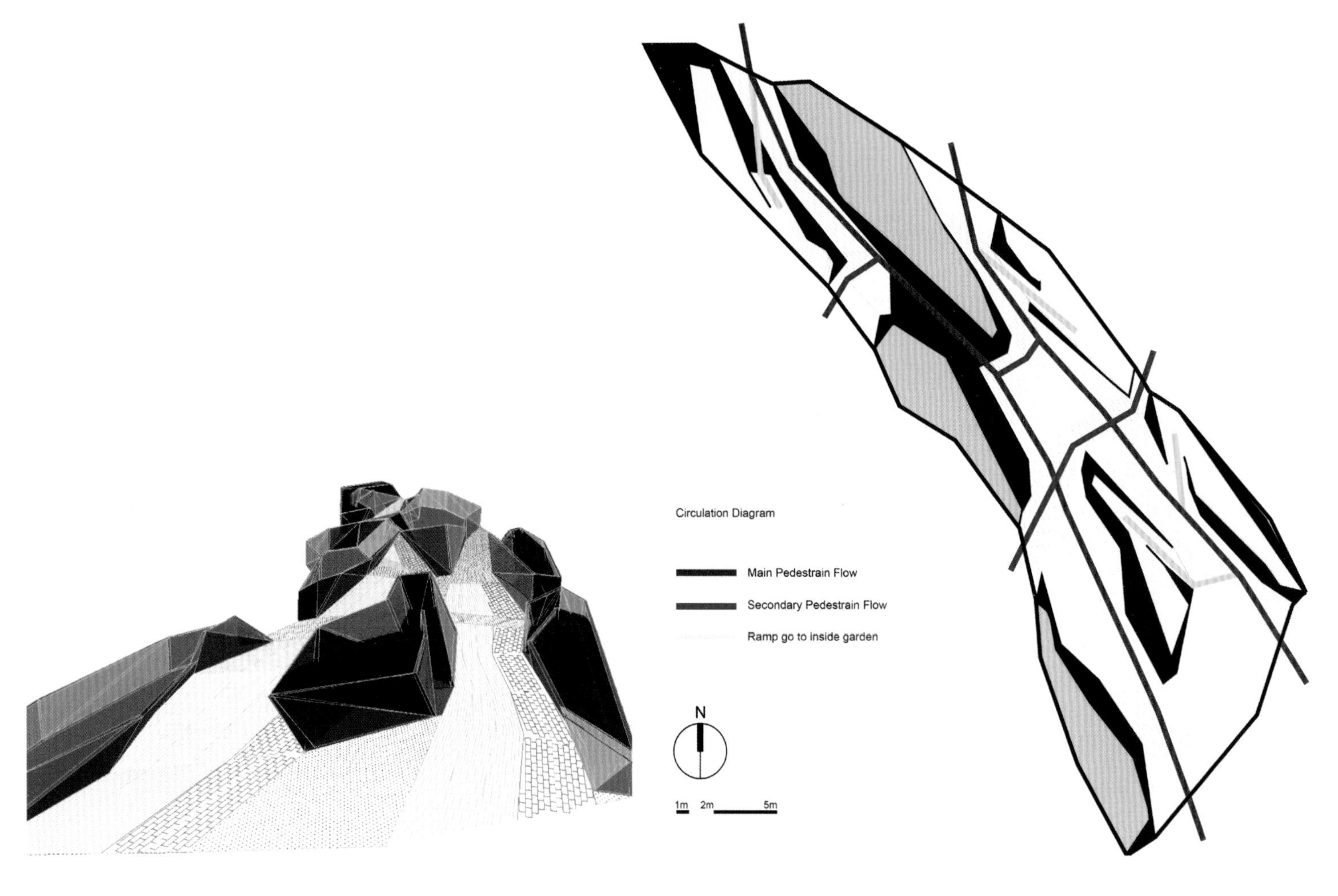

图 05 空间分析图
Fig 05 Space analysis

本设计主体思想可以追溯到苏州的狮子林，花园邀请游客内外行走，攀爬游憩，以最大限度邀请体验者思考沉思图（图 01）。这一传统的公园类型被转成当代混凝土倾斜墙壁版本，我们期望用周围人的水平基准流动性与沉重坚固的材料进行对比。

在自然界，地面的下沉产生多层的表面，植被生长形成的垂直绿地，自然环境中形成溶洞，为动物和植物创造新的生存环境。而设计灵感正是从这里捕获：三类口袋花园在其中交替——上升花园、下沉花园、封闭花园。在“上升花园”中植被从上方俯瞰游人，让人们产生一种浸在大自然中的感觉，游客们可以立刻对所看到的景观产生强烈的回应。而“下沉花园”则始于另一种相反的理念，使我们从俯瞰的角度体验自然。而“封闭花园”则用封闭的小空间容纳景观进行生态交流，人们可以从观察孔观察它的复杂性，以及平静的演变过程（图 02-03）。

想象我们在空间和时间上行走，可以感受到空中花园的古老思想。在整个花园中，从下面到上面，从里面到外面，从开敞到封闭，正是这些不同的景观体验。让游客了解一个多层次的从和谐与直观中产生的人与自然（图 04）。

由于设计地块具有很强的线性特征，最初的任务是创造两个平行走廊式的主轴结构。较长的廊道引导游客从一个由混凝土围合的空间中通过，最终沉浸在茂密的植被树冠中。这条主要通道引领体验者沿路进入“下沉花园”欣赏景观，以及从外部观赏“封闭花园”的生态

a microcosm in a series of pocket landscapes for reflection, relaxation and ultimate playfulness.

The first idea is that of the canyon, as a compressed landscape into the space of a corridor. Natural canyons both have the intense spatial quality of the enclosed space, but also inherit the layered characteristics of the geological formation, showing strata, cracks and faults of the matrix rock. We wanted to use this feeling of thickness and formational to explore how architectural elements can fabricate a passage where material, shade and enclosure would dictate the main spatial experience.

As if we were traveling both in space and time, we arrive to the ancient ideas of hanging gardens, where vegetation is viewed from the underneath. This generates a sense of immersion in nature, favoring retreat and reflection within the visitor. The garden also includes sunken gardens which start from a different point, enabling to experience nature from above, but also frame the skyline when we are inside them, closing the look of inside, outside of nature. It is perhaps this constant travel from inside to outside nature which shall let the visitor understand a multilayered relationship to nature and the environment, more in line with concepts of harmony and immediacy(Fig.02-03).

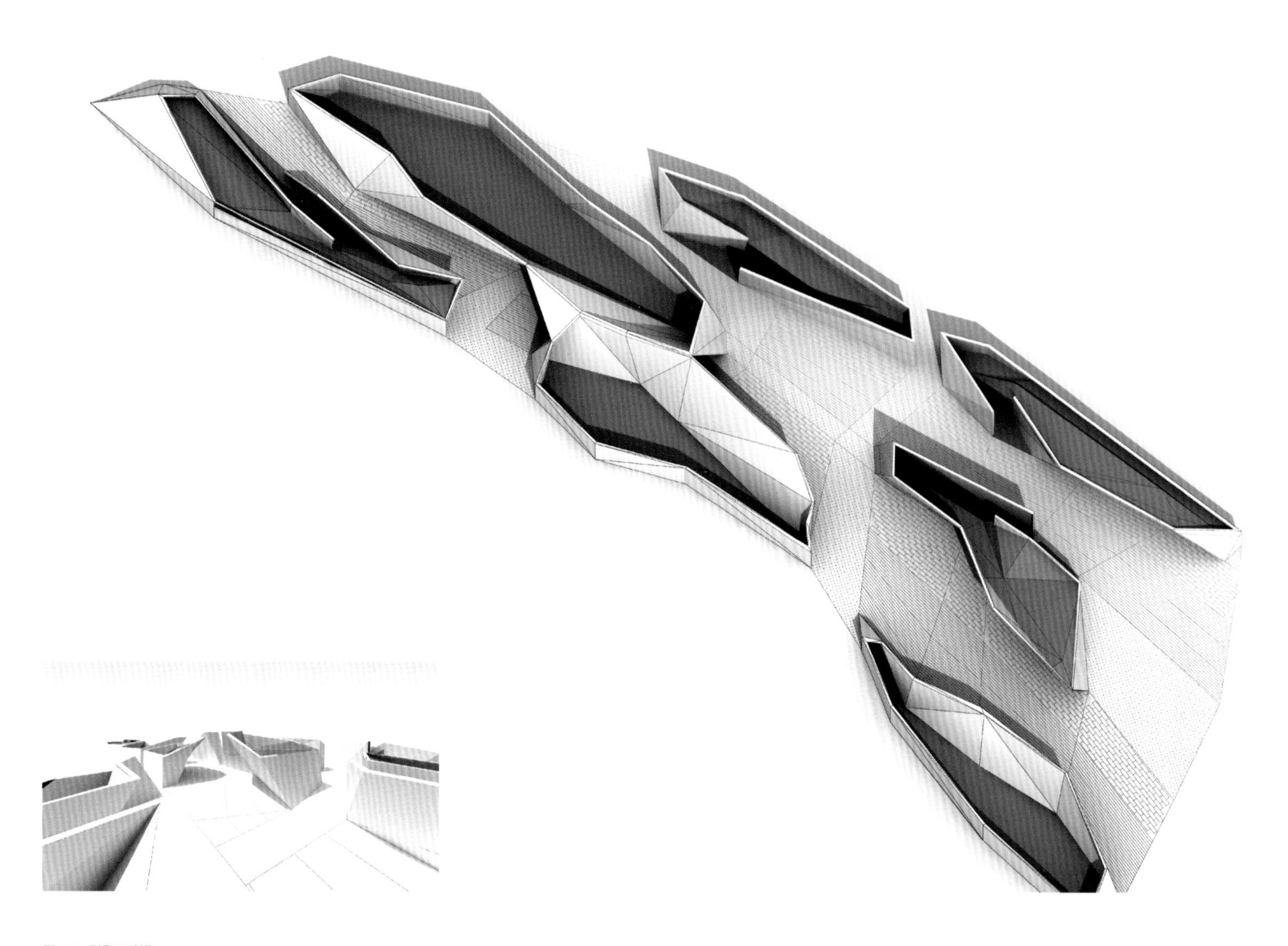

图 06 顶视及透视
Fig 06 Top view and perspective

系统演替，实现峡谷效应。第二个廊道与第一个的感觉截然相反，是一种游走于景观上方的体验，为了达到这种效果，第二个廊道连接了两个上升花园，人们可以进入上升花园，也可以在廊道上观测上升花园里的景观。两个走廊由一系列不同材料的铺装相连，形成不同的地面景观，从而产生景观多样性及不同的景观心灵体验（图 05–07）。

封闭花园作为大型景观的一种延伸，是传统苏州园林和一池三山的微型景观的现代版本。人们可以观察到每一个对象，想象自己在其间游走并与自然产生亲密的关系。一些传统的花园也纳入到这些混凝土的岩石元素，使它们成为山川大地的缩影。而凹陷地表的概念，正是为了让游客可以体验完整的三维公园，攀登到花园上方，亦可观赏其他花园。最后，道路网络的设计组织了整个花园的人流量，给予这个花园一种整体性、连续性与和谐感（图 08–15）。

后记：在院子建设的过程中，各种突发的情况（如地面标高的变化与场地出入口的改变）反而给设计带来了很多意想不到的乐趣。而最后的结果中，园子比之前想象的更加多地融合到了地表内部，新加的坡道和台阶也为空间的丰富性增添了更大的可能。这是一个设计师与体验者的互动，一个多维空间中的游戏（图 16）。□

This effects are also found in nature, where the ground sinks generating a spontaneous multiplication of the surface, vegetated structure and vertical green cover. Natural sinkholes happen in the landscape generated new conditions both in terms of scenery but also flora and fauna. This leads to the idea of the captured microcosm, which inspired to design enclosed pots or small landscape ecosystems which, while being inaccessible for the pedestrian, can be contemplated from the outside and where a small choreography of ecological activities can be observed in its complexity and calm evolution. We have looked at existing techniques of green wall and vertical planting combined with ponds, mist generators and scent dispensers. All these shall enable the fabrication of intense ecologies where the dripping of water, the freshness in the air and the subtle movements of fish shall transport the observer to an imagined pocket of alternative natures(Fig.04).

Given that the site has a strong linear character, the first design task is the structuring of two corridors parallel to the main axis. The longer corridor takes the user through a strong experience of being surrounded by concrete, End immersed in a dense

TEXTURES COMPILATION

FURNITURE

Material list

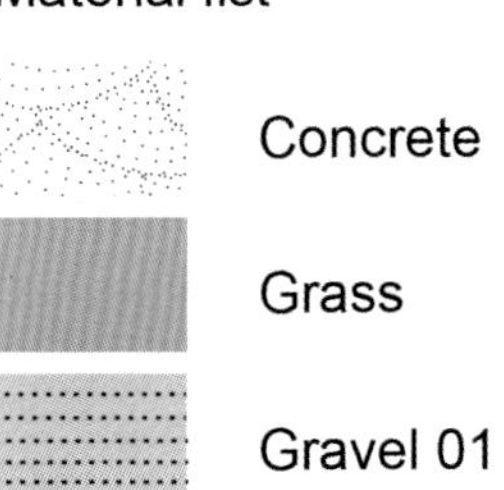

Concrete

Grass

Gravel 01

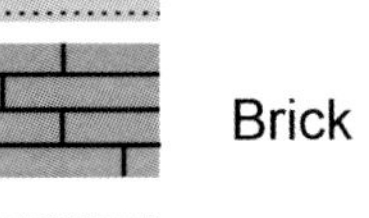

Brick

Gravel 02

N

1m 2m 5m

图 07 铺装策略
Fig 07 Pavement concept

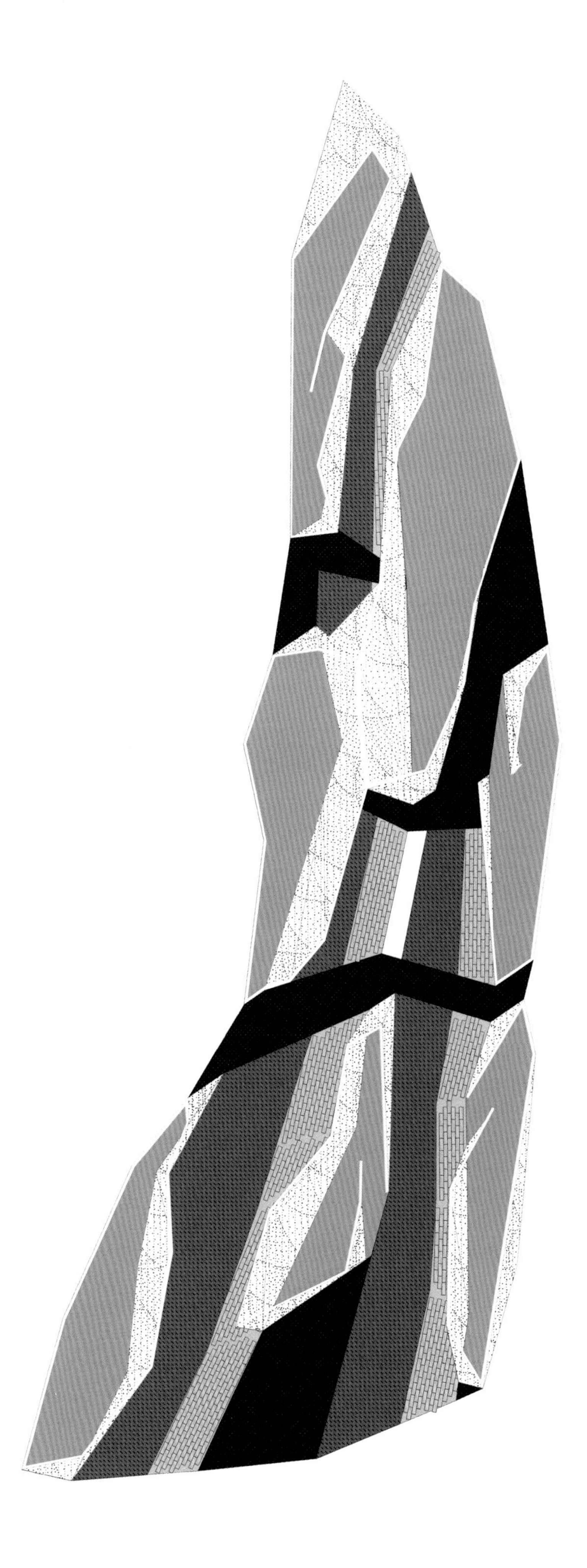

vegetation canopy. The canyon effect is achieved thought the incorporation of oversized pots which delineate the path and are both micro gardens or pocket landscapes which can be either entered or observed from the exterior. The second corridor, which stems from the first one, moves from the feeling of being under the Landscape, to that of being on top of the Landscape. In order to do this, a gentle slope places the spectator on a slightly upper position, while the pot is lowered slightly. The combine effect reverts the effect of the main axis in such a way that the hanging gardens have given way to a sunken garden east of the corridor. Both corridors are linked through a series of landscape terraces of different materials, generating a diverse ground which is the ultimate responsible for diversity and variety of experiences (Fig.05-07).

The pots serve as meandering elements within the larger landscape, a large scale version of traditional ShouZhou garden component of the rock boulder and the miniaturized landscape within the flow of the sea. One can can observe each object but also imagine oneself climbing into it in an act of intimate relationship with nature. Some of the traditional gardens also incorporate an element of circulation within these rocks, so that they become both miniature, but also belvedere and observation. Sunken Ground concept uses these traditions to generate a diverse experience where the levels inside the pots are thought so that visitors can experience the full three dimensionality of the park, climbing to the top of the pot, looking at other pots and the flow of people around. Finally, pavement and plan patterns flow along the entire site, giving it a sense of totality, continuity and ultimate harmony to the park(Fig.08-15).

Postscript: in the process of the construction, all kinds of emergency situations (such as changes in ground elevation and changes in exit or entrance) but to design comes with lots of fun. In the end, the garden got more and more integrated into the land than previously thought , new ramps and footstep also added more possibly to enrich the space. This is a designer and interactive experience, a multidimensional game(Fig.16). ■

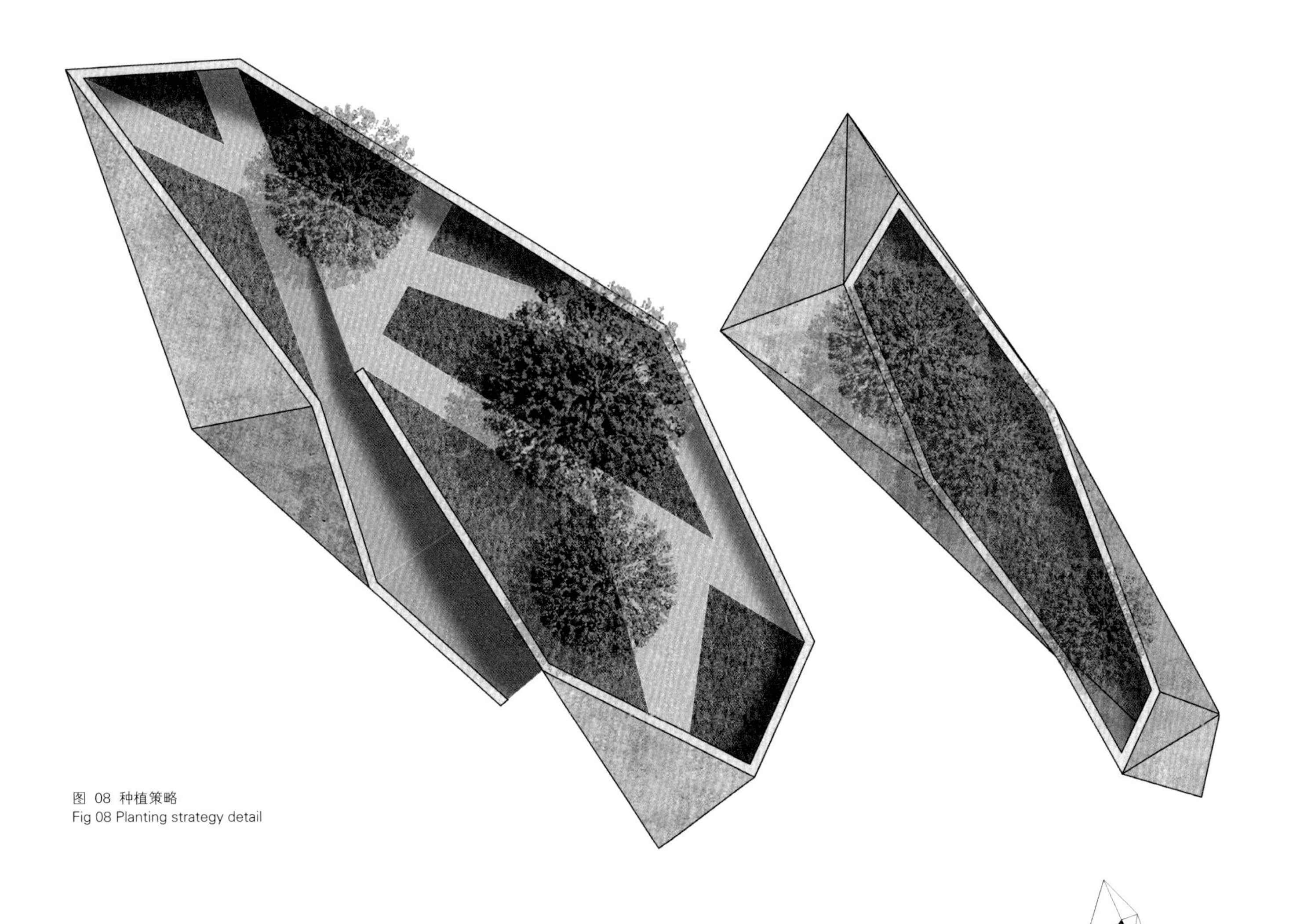

图 08 种植策略
Fig 08 Planting strategy detail

Big trees 大型樹
Medium trees 中型樹
Small trees 小型樹
Shrubs 灌木
Ground Cover 中型樹
Flowes 花卉

N

1m 2m 5m

图 09 植物搭配
Fig 09 Plants

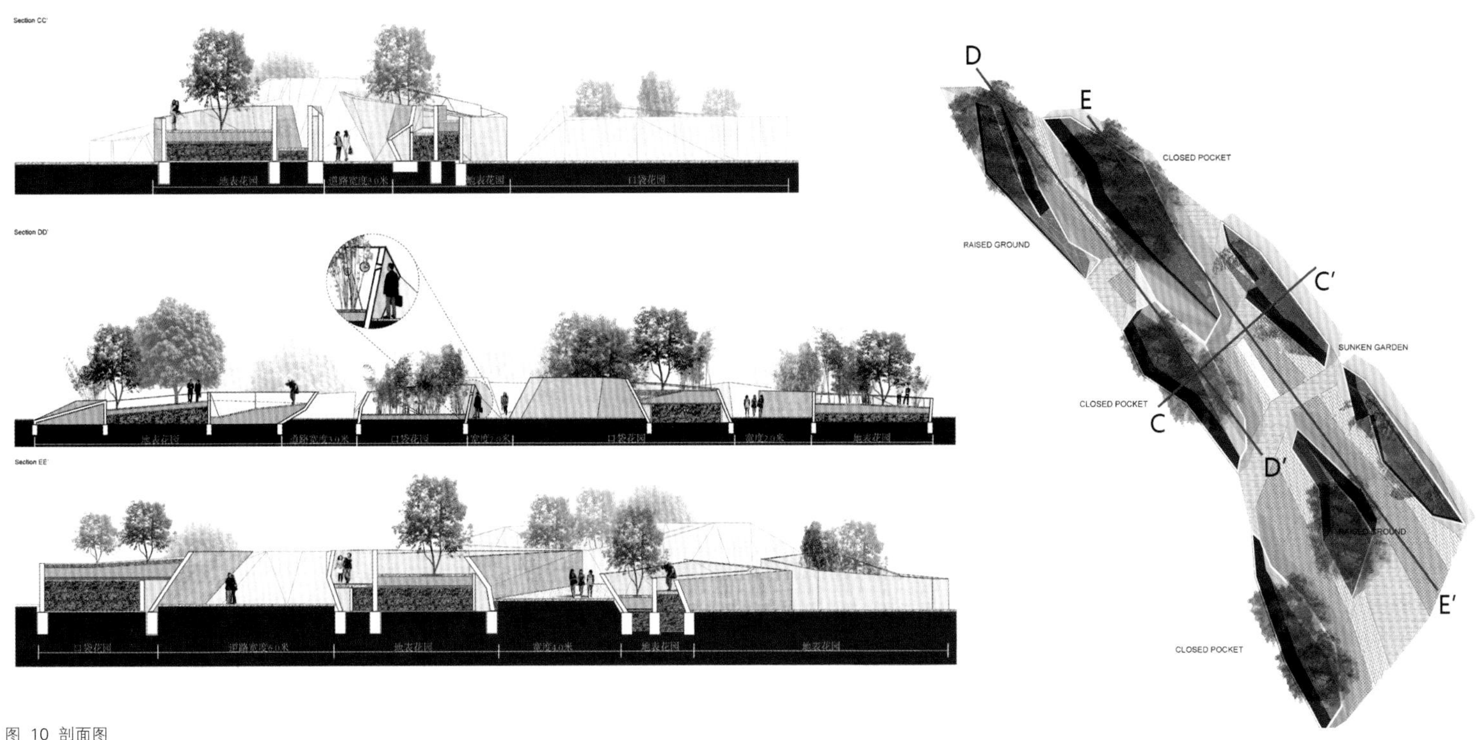

图 10 剖面图
Fig 10 Sectional view

图 11 实景图 1
Fig 11 Real map 1

图 12 实景图 2
Fig 12 Real map 2

图 13 实景图 3
Fig 13 Real map 3

图 14 实景图 4
Fig 14 Real map 4

图 15 实景图 5
Fig 15 Real map 5

图 16 效果图
Fig 16 Rendering

作者简介：

伊娃·卡斯特罗 / 女 / 景观建筑师 / 普玛建筑设计事务所合伙人
王　川 / 男 / 景观建筑师 / 普玛建筑设计事务所合伙人

Biography:

Eva Castro / female / landscape architect / Plasma St udio partner
Chuan Wang / male / landscape architect / Plasma S tudio partner

“流水印 2013”
“METAL LABYRINTH 2013”

朱育帆　孟凡玉　崔师尧

Yufan Zhu　Fanyu

项目位置：中国，北京，第九届北京园博会
项目面积：1,000m²
委托单位：第九届中国（北京）国际园林博览会组委会
设计单位：朱育帆一语景观工作室
景观设计：朱育帆、孟凡玉、崔师尧、吕回、翟薇薇、
　　　　　常钰琳、魏方、姚玉君、贾萌飞、杨思
完成时间：2013 年 5 月

Location: The 9th China (Beijing) Inter
Area: 1,000m²
Client: The 9th China (Beijing) International Gar
Designer: Zhu Yufan Y³C Landscape Studio
Landscape Design: Yufan Zhu, Fanyu Meng, Shiyao Cui,
Yulin Chang, Fang Wei, Yujun Yao, Mengfei Jia, Si Yang
Completion: May, 2013

作为中国风景园林行业的又一次国际性盛会，第九届中国（北京）国际园林博览会（后简称"北京园博会"）将于2013年5月在丰台区永定河畔举办，本工作室设计作品“流水印2013”有幸成为园区“设计师广场”6个实建项目之一参与到园博会的建设实践当中。

设计师园与中国式风景园林博览会

大型博览会发轫于1851年的伦敦水晶宫国际工业博览会，而当1999年世界园艺博览会首次落户中国昆明，也从此揭开了中国风景园林行业登上“城市事件”大舞台的序幕。此后，中国国际园林花卉博览会（简称园博会）和世界园艺博览会（简称世园会）在国内各大城市几乎是交相举办，至规划中的2016年唐山世园会在不到20年的时间内已达16次，堪称轰轰烈烈，举世罕见，也逐渐形成了风景园林博览会的政府直接推动、追求规模宏大、内容丰富全面等中国特色。

2007年由王向荣教授编制的厦门园博会总体规划中首次将“设计师园”内容引入博览会，这个专门为风景园林设计师个体提供的小型创作舞台立刻在行业内得到了较高的关注，其后2009年济南园博会和2011年西安世园会分别设立“设计师展园”和“大师园”，至北京园博会的“设计师广场”已是第4次，并呈现出越发明显的国际化趋势，中外设计师的同台竞技也引发了越来越显著的社会效应。

本次园博会“设计师广场”所规定的展示主题是"园林文化的传承与创新"，作为中国风景园林师中国园林传统文化的传承表达无疑是创作概念的第一个支点。

博览会的本质是个大型秀场（SHOW），它既有展时光鲜的一面，也有展后萧条的一面，这种巨大反差也是困扰所有事件型举办方的世界性难题。因此通常而言博览会的多数建设项目的属性带有很强的临时性，即使像1929年巴塞罗那博览会德国馆这样的旷世之作会后也被拆除。而事实证明中国风景园林博览会的一个特点就是室外展场的建设基本是永久性的，这就带来一个问题，即设计作品在展后是否能够延续展时的养护管理力度，如果不能，那么将对设计作品产生多大的不利影响？以往中国风景园林博览会的经验教训一再告诉我们：如果设计师在设计阶段的创作理念只图一时之快而完全不顾及后期的养护管理，那么这个设计前期将为其后期的“生存状态”埋下无穷的隐患，高关注度下作品设计的精细复杂很可能对应于越发糟糕的展后结果，这一点其实是设计师和承接方都不愿意看到的。

某种程度上中国式风景园林博览会展园的高品质取决于设计构思的“一前一后”：既要注重设计理念的前瞻性表达和适合国情的高品质的可实现性（对应于展时），也要注意应对后期养护管理特点的设计策略（对应于展后）。因此，这种思考也产生了我们对于设计师广场创作的另外一个支点，即如何选用合理的设计模式以降低后期的管理成本和难度。

设计策略、模式与原则

1. 开放型空间设计模式

博览园由众多园中园集群组成，各展园相对独立，摊位式的布局使得相互间的关系不佳且不测，所以设计师更愿意使用内向式的布局保证作品的独立完整。城市公共空间的经验告诉我们在开放空间中形成和制造封闭空间（阴角）会提高管理难度，内向空间可以降低使用频度但却提高了形成更大破坏的可能性，从这个意义上讲开放型空间要比封闭型空间易于管理。

2.“瘦身”设计

设计师展园在园博会中社会关注度较高，投资一般比园区其他的户外场所要高，北京园博会设计师广场投资估算达每平方米1,000元，总投资希望控制在100万元以内，这对于国内工程而言也已经是一个不错的投入，对于设计师广场工程品质而言是一个基本保证，但如果能够进一步根据展园的特殊性进行投资分配调整，果断削减不必要的设计分项，将资金集中投入是明智的选择，我们认为照明系统不是必需的，维护难度大的水景系统和灌溉系统可以取消。

The 9th China (Beijing) International Garden Expo as an international event in Chinese landscape garden industry, will be held by the Yongding River in Fengtai District in May, 2013. It is an honor for our studio that our "Metal Labyrinth 2013" will be adopted by the Beijing Garden Expo as one of the 6 construction projects.

Designer Garden & Chinese landscape garden expo

The large-scale expo derives its origin from the Great Crystal Palace Exhibition held in London in 1851. Since the International Garden Expo in Kunming, China, the Chinese landscape garden industry has been striking poses on the "City Event" stage. Since that event, the China International Garden Expo and the International Horticultural Expo have been held alternatively in big cities of China. Including the upcoming Tangshan Expo in 2016, the horticultural expo will have been held only 16 times in two decades, and so can be deemed an important and rare event. Due to these events, the landscape garden expo in China has developed its own characteristics, such as government promotion, large-scale projects and comprehensive exhibitions.

The "Designer Garden", as a specified small creation stage for landscape garden designers, was first introduced in the Overall Planning of the Xiamen Garden Expo by Professor Wang Xiangrong in 2007, and soon became popular within the industry. Since then, "Designer Exhibition Garden" and "Master Garden" were established respectively for the Jinan Garden Expo (2009) and the Xi'an horticultural Expo (2011). The "Designer Square" of the Beijing Garden Expo is the 4th instance of the inclusion of these gardens, and it is following a distinct and increasingly international trend. Moreover, the competition between Chinese and foreign designers also brings about visible social effects.

The exhibition theme of "Designer Square" for this Garden Expo is "Inheritance and Innovation of Garden Culture". Thus the launching point for the designs is an emphasis on garden designers inheriting traditional culture.

An expo, by nature, is a large show that is not only aimed at displaying beauty and greatness, but is also to a degree responsible for considering the short and long term economic and environmental effects it has on a city. This situation has bothered organizers across the world. Generally, most of the construction projects for expos are temporary, even masterpieces like the German Pavilion at the Barcelona Expo in 1929 was removed after the event. However, the outdoor exhibition fields constructed for the China landscape garden expo are permanent; thus a problem occurs: Whether the maintenance and management level during the Expo will be maintained after the event, if not, what adverse effects it will cause for the design works? The experience of previous Chinese landscape garden expos tells us that if a designer only considers the temporary effect and pay no attention to post-maintenance and management during the design stage, such designs will have problems in the future. A delicate and complex design is likely to deteriorate to some degree, and this is often the last thing that designers and contractors consider.

To a degree, the number of exhibition gardens in the Chinese landscape garden expo should depend on the current national economic conditions (during the expo) as well as the prospect

图 01 北京潭柘寺猗千亭（西晋始建寺，元、明、清历经修建，乾隆亲笔书写亭匾）
Fig01 Yixuan Pavilion of Beijing Tanzhe Temple (the temple, built in the Western Jin Dynasty and renovated in Yuan, Ming and Qing dynasties respectively, was named by Qianlong Emperor, the sixth emperor of the Manchu-led Qing Dynasty)

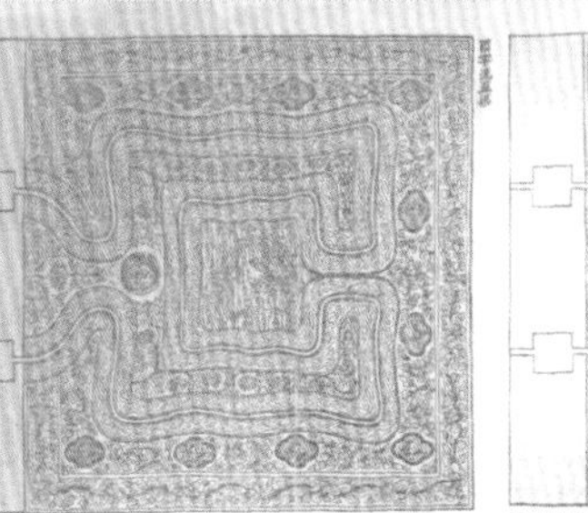

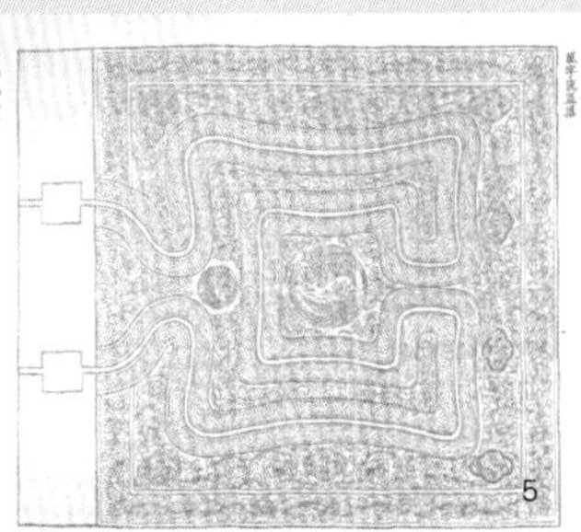

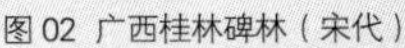

图 02 广西桂林碑林（宋代）
Fig02 Forest of Steles, Guilin, Guangxi (the Song Dynasty)

图 03 北京恭王府（清代 1776）（图 1-8 引自网络）
Fig03 Prince Gong Mansion, Beijing (1776, the Qing Dynasty; Figures 1-8 are sourced from the internet)

图 04 四川宜宾流杯池（北宋 1098，黄庭坚谪居戎州（今宜宾）时修建）
Fig04 Flow Cup Pool, Yibin, Sichuan (1098, the Northern Song Dynasty; built when Huang Tingjian was exiled to Rongzhou (now Yibin)

图 05《营造法式》中的水渠形制（宋 1100）
Fig05 Canal Form in "Yingzao Fashi" (also the Rules of Architecture) (1100, the Song Dynasty)

图 06 北京故宫禊赏亭（清乾隆 1772 年）
Fig06 Xishang Pavilion in the Imperial Palace, Beijing (1772, during Qianlong Emperor Period in the Qing Dynasty)

图 07 重庆大足石刻（晚唐 892 年 – 南宋 1162 年）
Fig07 The Dazu Rock Carvings, Chongqing (892~1162, from the late Tang Dynasty to the Southern Song Dynasty)

图 08 东京（今河南开封）金明池（北宋 976–981）
Fig08 Jinming Pool, Dongjing (now Kaifeng), Henan (976–981, the Northern Song Dynasty)

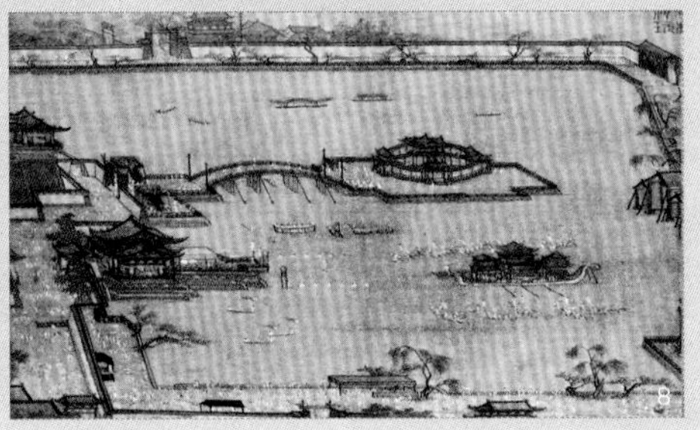

3. 无缝（少缝）设计

一件景观设计作品是由不同和相同设计材料拼接组装而成的。设计上对表皮层“缝”的处理和表皮层与结构层的连接处理是基本功，相对而言，表皮层拼接越少的设计整体性越强，拼接越多的设计显得越精细，而这种精细设计的维持尤其需要后期的精心养护，因为损毁通常从拼接的缝开始。

4. 高强度材料

设计主要材料的选取应相当慎重，材料本身的强度是要点，在可行的情况下尽量避免使用容易损坏或磨损的材料，如板型的玻璃、塑料、木材和竹材等。当然，所有针对后期维护管理的设计考虑均不能降低一个设计标准，即作品本身的概念性和艺术性。

曲水流觞、“流水印”系列与中国园林文化

1660 年前[1] 发生在会稽的一次文人聚会成为中国历史上最著名的一次聚会，“曲水流觞”的典故也自此传咏千古，而这次聚会与园林密切相关。此后中国园林中的“曲水流觞”沿着两条线并行发展，一条是继续呈现出自然的蜿蜒辗转形态的水道，第二条是演变为抽象几何化的流杯渠。

of long-term maintenance and management (post- expo). Thus, an important consideration for the gardens in "Designer Square" is how to select a rational design that reduces the cost and difficulty of post-management.

Design strategy, mode and principle

1.Open space layout

The expo site consists of numerous "garden within garden" assemblies; each garden is conceived in relative isolation from the others. Therefore relationship between such gardens is often poor, and designers generally prefer an inward oriented layout to ensure the independence and completeness of a work. The history of urban public space tells us that enclosed spaces increase management difficulty and that open spatial layouts are preferable.

2. "Low-cost" Design

The designer exhibition garden of the Garden Expo enjoys more

[1] 东晋永和九年（公元 353 年）。
[1] The 9th year of Yonghe in the Eastern Jin Dynasty (AD 353).

从图形设计的角度，“曲水流觞”属于一条线的艺术，1100 年刊行的《营造法式》中已经出现了“国字流杯渠”和“风字流杯渠”式样，“曲水流觞”呈如盘结在一个正方形中的蛇形叠篆线，说明至少发展到北宋中期“曲水流觞”分支“流杯渠”已获得官方认证，这条线也已经高度图示化和符号化，并凝练成具备特定形式语言功能的文化线，而它本身代表的正是中国园林和传统文化精神。清代流杯渠的诸多遗存基本都是些小尺度的强调趣味性的流杯亭，流杯渠纹也演变为一条在一定几何形状内容纳尽可能长尺度的水线，图式上出现了如潭柘寺猗玕亭中“龙虎符”等诸多变化（图 01）。

由一个事件演化为一种代表性图式的现象即便在如此博大精深的中国园林文化中也绝无仅有，“流觞曲水”就是一条风格化的流水印迹线（图 02–08），我们要做的就是如何再度演绎和拓展这条线的艺术表现力。

“流水印 2013”（图 09–10）延续了我们在 2010 年意大利威尼斯建筑双年展中国馆作品“流水印”的设计概念（图 11），并将其发展成为一个系列性作品，它用“锈钢板铸造的并非一件理查德·塞拉式的雕塑，而是一个广场式开放空间的‘路引’，其中糅合了东方人对于景观的认知，规范和提示着人在环境中的行为。”[2]

attention and higher funding than other gardens, for example, the investment in "Designer Square" at the Beijing Garden Expo is estimated at RMB1000/m^2 while the total investment is expected to be contained to RMB100 million. As for the Chinese projects, there is a considerable investment in guaranteeing the quality of "Designer Square". Thus by distributing and adjusting the investment according to the specific characteristics of the exhibition gardens, and focusing primarily on essential elements, funding should be sufficient.

3. Design with fewer seams

A landscape design is made up of different or similar materials, so the treatment of "seams" on surface layers and the connections between the surface layer and the structural layer are primary concerns. Comparatively speaking, fewer seams on the surface layer will lead to a more resilient construction. Since the seams are where damage is likely to occur, these areas require particular attention and good maintenance (especially over time).

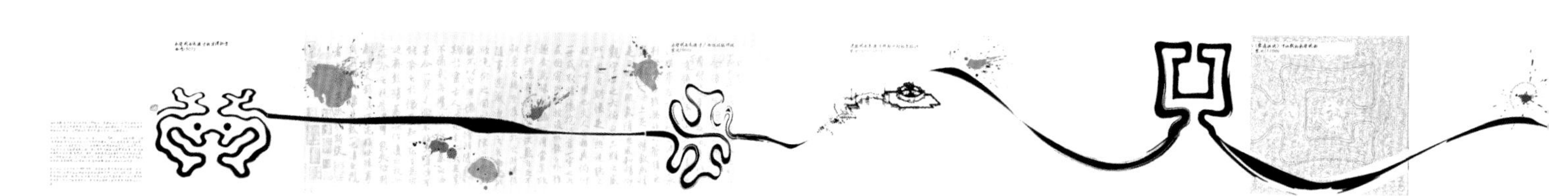

[2] 引自策展人唐克扬先生关于第十二届威尼斯建筑双年展中国馆作品‘流水印’的评述。

[2] Refer to the comments of Curator Tang Keyang on the “flow watermarking” of China Pavilion in the 12th Venice Architectural Biennale in Italy.

图 09-10 威尼斯双年展装置艺术展品“流水印”（朱育帆摄）
Fig 09-10 " Metal Labyrinth ", an Installation Art Exhibit at the Venice Biennale Exhibition 2010 (Taken by Yufan Zhu)

无疑，“流水印”抽象和发展自传统的“流杯渠”纹样，并试图将原型中的平面化向第三维竖向空间进行拓展，镜面不锈钢的顶面取代了真实的水流同时又保持了水空间的延展性和渗透性。由于当时中国馆室外场地及其施工条件的苛刻限制，尤其是没有重塑地形的可能，9m 见方的“流水印”与其环境并不完全匹配，细节体验性的处理上还是比较粗燥的，远没有达到美学意义上的完形，而双年展上所有出现的不利客场因素在北京园博会主场都将不存在，设计品质的升级也完全成为可能。

园博会设计师广场 4# 场地与“流水印 2013”

北京园博会设计师广场 4# 地块（图 12）位于园区“锦绣花谷”东南方，与大师园与设计师广场的南入口区域，平面呈长条梯形，长轴为东北西南方向，地块东临主园路，短轴向（西界）为西偏北面对谷地，北接美国风景园林师戴安娜·巴尔莫里（Diana Balmori）女士的作品“风波园”，占地面积约 1,000 平方米。

地块朝向谷地，理论上完全可以藉借“锦绣花谷”的景观创造更

4.Strong materials

Special care should be taken with material selection and emphasis should be attached to durability. Easily damageable and degradable materials such as glass, plastic, timber and bamboo should be avoided. These concerns can all be addressed without adversely affecting the integrity of the design.

Floating wine cups along winding water, "flow watermarking" series and Chinese garden culture

The elegant "floating wine cups along winding water" custom is derived from a meeting of literati held in Kuaiji 1660 years ago. This meeting addressed the art of garden design and developed the concept of "floating wine cups along winding water" which describes two parallel paths in Chinese garden: one is to display a naturally winding waterway, and the other to show an abstract and geometrical flowing 'cup' canal.

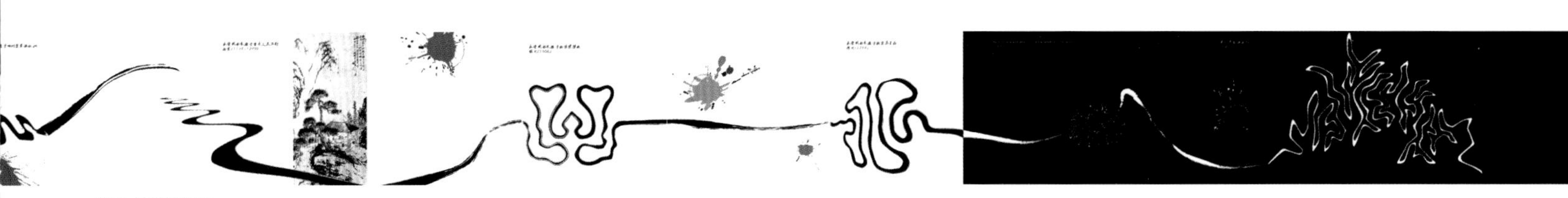

图 11 设计概念图
Fig11 Design Concept

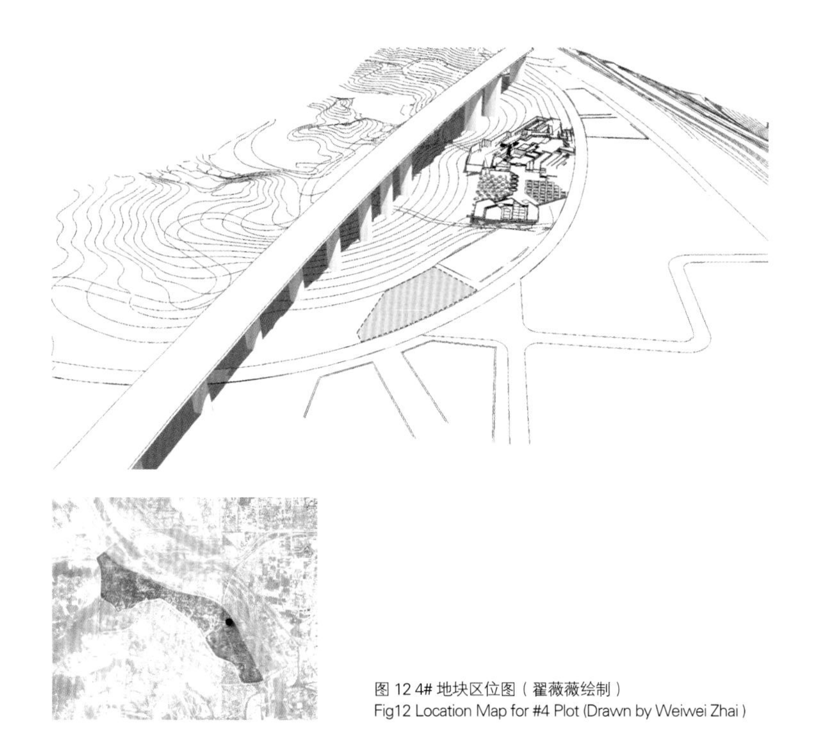

图 12 4# 地块区位图（翟薇薇绘制）
Fig12 Location Map for #4 Plot (Drawn by Weiwei Zhai)

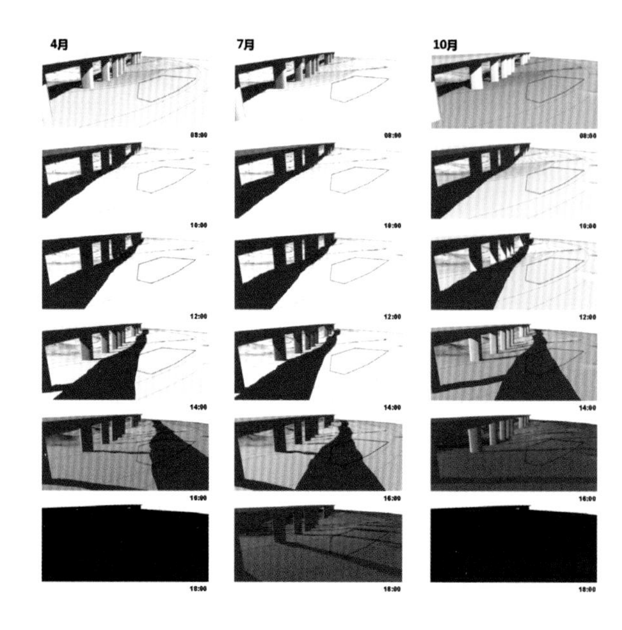

图 13 4 月、7 月、10 月一天中典型的阴影变化（翟薇薇绘制）
Fig13 Daily Shadow Change in April, July and October (Drawn by Weiwei Zhai)

大的景深，但实际空间并非如此。首先地块是斜对花谷，其次距离地块西界平均 20m、高达 20m 的铁路高架桥（京石高铁）在场地西斜上空南北向腾空而过，与桥体下方阵列的支撑柱组成了巨大的工程体量，对场地直接形成了视高比大于 1:1 的强势影响，同时阵列的桥墩体明显减小了观看花谷的可视面，从而进一步削弱了地块和谷地之间的良性关系。这种地块的外部条件使得设计空间型的选择陷入两难：如果采用内向的空间型，那么将难以屏蔽高架桥的身影与园内景观的冲突，如果采用直接外向借景花谷的空间型，那么将同样要面对难以消解桥墩负面影响的境地。我们认为这时设计策略调整的重点应该从到“划地为牢”转移到如何制造新的趣味中心，同时开放的空间形式有利于让诸多不利的外围因素之间进行自我消解，使游人不再特别关注它们（图 13)。

4# 地块的设计采用了简明的开放式空间型，我们首先将沿主路一侧的步道放大使其转变成为一条观看全园的全景线，地块内并没有再设置常规的园路，而是让观者直接面对一整块带有地形起伏的草地，长达 350m 的“流水印”[3] 盘踞在它的中心并覆盖了 60% 的区域，成为具有绝对意义的焦点。这种设计貌似将 4# 地块转换为一处纯粹的观演空间，然而这种“观演”不会是静态的，点状离散的步道铺装暗示了观者进一步运动的方向，走近和进入“流水印”本体是一种对无穷变化空间的过程体验。

“流水印 2013”的整体形式由一条曲线演化而来，平面设计上在流杯渠纹样基本特点的基础上向白描[4] 风格进行变化以形成一种复杂的图式；形体上流水印是一个由变截面金属箱体构成的三维曲线体，与传统流杯渠空间处理平面化不同的是，“流水印 2013”在顶面和底面层均拓展了竖向变化的可能性（图 14）。

场地在设计红线边缘已经开始坡向锦绣谷，1/7 的坡度有明显的向下滑落感，长达 100m 的稳定边坡使人强烈感受到负向空间的吸引，因此设计中需要调整空间感缓解这种紧迫压力。

我们设计了一个凸起的地形（最高点近 1.2m），一方面把人与强大的负向空间相对隔离开来形成稳定感；另一方面也便于“流水印”

From a graphic design perspective, "floating wine cups along winding water" is a linear art. Flowing cup canals, which takes the shape of the Chinese characters " 国 " or " 风 ", is recorded firstly in the "Yingzao Fashi" (namely The Rules of Architecture) published in 1100. The "Floating wine cups along winding water" looks like an official seal in the figure of snake winded in a square, this shows that the "Floating wine cups along winding water" was officially approved as early as the middle of the Northern Song. It is a highly graphic and symbolic cultural line equipped with inherent linguistic meanings representing the spirit of Chinese gardens and traditional culture. Most of the flow cup canals in the Qing Dynasty were small flow cup pavilions used for entertainment, thereof the flow cup canal has evolved into a long waterline specified into a geometrical shape. The picture shows the changes of "Longhufu" of Yixuan Pavilion in Tanzhe Temple(Fig.01) .

A representative cultural pattern evolving from an event is very unique, even in a culture as deep as that of Chinese garden culture. "Floating wine cups along winding water" is a stylized ground-level watermarking trace (Fig.02-08), which forms the basis for the development of our artistic expression.

The "Metal Labyrinth 2013" (Fig.09-10)continues the design concept (Fig.11)of " Metal Labyrinth " constructed at the China pavilion of the 2010 Venice Architectural Biennale in Italy, developing it into a series. The construction of rusted cor-ten steel plates is not a Richard Serra type sculpture, but rather an open square type of "Roadmarking", which represents the cognition and code of landscape and reflects human behavior in the environment."

It is of no doubt that the " Metal Labyrinth " is abstracted and developed from the traditional "flow cup canal" pattern and

[3] 双侧面和顶面的总延长达 1,000m，钢板总面积达 320m^2。
[3] The sides and top surface are extended by 1000m in total, and the gross area of steel plate is up to 320m^2.

[4] 中国传统国画“线勾”技法。
[4] A "line drawing" skill in Chinese traditional painting.

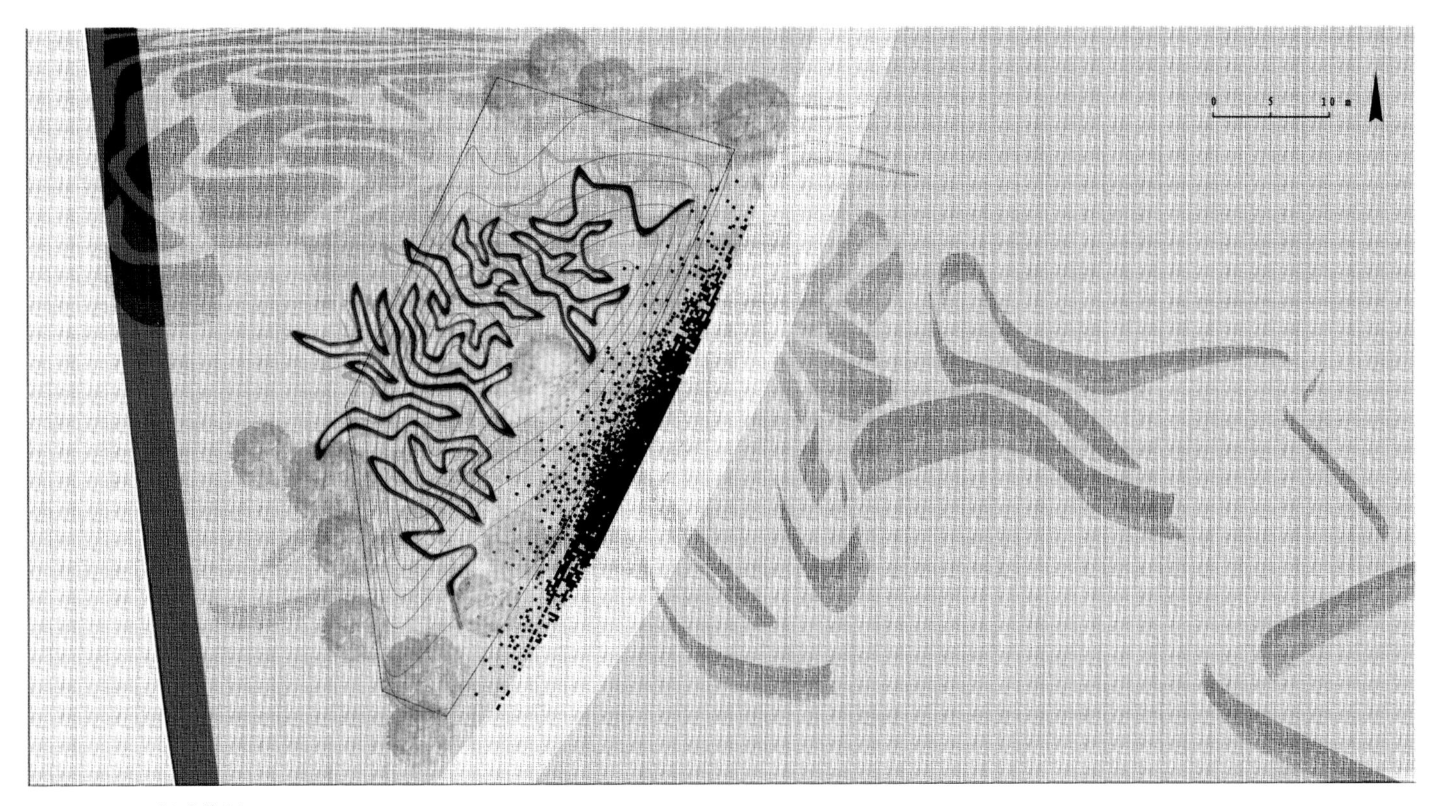

图 14 总平面图（崔师尧绘制）
Fig14 General Layout (Drawn by Shiyao Cui)

的展示。人站在场地内无法全揽，促使其进入体验。在这样的处理下，消极的空间被正面化，更加明确的向主路上的三角形集散空间打开。地形的隆起变化增加了全景线与锦绣花谷之间的空间层次，提升了空间的戏剧性，而“流水印”顶面和底面起伏与地形处理的相互协同呼应大大增强了这件作品的艺术表现力。流水印的核心区局部在形体上脱离了地面形成了更大的趣味空间，像飞奔中的水流暂时摆脱了地心重力的吸引，孩子们可游走上下。到此游戏并没有结束，在全景线视线的背处流水印沿地形向西流向谷地，使得空间变得不测，2m 的落差使得流水印回归重力并给予游者一份惊喜，暗示了空间的延展性和无限性（图 15–18）。

至此，流水印已经成为一个可观、可坐、可站、可躺、可在表面滑移的公共参与的巨型装置，镜面不锈钢的顶面映射变化着的天空，也将人的活动纳入其中，此时人的角色转换成为“流觞”。

非常有趣的是，4# 地块设计的这种空间型与苏州留园五峰仙馆及其前院空间（图 19–20）关系有异曲同工之处：游人站在全景线上观看流水印如同在五峰仙馆中观看庭院中的五老峰，进入流水印的兴致有如在五老峰山石间游走，而山后流水印的延展下坠还原为在五老峰和院墙之间获得的那份惊喜，但与之不同又具有当代意义的是：“流水印 2013”的空间是开放的，并非中国传统园林的封闭空间。

“流水印 2013”是一件由考登钢和不锈钢通焊而成的整体，它可以划入大地艺术的范畴，强调了景观学水平向的基本意义，也产生了地形学意义上的专业性拓展（图 21–23）。作为 2013 第九届中国（北京）国际园林博览会的设计师展园之一，“流水印 2013”（图 24–27）是中国风景园林师传承园林文化精神的当代表达，验证了文化经典仍具有永恒的生命力，同时流水印也是对中国现阶段风景园林行业设计技术和施工技术的一种展现。无论展时展后，花开花落，“流水印 2013”都将静待这片大地林木绿草如荫，并默默地记录下这份时代的变迁和感动。□

attempts to expand the original planar pattern into the third dimension of vertical space. The mirror is finished in stainless steel, which stands for the flow of water, and is designed to maintain the ductility and permeability required for a water space pattern. Due to the unfavorable outdoor site and construction conditions of the China pavilion, especially the impossibility to reshaping the land, the 9m^2 " Metal Labyrinth " was not perfectly matched to its environment, and the aesthetic experience of the details was far from sublime. However, the unfavorable factors of the Biennale do not exist at the Beijing Garden Expo, and the design quality will be upgraded.

The #4 plot in the designer square of Beijing Garden Expo - "Metal Labyrinth 2013"

The #4 plot in the Designer Square of Beijing Garden Expo (Fig.12) is located in the southeast of "Jinxiu Flower Valley". This plot, together with the Master Garden and the south entrance of Designer Square, form a trapezoid, of which the major axis is southwest by northeast (with the main garden path to the east) while the minor axis is northwest by southeast (confronted with the valley). Located in the north area is the 1000m^2 "Rimple Garden" designed by Ms. Diana Balmori, an American landscape gardener.

Firstly, the valley runs parallel to the plot; secondly, a 20m high railway viaduct (from Beijing to Shijiazhuang) crosses through the air in a south-north direction 20m away from the plot. The bridge and piers below forms a massive engineering volume, which leads to a strong impact (with ratio of height of eye as 1:1) on the plot. Moreover, the piers greatly reduce the visual surface of the valley,

15

16

图 15-18 渲染效果图（吕回绘制）
Fig15-18 Rendering (Drawn by Hui Lv)

图 19 五峰仙馆内部透视与前院对景 (引自 –< 江南园林图录：庭院 . 景观建筑 > 作者：刘先觉，潘谷西编出版社：东南大学出版社出版时间：2007–9–1)
Fig19 Internal Perspective of Wufeng Xiangguan (Opposite to Its Front Yard) (Quoted from "Catalogue of South China Garden Pictures: Yard & Landscape Architecture", jointly prepared by Xianjue Liu and Guxi Pan , published by Southeast University on September 1, 2007)

which further weakens the proper relationship between the plot and valley. The pattern selection of design space is caught in a dilemma due to such external conditions: If an inward space pattern is adopted, it will be difficult to avoid the viaduct shadow and its conflict with the landscape; if an open space pattern directly borrowing space from the valley is adopted, it is also difficult to eliminate the negative impact imposed by the piers. Therefore, emphasis shall be transferred from "restrained area" to new interest center; meanwhile, an open space pattern provides an environment where unfavorable external factors may negate each other. Thus, the visitors will pay less attention to the negative impact(Fig.13).

A concise open space pattern is adopted for the #4 plot: the footpath along one side of the main road is developed into a panorama of the garden, and no other regular path is arranged. Visitors will see big undulating grassland with the 350m-long " Metal Labyrinth " located in the center, accounting for 60% of the total area. This will become a significant focus. It seems that the #4 plot has been converted to a spatial pattern for "viewing and performing" only. However, such "viewing and performing" is dynamic, and the scattered dot pavement will direct your next step. The process to enter the "Metal Labyrinth " is an experience in eternally changing space.

The overall pattern of the "Metal Labyrinth 2013" is derived from a curve. In the planar design, it changes into a “line drawing” style on the basis of the basic characteristics of the flow cup canal; while in shape, it is a three-dimensional curve construction made of variable cross-section metal boxes. The difference between it and the tradition flow cup canal is that the spatial treatment of "Metal Labyrinth 2013" has developed the vertical change possibility in the top and bottom surfaces (Fig.14).

The red line edge of the site is extended towards the Jinxiu Valley. The 1/7 gradient gives a visible falling sense and the 100m stable side slope provides a strong attraction of negative space. Thus the spatial sensation shall be adjusted to mitigate this tension.

We have developed a raised position (up to 1.2m). On the one hand, this provides a sense of separation between people and the negative space to form a sense of stability; on the other hand, it is convenient to display the " Metal Labyrinth ". No panorama is

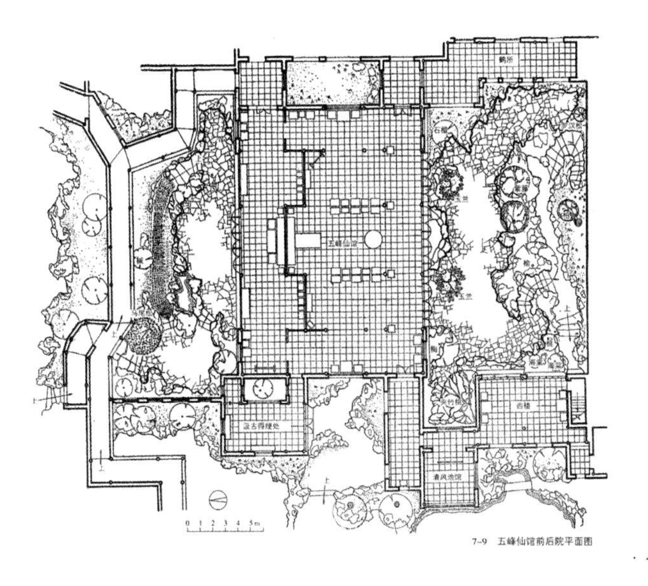

图 20 五峰仙馆前后院平面图（引自 –< 江南园林图录：庭院．景观 建筑 > 作者：刘先觉，潘谷西编 出版社：东南大学出版社出版时间：2007–9–1）
Fig20 Plan View of the Front and Back Yards of Wufeng Xianguan (a famous hall in Suzhou) (Quoted from "Catalogue of SouthChina Garden Pictures: Yard &Landscape prepared by Xianjue Liu and Guxi Pan, published by Southeast Universityon September 1, 2007)

available within the site, so visitors must enter it to experience all of the scenes. In this way, the negative space becomes positive, and the unfolding towards the triangular distribution space in the main road is specified. The changes in grading have enriched the spatial relationship between the panorama path and Jinxiu Flower Valley and added color to the space. Moreover, the undulating top and bottom of the " Metal Labyrinth " is echoed in the spatial treatment. Thus, the artistic expression of " Metal Labyrinth " is strengthened. Part of the Metal Labyrinth core area is breaking away from the ground to form a more interesting space, just as the flowing water is free from gravity. What's more, on the other side of the panorama path view, the Metal Labyrinth is flowing westwards into the valley. Thus, the ductility and infinity of space is fully displayed, the space becomes more varied and the 2m fall brings gravity back to the Metal Labyrinth and also brings a big surprise to visitors (Fig.15-18).

So far, the Metal Labyrinth has become a large-scale construction of public participation which can be used to sit, stand, lie and slide. The mirror finished stainless steel surface reflects the sky and visitors so that the visitors become "flow cups".

What is interesting is that, the spatial design of #4 plot can match that of Wufeng Xianguan in Suzhou Liu Garden and its

图 21 制作过程照片
Fig21 Fabrication Picture

图 22 施工安装过程照片 1
Fig22 No.1 Installation Picture

图 23 施工实景照片之 2
Fig23 No.2 Installation Picture

图 24 流水印全景图 1
Fig24 No.1 Overall View "Metal Labyrinth"

图 25 流水印全景图 2
Fig25 No.2 Overall View "Metal Labyrinth"

图27 流水印全景图4
Fig27 No.4 Overall View "Metal Labyrinth"

图26 流水印全景图3
Fig26 No.3 Overall View "Metal Labyrinth"

front yard (Fig.19-20). Namely, when visitors view the Metal Labyrinth standing on the panorama path, it is like viewing Wulao Peak in the yard from Wufeng Xianguan; and when they enter the flow watermarking, it is like walking among the hillstones of Wulao Peak. Moreover, the downward extension of the Metal Labyrinth behind the hill can give you the same surprise as received from between Wulao Peak and the yard wall. However, the difference lies in that the former has contemporary significance; "Metal Labyrinth 2013" is open in space, rather than closed like traditional Chinese gardens.

"Metal Labyrinth 2013" is wholly welded of corten steel and stainless steel and can be regarded as earth art. It highlights the basic meaning of horizontal landscape and also brings a special extension to the topography (Fig.21-23). As an exhibition garden of the 9th China (Beijing) International Garden Expo, "Metal Labyrinth 2013"(Fig.24-27) represents the modern inheritance of the garden culture and spirit by Chinese landscape garden designers and verifies the eternal vitality of classic culture. In addition, the Metal Labyrinth is also a demonstration of the design and construction technologies in China's current landscape garden industry. Into the future, "Metal Labyrinth 2013" will continuously add color to our lives and witness the changes in history. ■

作者简介：

朱育帆／男／教授／博士生导师／清华大学建筑学院景观学系副系主任／中国北京

孟凡玉／女／硕士／项目经理／朱育帆景观工作室／中国北京

崔师尧／男／硕士／设计师／朱育帆景观工作室／中国北京

Biography:

Yufan Zhu / male / Professor / PHD Supervisor / Deputy Head of Department of Landscape Architecture, School of Architecture, Tsinghua University / Beijing, China

Fanyu Meng / female / Master / Project Manager / Zhuyufan Landscape Atelier / Beijing, China

Shiyao Cui / male / Master / Designer / Zhu Yufan Landscape Studio / Beijing, China

查尔斯·沙（校订）

English reviewed by Charles Sands

图 01 概念图
Fig01 Concept Diagrams

小径花园
PATH GARDEN

克里斯托弗·康茨工作室

Christopher Counts Studio

项目位置：中国，北京，第九届北京园博会
项目面积：1,000m²
委托单位：第九届中国（北京）国际园林博览会组委会
设计单位：克里斯托弗·康茨工作室（CCS）
景观设计：克里斯托弗·康茨工作室（CCS）
完成时间：2013 年 5 月

Location: The 9th China (Beijing) International Garden Expo, Beijing, China
Area: 1,000m²
Client: The 9th China (Beijing) International Garden Expo committee
Designer: Christopher Counts Studio (CCS)
Landscape Design: Christopher Counts Studio (CCS)
Completion: May, 2013

图 02 悬浮的 Axon：桥为游人提供了新的视角，在道路的不同高度，游客可以眺望到园内和园外不同的景象

Fig02 Aerial Axon Diagram: The Bridge provides a new perspective of inside and outside the garden as viewed from various overlooks at varying heights along the path

小径花园的设计为参观者带来激发他们好奇、兴奋与探求欲的有趣体验。以运动的研究为构想，设计者渴望提供一个与内向的、以内部物体为导向的传统园林不同的作品。设计的主要构成是一条垂直的、视觉效果夸张的道路系统，它将引导游人在 1,000m^2 的三角形公园中穿越 3.5m 的垂直空间。这种抽象的效果是带来了一系列可以在移动中体验到的流动的空间（图 01）。逐步展开的景色在花园的内部景观、临近的花园和宽广的外部景观间不断变化，带来了多样的对花园和其内涵的观赏、理解（图 02）。这条富于表现力的园路开始于花园底层，通过缓慢抬升的雕塑化的地形将人们逐渐引向天空，在这里人们不知不觉地走入了树冠区域。这样的设计特色为游人提供了行走在树荫中，并且能够从高处欣赏花园和外部景色的机会（图 03–05）。

“融入环境”的概念同样也可以通过对道路和树木的延续以及越过树冠对公共区域的眺望来得以表现。在临近的公共道路区域，横跨的桥梁变成了一个遮蔽的结构，为游人提供遮阳和休息的便利场所。树冠通道的下方被涂成了品红色来加强花园游乐的氛围（图 06），并且通过强化入口来吸引人们进入花园（图 07–08）。

花园动态地设置了运用大胆色彩的道路、地形和景观，创造了一幅生动的画作，这里有盛开的薰衣草做的地毯，整洁的草坪以及会因人们到来而激活的流动的混凝土彩虹（图 09–11）。塑造的地形也成为了生动地展示攀缘植物的设施，这些植物攀爬在斜坡上，吸引游人来细致地观察不断变化的细节材质。波动的、流动的种植形态被带状种植的白杨树打断，它们的竖向线条和特别的白色树干增添了空间的层次，确定了花园的竖向结构（图 12）。对花园中各种元素进行相互关联的构成并且表现它们在材质上的联系为人们含蓄地展现了一个令人兴奋的 21 世纪现代花园（图 13–15）。

克里斯托弗•康茨工作室是一个获得过很多奖项并且因其对建筑、

The ‘Path Garden’ is designed to inspire visitors with a playful sense of wonder, excitement, and discovery. Imagined as an investigation in movement, the work aspires to offer an alternative to traditional interior or object-oriented gardens. The engine of the design is a vertically and horizontally exaggerated path system that guides visitors through 3.5 meters of elevation change within a triangular shaped 1,000 square meter garden. The effect of this compression is a garden defined by fluid spatial compositions understood through experiential movement(Fig.01). The unfolding views constantly fluctuate between interior views of the garden, perspectives of adjacent gardens, and broad overlooks of the expo providing a multitude of ways to see and understand the garden and its context(Fig.02). The expressive walkway begins at-grade and gently lifts visitors into the sky through sculptural earthworks that seamlessly transitions into an elevated canopy walk. This exciting design feature creates opportunities for visitors to walk amongst tree canopies and offers an elevated perspective to view the garden and expo from above(Fig.03-05).

The concept of engaging the context is also expressed through the extension of the pavement, trees, and canopy overlook into the public space. Within the adjacent public walks, the overlook bridge structure becomes a shade structure with benches offering visitors public amenities of respite from the sun and a resting place. The underside of the canopy walk is painted deep magenta(Fig.06) to enhance the playful nature of the garden and frame entrances inviting visitors into the garden(Fig.07-08).

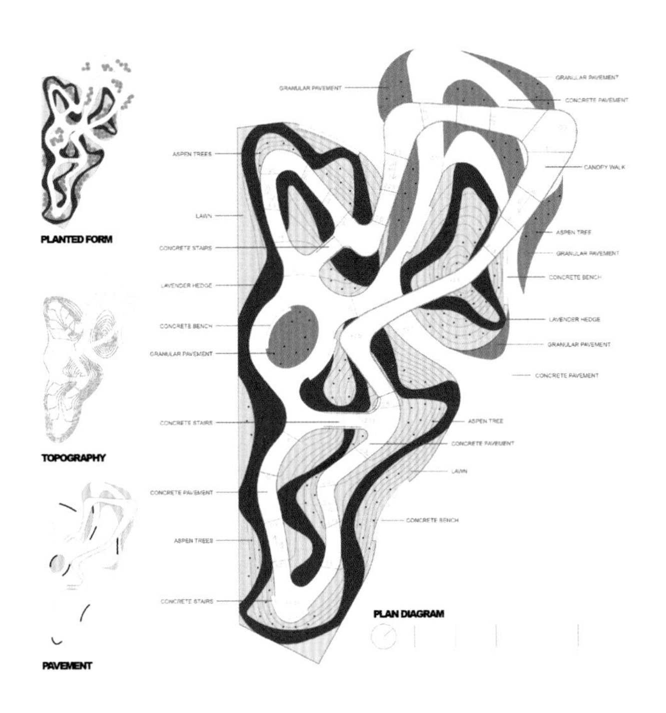

图 03 设计图：4 张图示共同阐释了种植规划，地形随着道路的变化以及按条理放置的游客坐等
Fig 03 Plan Diagrams: All four (4) of these plans are intended to illustrate the planting scheme, topographic change along the path, and strategically placed custom benches

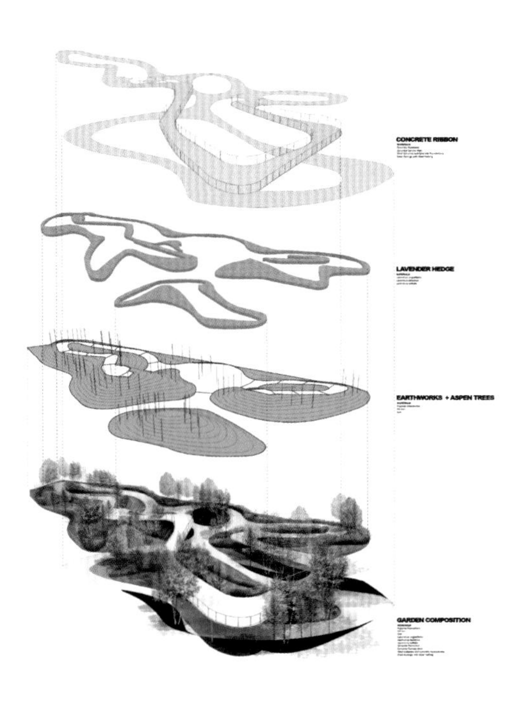

图 04 Axon 分析：说明了 Axon 从地面开始的不同层面的设计
Fig04 Exploded Axon Diagram: Illustrating the various layers of the design from the ground up

图 05 鸟瞰图
Fig05 Bird's eye view

图 06 仰望小径
Fig06 Upward view

图 07 公共入口：从三角形花园的南面进入，可以看到富有表现力的植物种植展现的强烈动感
Fig 07 Public Entrance: Entering the design viewing south into the inner triangular garden reveals the powerful movement of the expressive planting swaths

图 08　入口夜景：在夜间游览花园带来了令人激动的、独特的体验，波动的植物种植形态和地形带来的丰富变化的投影创造了丰富的观感体验的惊喜
Fig08　Entry View at Night: experiencing the garden at night poses an exciting and unique opportunity in which the undulating planting forms and topography cast unexpected and rich shadows to create experiential richness and an element of surprise

图 09 设计过程：在整个设计过程中制作了不同规模的，不同颜色，不同材质的模型来探究，完善设计构思
Fig09 Process Collage: throughout the entire process, models at various scales, colors, and with different materials were created in order to explore and refine design ideas

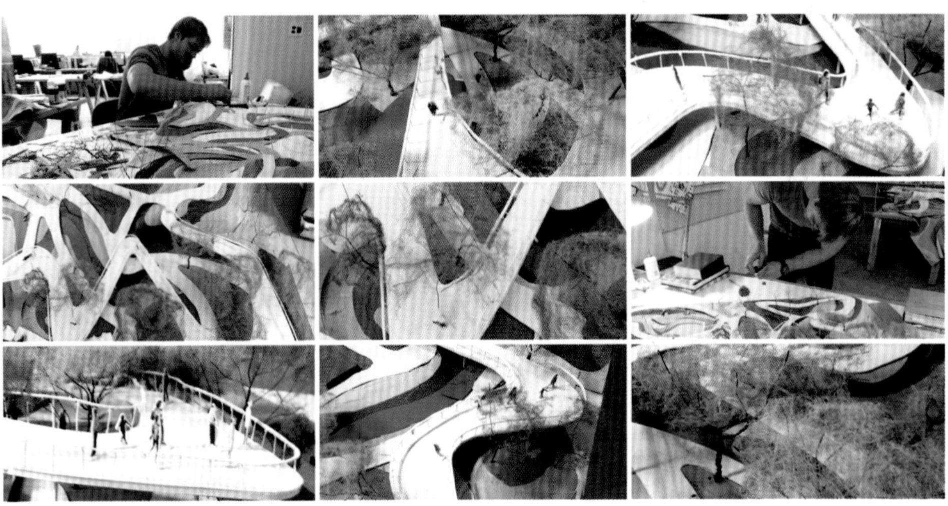

图 10　纸质模型：对颜色的探索是设计过程中必要的部分，这决定了对植物材料的选择并且帮助形成新的想法和设计形式。这个尝试带来了选择一个充满活力的三维种植配色的灵感
Fig10 Paper Mache Model: this colorful exploration was an integral part of the design process, which lead to the selection of the plant materials as well as helped to generate exciting new ideas and forms. This exercise inspired the selection of a robust and three-dimensional planting palette

图 11 鸟瞰图：场地中不同的地形高度鼓励游人寻找机会内部，外部欣赏花园以及整个博览会园区
Fig11 Prospect Overlook: From various elevations throughout the site, visitors are encouraged to seek opportunities for viewing the garden both in its interior, but outwards and beyond to the rest of the expo as well

图 12 杨树林：中心花园区域试图带来永恒的氛围并且引导游人通过多条道路向上方游览，体验在高处的视角和新的景观
Fig12 Aspen Tree Grove: the inner garden area is intended to be as much about reflecting internally as well as being inspired to move beyond and explore the several other pathways that lead to elevated viewpoints and new variations

图 13 中央白杨树丛和薰衣草种植
Fig 13 Central Aspen Grove and Lavender Planting

图 14 从展园道路上看小径花园
Fig14 View from Expo Path

地形和城市空间熟练的规划设计而得到纽约市认可的设计公司。设计室对于城市规划和景观设计的热情体现在他们对于塑造富有丰富的体验，理性的逻辑和融入可持续性发展的动人环境的追求。克里斯托弗·康茨工作室的员工有城市规划师、建筑师和景观师，他们一起组成了一个多学科综合的设计团队，以此来致力于创造丰富市民日常生活的创新景观。工作室宽泛的设计手法因其富有经验的设计团队的知名城市设计而被众所周知。这样的经历塑造了公司合作的精神，优先考虑客户、顾问和公众参与，并且将设计中遇到的矛盾与挑战当作令人兴奋的创作体验。公司通过严格地关注项目的管理，卓越的技术和对预算的负责来确保达到既定的目标和客户的期望。现在，克里斯托弗·康茨工作室的项目包含了从预算14.8万美元的有22年历史而未被好好利用的罗列城市广场（美国北卡罗来纳州）的竞赛项目到55hm^{2}在建的城市景观设计，这个设计以其为15,000人创造多样的活动场地而被中国视为现期最大的单体项目。其他作品包括了赢得2013北京园博会竞赛的作品。公司也因其出色的合作伙伴而知名，其中包括了皮特·沃克、黛安娜·巴默瑞、彼得·拉茨以及伊娃·卡斯特罗。□

The cinematic arrangement of bold chromatic juxtapositions of plants, topography, and views creates a living painting expressed as a flourishing carpet of thick lavender, taut lawn, and flowing concrete ribbons that come to life as each visitor moves through the garden(Fig.09-11). Earthworks become dramatic presentation devices for the tapestries of plants as they engage the steep slopes and tilt towards the visitor for detailed viewing of the shifting compositions. These undulating drifts of planting are punctuated by strands of aspen trees whose vertical character and distinctive white trunks create spatial complexity and become a vertical datum within the garden(Fig.12). The interrelated composition and compression of garden elements and its relationship to its context define its contemporary design expression that modestly aspires to contribute to the exciting possibilities of 21st century garden design(Fig.13-15).

Christopher Counts Studio (CCS) is an award-winning design firm based in New York City recognized for its fluid compositions of architecture, topography, and urban space. The firm's passion for urban design and landscape is reflected in its commitment to creating inspiring environments distinguished by experiential richness, pragmatic logic, and integrated approach to sustainability. The CCS staff is comprised of urban designers, architects, and landscape architects that form a multi-disciplinary studio environment dedicated to innovative design solutions that enhance the experience of daily life in cities. The firm's inclusive approach to design is informed by experience in leading design teams through complex high-profile urban projects. This experience has shaped a collaborative spirit that prioritizes client, consultant, and public participation and views design contradictions and challenges as exciting creative opportunities. The firm is committed to ensuring the achievement of the goals and aspirations of our clients through rigorous attention to project management, technical excellence, and budgetary responsibility. Current CCS projects range from a competition awarded $14.8 million improvement of a 220-year-old under- performing square in downtown Raleigh, NC, to a 55-acre on-structure urban landscape design and mixed-use development for 15,000 people currently recognized as the largest single-phase project in China. Other projects include the design for the competition awarded 2013 Beijing International Garden Expo. The firm is honored to be included with the distinguished participants that include Peter Walker, Diana Balmori, Peter Latz, and Eva Castro.■

图15 从上部的路径俯视花园
Fig15 View from Upper Path

作者简介：

克里斯托弗·康茨工作室 / 国际城市设计和景观设计 / 美国纽约市布鲁克林区

Biography:

Christopher Counts Studio / International Urban Design and Landscape Architecture / Brooklyn, New York, USA

HUAYU
华宇园林
住建部园林工程施工一级企业
设计.施工.苗木.养护

HUAYU
华宇园林

Ming Garden

章俊华　Junhua Zhang

场地呈近似梯形又四边不等的斜方形，设计采用让出外围空间，简化一切异质的细部变化来衬托主景（图01）。并尝试引喻中国传统思想文化——“天圆地方”，将北京特有的场域特征（图02）通过“圆”与“方”的形式组合划分为“内·圆”、“外·方”两组空间，借以表达人与自然的宇宙观。条带状肌理象征着养育人类的母亲——大地；黄色琉璃脊红色彩叶草的神韵洋溢着皇城的秩序；周边的沟带界定了城池，烘托了场所的精神。直白、整然、理性、日常是“外·方”空间的特质。围合式下沉空间营造无限的冥思；垂映水池中的景象梦幻着太空的偶遇；高耸的中央圆形场所再现小宇宙空间的存在。厚重、神秘、超然、非日常是“内·圆”空间的特质。“日月丽乎天，百谷草木丽乎土。”故择之：“明园”也（图03-09）。

项目位置：中国，北京，第九届北京园博会
项目面积：1，275 m^2
委托单位：第九届中国（北京）国际园林博览会组委会
设计单位：R-land 北京源树景观规划设计事务所
景观设计：章俊华
参与人员：白祖华　胡海波　杨珂　于沣　夏强　程涛　陈一心
余磊　高侃　马爱武　李松平　徐飞飞
施工单位：北京金五环风景园林工程责任有限公司
完成时间：2013年5月

Location: The 9th China (Beijing) International Garden Expo, Beijing, China
Area: 1,275m^2
Client: The 9th China (Beijing) International Garden Expo committee
Designer: R-land Beijing Ltd
Landscape Design: Junhua Zhang
Completion: May, 2013

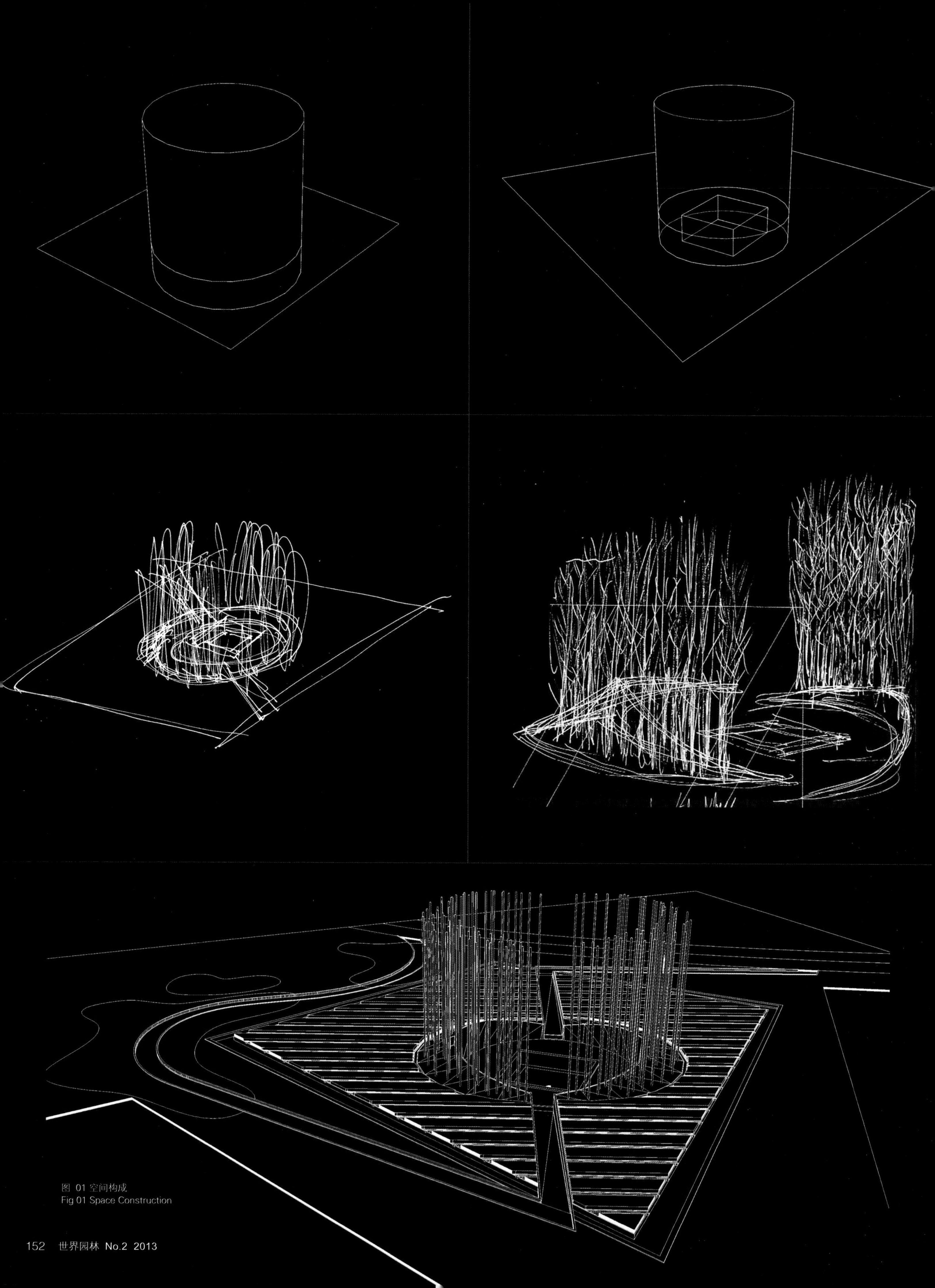

图 01 空间构成
Fig 01 Space Construction

在 13 亿中国人民的首都——全世界关注的北京，如何创作一个“园林文化的传承与创新”为主题的场所？如何让“变”与“不变”互相交融？如何让“无形”变成“有形”？如何让“非日常”再回到“日常”？如何诞生看似平凡但又不平凡的作品？在这些永远纠缠不清的话题中开始了我们的工作。

In Beijing, the capital city of 13 million Chinese people that has caused worldwide concern, how to create a site with the spirit of "inherit and innovate landscape culture", how to bring together "change" and "remain", how to make "invisible" into "visible", how to bring "unusual" back to "usual", and how to generate a common work that is in fact uncommon? In all of these questions without answers we started our job .

图 02 北京印象
Fig 02 Beijing Impression

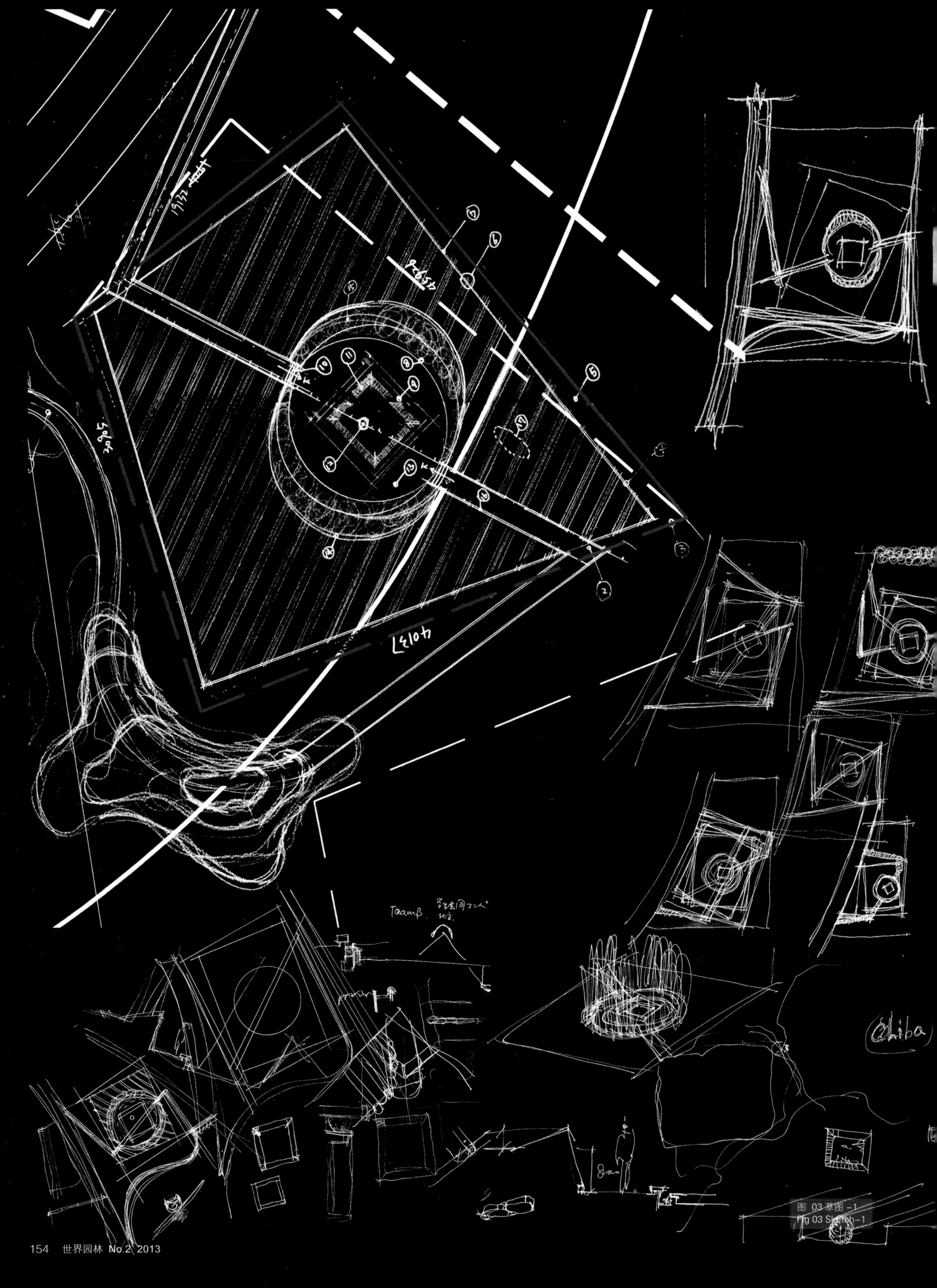

图 03 草图－1
Fig 03 Sketch－1

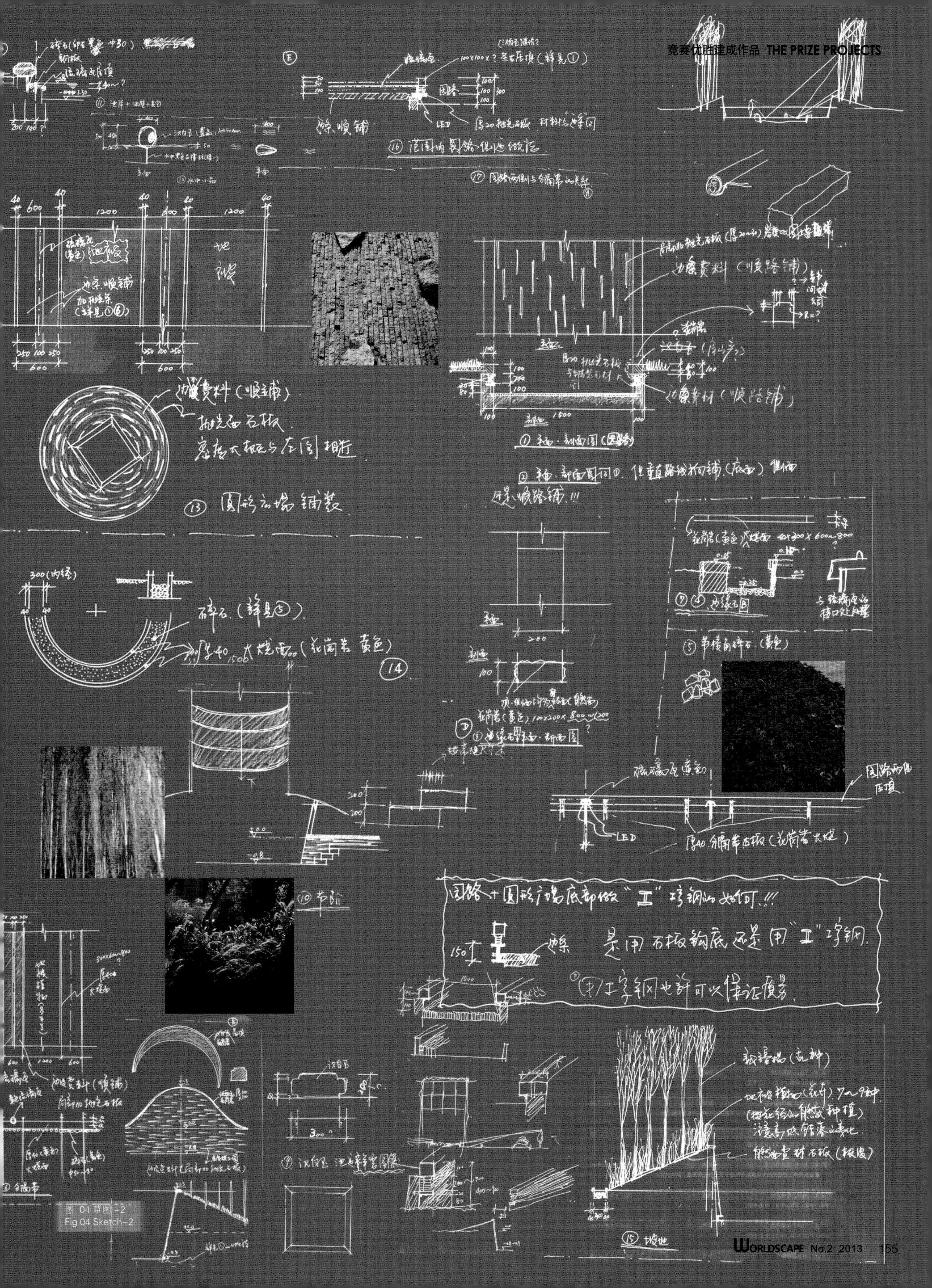

图 04 草图-2
Fig 04 Sketch-2

The field is similar to a rhombus with four unequal sides. After the removal of the peripheral room, the design is simplified and all the changes in the details highlight the main scene (Fig.01). Traditional Chinese thought and culture, specifically "orbital sky and rectangular earth," is employed. The open spaces in Beijing city (Fig.02) have been divided into two spaces—"inside round" and "outside square." The combination of circles and squares expresses the cosmology between humans and nature. The ribbon pattern represents Mother Nature nurturing humankind and the land; the romantic charm of the yellow glaze and the ridge-red Coleus blumei illustrate the order of the imperial city; the surrounding ditch belt defines the city and reflects the spirit of the place. The "outside square" is characterized by straightness, plainness, wholeness, reason, and commonness. The surrounding sinking space creates an aura of limitless meditation; the scene of "bending down in the pond" produces the illusion of an unexpected meeting; the towering central circumferential site presents the existence of the microcosm. The "inside round" has special, dignified, mysterious, supernatural, and uncommon traits. The name Ming Garden is adopted based on the passage, "The beautiful rhythm of the sun and the moon comes from the sky; the fragrance of grains and vegetation originates from the earth(Fig.03-09)."

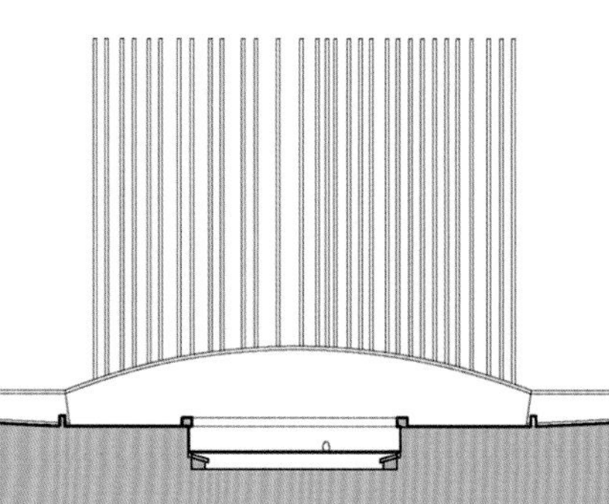

图 05 剖面图
Fig 05 Cutaway view-1

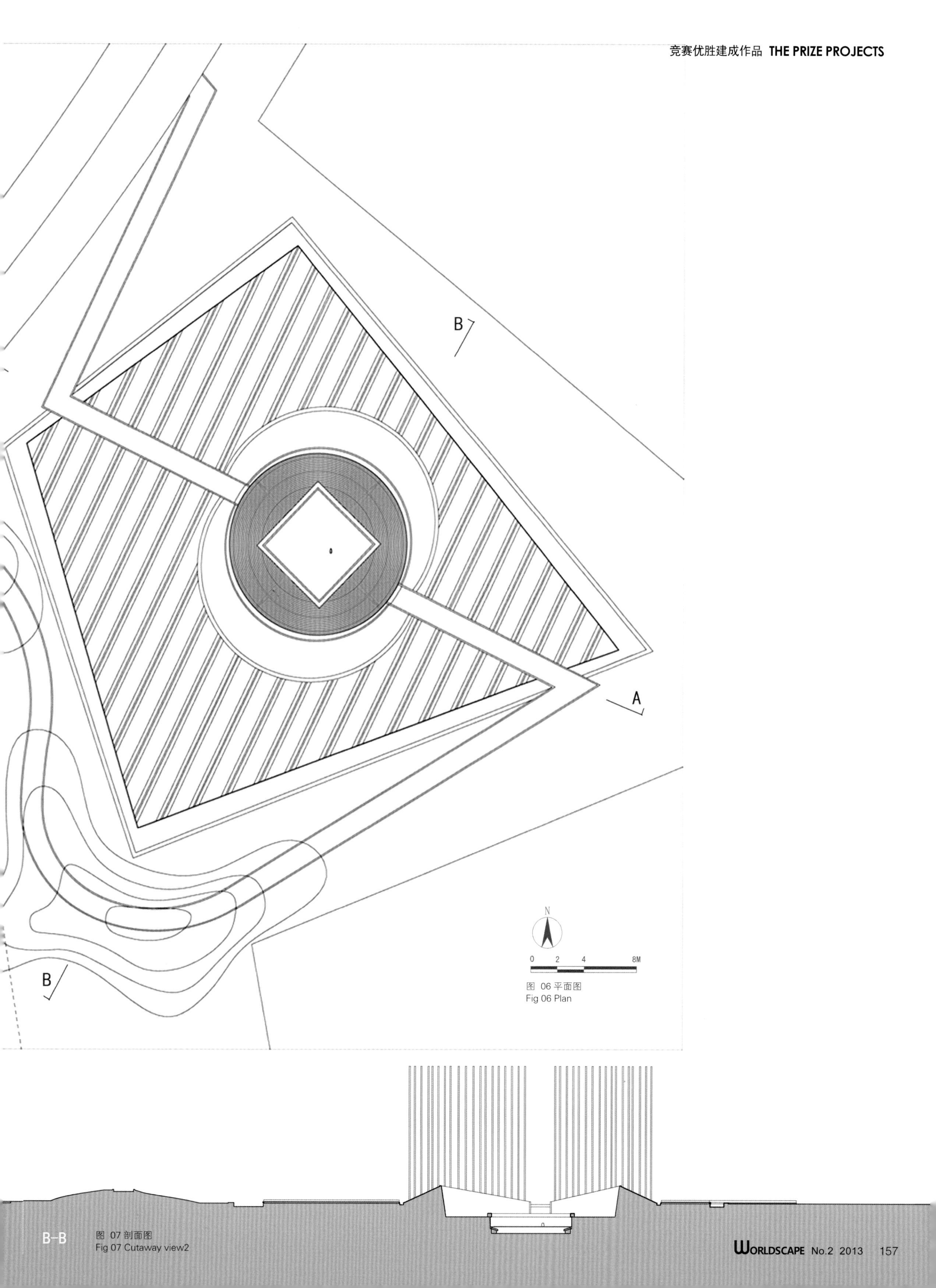

图 06 平面图
Fig 06 Plan

图 07 剖面图
Fig 07 Cutaway view2

图 08-09 模型
Fig 08-09 Model

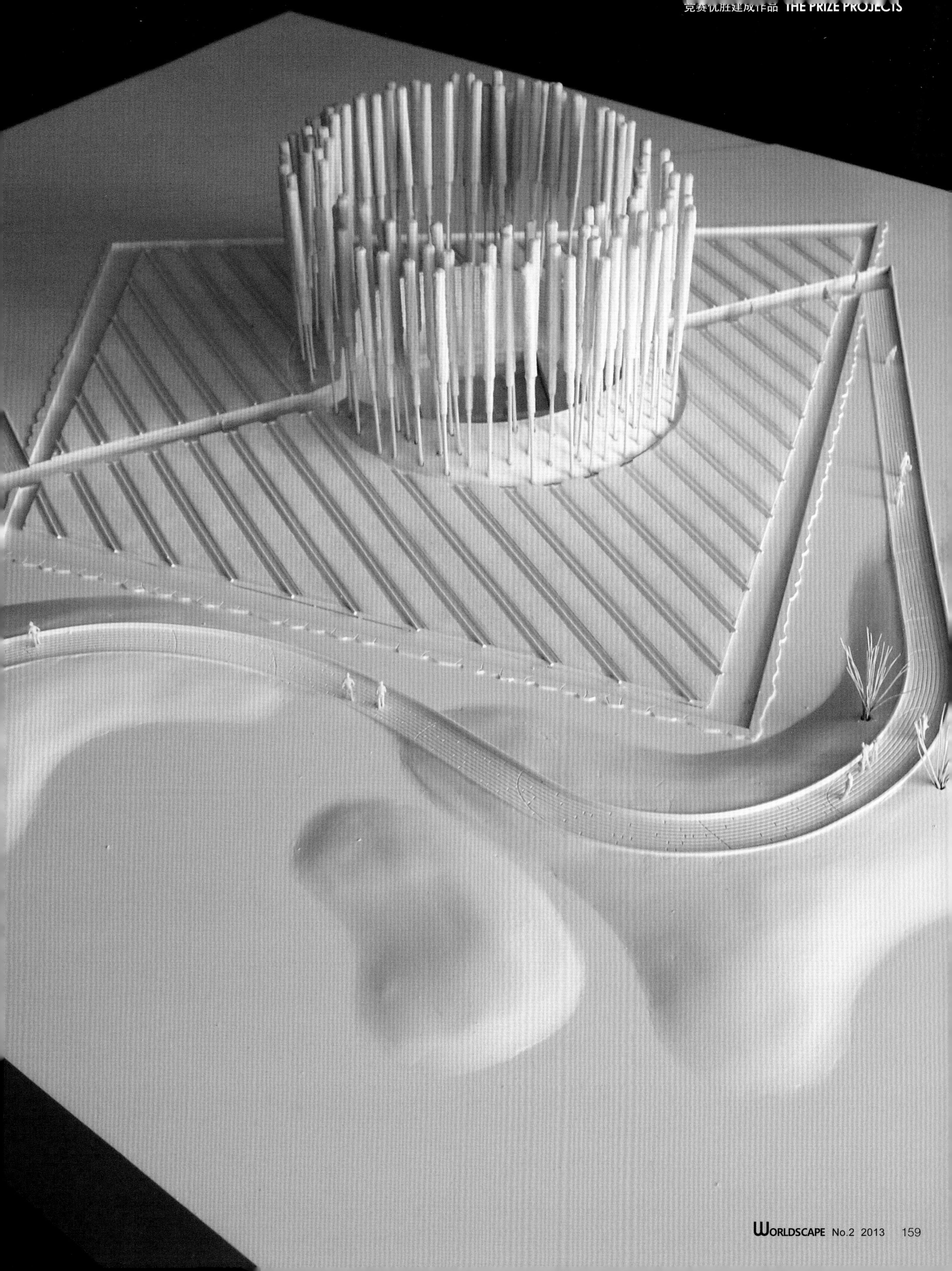

图 10 整然中的肌理，孕育着大地的生机
Fig10 The tidy texture pregnant with the vitality of the earth.

图 11 伏石沟强调“庭”的存在，演绎城池的精神
Fig11 Sunken stone ditches emphasize the presence of “yard”, expressing the spirit of city.

图 12 116 棵杨树塑造着“非日常性”的纵向空间
Fig12 116 poplars mold a vertical space of “unconventionality”.

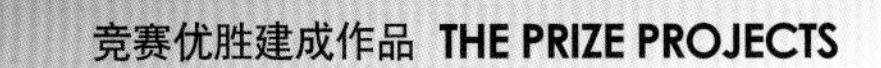

10

11

12

14

15

图 13 雾中的水池，洋溢着小宇宙的遐思
Fig13 The pool in the mist brims with the revery of microcosm.

图 14 由高渐低的弧形墙，强化空间的延伸
Fig14 The arc-shaped wall, gradually from high to low, intensifies the extension of space.

图 15 洒落墙壁的树影，勾画着时空的悟语
Fig15 The shadow of trees cast on the wall draws the apperception of space-time.

图 16 条凳的石材纹理与树影遥相呼应
Fig16 The stone texture of benches corresponds to the shadow of trees at a distance

图 17 渐变的入径，彰显着内外空间的“异”
Fig17 The gradient of the entrance trail manifests the “distinction” of inner and outer space.

16

17

图 18 高低起伏的园路，提供了场所的不同视角
Fig18 The undulating garden paths offer the diverse perspectives of the site.

图 19 夜景中的内圆空间
Fig19 Inner circle space in the night scene

图 20 LED 光带，烘托着中央（纵向）场所的存在
Fig 20 A LED light band highlights the central (longitudinal) place.

环绕内外的园路提供了俯视、平视、仰视不同角度的场域观赏；硬质外露面侧墙与地面铺装均采用边条石材（图 10），追求材料自身的内在质感和统一性；10cm 的路缘石及侧壁与地面接壤处的凹槽强化了线形的规整（图 11）；圆弧状分布的 116 棵杨树扮演着疏密变化的场景（图 12）；顶高 1.3m 的弧形坡面将横向线条延续至纵向（图 13–14）；1.5m–2.8m 的挡墙保持内•外场所空间的合与分•透与隔；7、8m 高的杨树寄托着乡土的情怀（图 15）；透过水中的琉璃瓦屋顶好似仰望星空的体感。此外，林下错落有致的地被和石材板皮（废料）、进入下沉广场的过门石、时序性的水面雾喷、模拟太空场景的背景音乐、强化线形的 LED 灯带、同心圆均一式的铺砌、下沉广场的下沉水面、祥云图案的汉白玉池壁、水面上漂浮的雕塑（图 16–29）……均传递了这样一种信息：景观设计不仅仅在于将自然界的存在进行再现，而是让这些存在能够被“看得见”。□

The path, surrounding the outside and inside of the garden, offers different perspectives form high and low elevations to create different scenery. The side wall, with its rough texture and the pavement are all constructed with strips of stone (Fig.10) which produces a material unity. The 10cm curb and the boundary channel between the side wall and the ground enhance the linear order (Fig.11). 116 aspens planted in a curve present a changing picture from thick to thin (Fig.12). The curved slopes with a height of 1.3 m extending across the line lengthwise (Fig.13-14); the retaining wall with a height of 1.5m to 2.8m keeps the connection between inside and outside, both separate and combined in a state of penetration and compartment. 8m high aspens express a sense of homeland (Fig.15); the glazed tile under the water can be seen in the sense of looking at the stars. In addition, the groundcover and stone panels with the special rhythm under the grove, the stone frame at the entrance of sinking square, the water spraying in a timed sequence, the water tank made of white jade in the shape of a cloud, and the sculpture floating on water (Fig.16-29) all express the message that landscape design not only represents nature to people but also exists in and of itself. ■

图 21 方・圆组合的空间，编制着夜空中的神秘
Fig21 The space that combines square and round symbolizes the mystery of night sky.

图 22–23 黄与红的神韵洋溢着皇城的秩序
Fig22–23 The verve of yellow and red implies the order of the imperial city.

图 24 设备箱围挡的成景
Fig24 View created by enclosing equipments.

图 25 俯视方池中的屋檐（大地），垂映天空与日月
Fig25 Overlooking the square pond that reflects eaves, sky, the sun and the moon.

26
27
28
29

图 26 夜景下的墙脚线，描绘着宇宙中的星体
Fig26 The night scene of the corner line depicts stars of the universe.

图 27 仰望夜空，好似繁星满月
Fig27 The leaves shining with silver light looks Like stars when you looking up at the night sky.

图 28 石材的肌理与祥云雕刻，寄托着京城的厚重与神秘
Fig 28 the stone texture and propitious cloud carvings demonstrate solemnity and mystery of Beijing.

图 29 围合式下沉内圆空间，营造着无限的冥思
Fig29 The enclosed and sunken inner circle space inspires the boundless meditation.

作者简介：

章俊华 / 男 / 日本千葉大学教授 / R-land 北京源树景观规划设计事务所合伙人 / 中国北京

Biography:

Junhua Zhang / Male / professor of Japan Chiba University / parterner R-land Beijing Ltd / Beijing, China

感谢：中国风景园林网
北京金五环风景园林工程有限责任公司
Appreciation：Chinese landscape architecture network
Beijing gold rings Landscape Engineering Co., Ltd.

查尔斯·沙（校订）
English reviewed by Charles Sands

声波
SOUND WAVES

巴默瑞联合设计事务所

Balmori Associates

项目位置：中国，北京，第九届北京园博会
项目面积：1,000m^2
委托单位：第九届中国（北京）国际园林博览会组委会
设计单位：巴默瑞联合设计事务所
景观设计：黛安娜·巴默瑞、诺薇密·赖福瑞蒂巴尼、凌皓欣、莫亚·卡森、马克·索曼、方晨露、杨逸伦、裴成真
完成时间：2013 年 5 月

Location: The 9th China (Beijing) International Garden Expo, Beijing, China
Area: 1,000m^2
Client: The 9th China (Beijing) International Garden Expo committee
Designer: Balmori Associates
Landscape Design: Diana Balmori, Noemie Lafaurie-Debany, Hao-Hsin Ling, Moa Karolina Carlsson, Mark Thomann, Chenlu Fang, Yi Lun Yang, Sungjin Na.
Completion: May, 2013

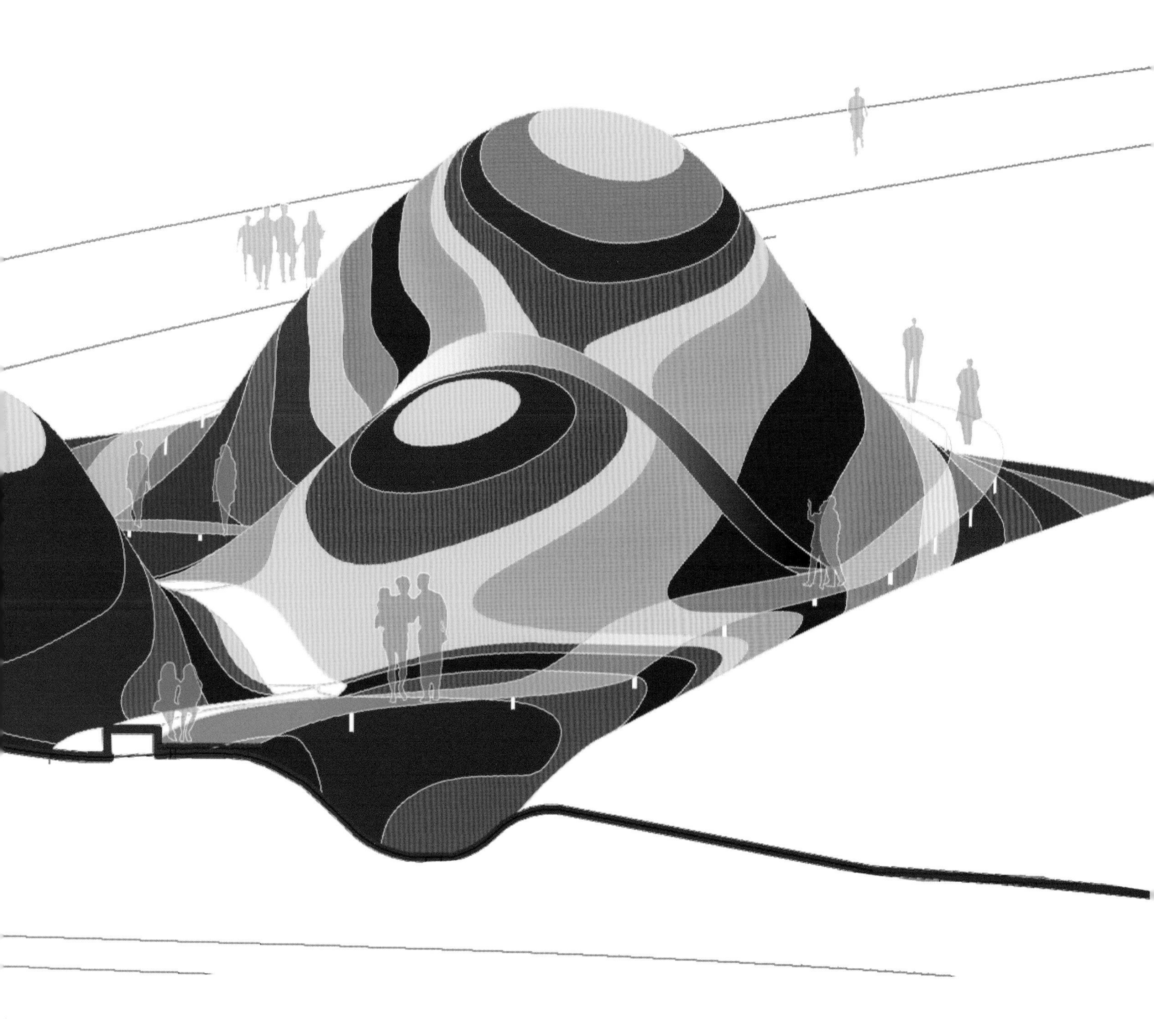

图 01 声波园的概念图
Fig 01 Conceptual painting for the garden

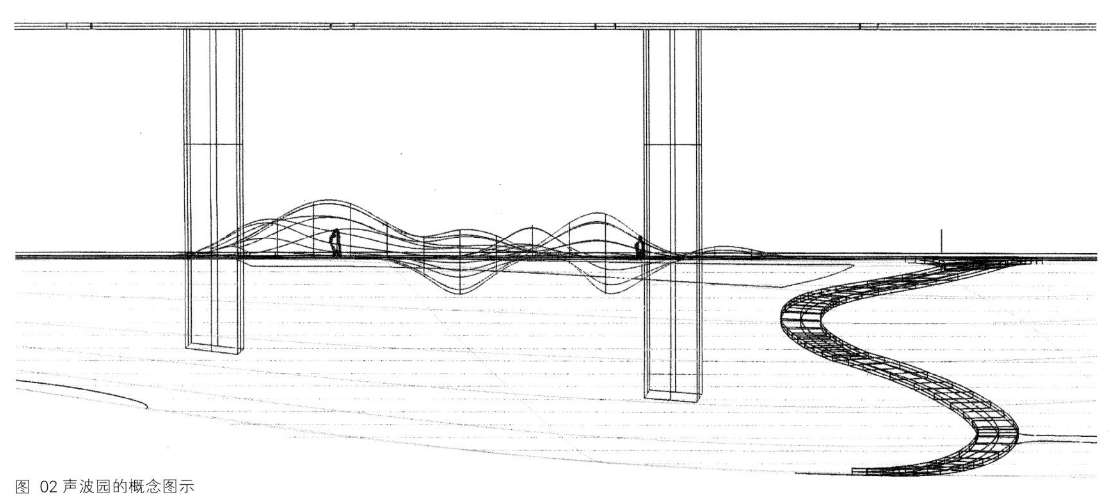

图 02 声波园的概念图示
Fig 02 Concept for the garden

在 2012 年，巴默瑞联合设计事务所受邀参与了第九届中国（北京）国际园林博览会，此一盛会展示传统与现代并重的园林设计。Sound Waves(声波）坐落于由各国知名景观规划师所设计的一系列不同的花园地块之间。它掌握了中国山水画描绘自然所激发的情感，模拟了桂林山水乐园与丽江的景象（图 01-03)。带状的植栽设计像是立体的挥毫，重新诠释了传统上对等高线的定义。造园手法不是简单地把相同高度的点连接起来，而是连接映射出园中自然环境（微气候）相同的各区块（图 04）。

为了捕捉基地上超过 140 种不同的自然环境特质（图 05-06），我们建立了一套计算模型软体来模拟花园环境中变动的信息流。先进的程式模拟技术可以分析任意特定地点在全年中的自然环境参数，包

In 2012, Balmori Associates was invited to participate in the 9th China (Beijing) International Garden Expo, which showcases both traditional and modern gardening. Situated among a series of gardens by internationally acclaimed landscape and urban practitioners, Sound Waves embodies the feelings triggered by viewing nature as depicted in Chinese landscape painting, reproducing the appearance of the magical Guilin's mountains of the Li River (Fig. 01-03). Bands of planting, like three-dimensional brushstrokes, play on the conventional reading of topographic contours, not connecting points of equal elevation, but instead mapping areas of similar conditions (Fig. 04).

图 03 早期的效果图
Fig 03 Early drawing

图 04 感观情感触动的概念图
Fig 04 Explosion of feelings

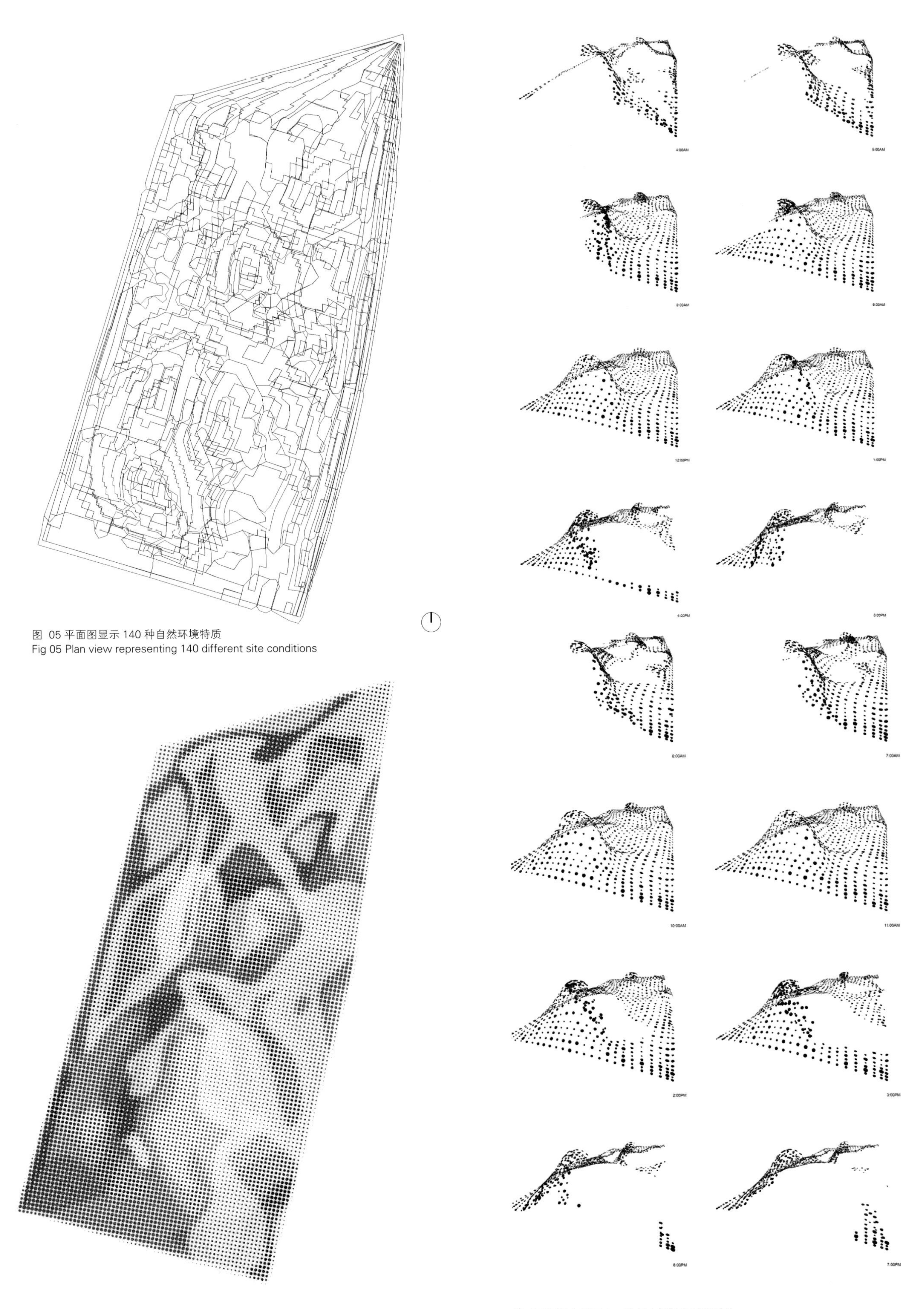

图 05 平面图显示 140 种自然环境特质
Fig 05 Plan view representing 140 different site conditions

图 06 平面图显示数据转换成植物配置带状区块
Fig 06 Plan view with data interpreted as planting bands

图 07 夏至六月二十一日每小时当日日照时数
Fig 07 Sun each hour of the day (calculated at solstice Beijing, June 21st)

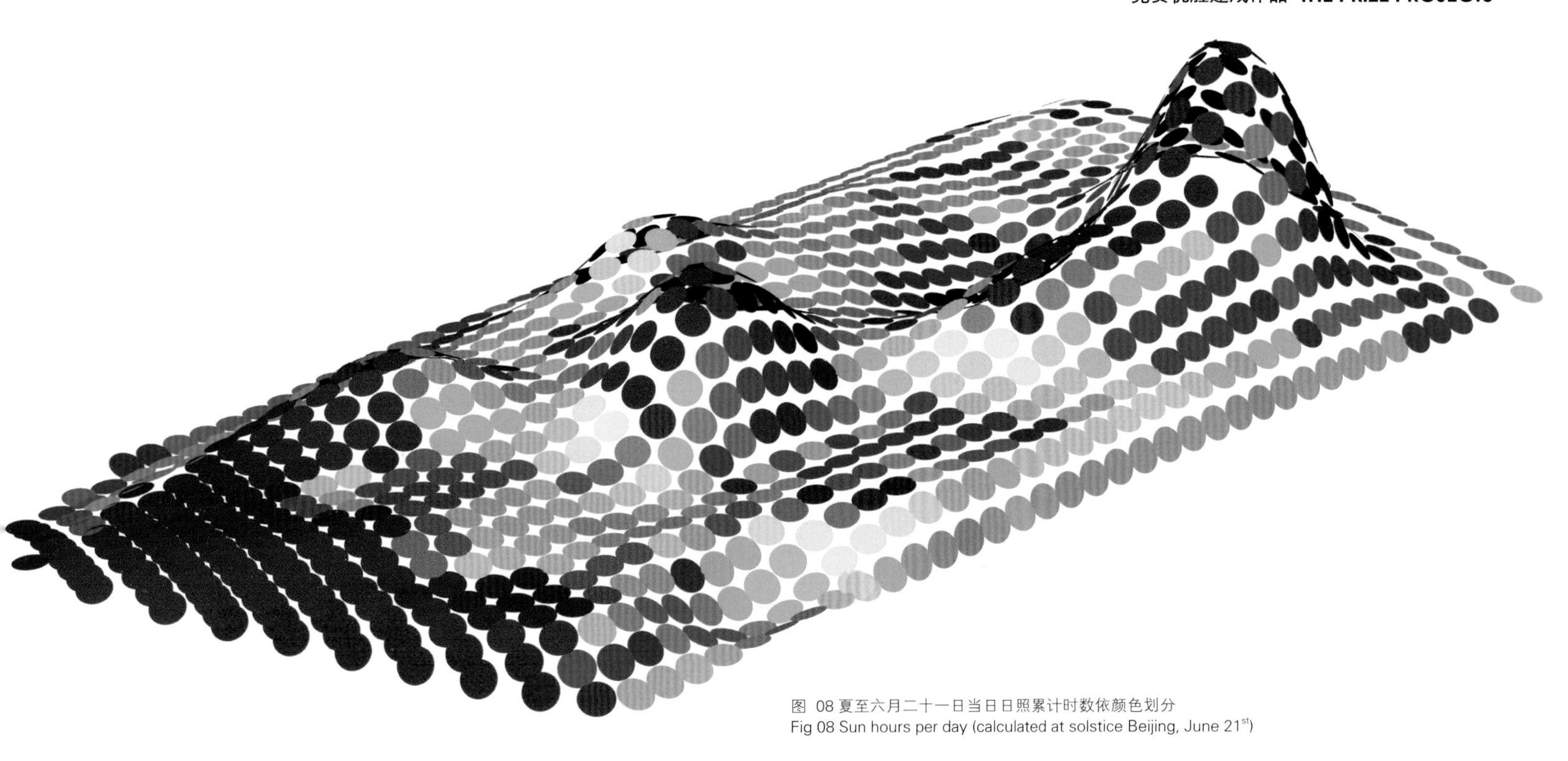

图 08 夏至六月二十一日当日日照累计时数依颜色划分
Fig 08 Sun hours per day (calculated at solstice Beijing, June 21st)

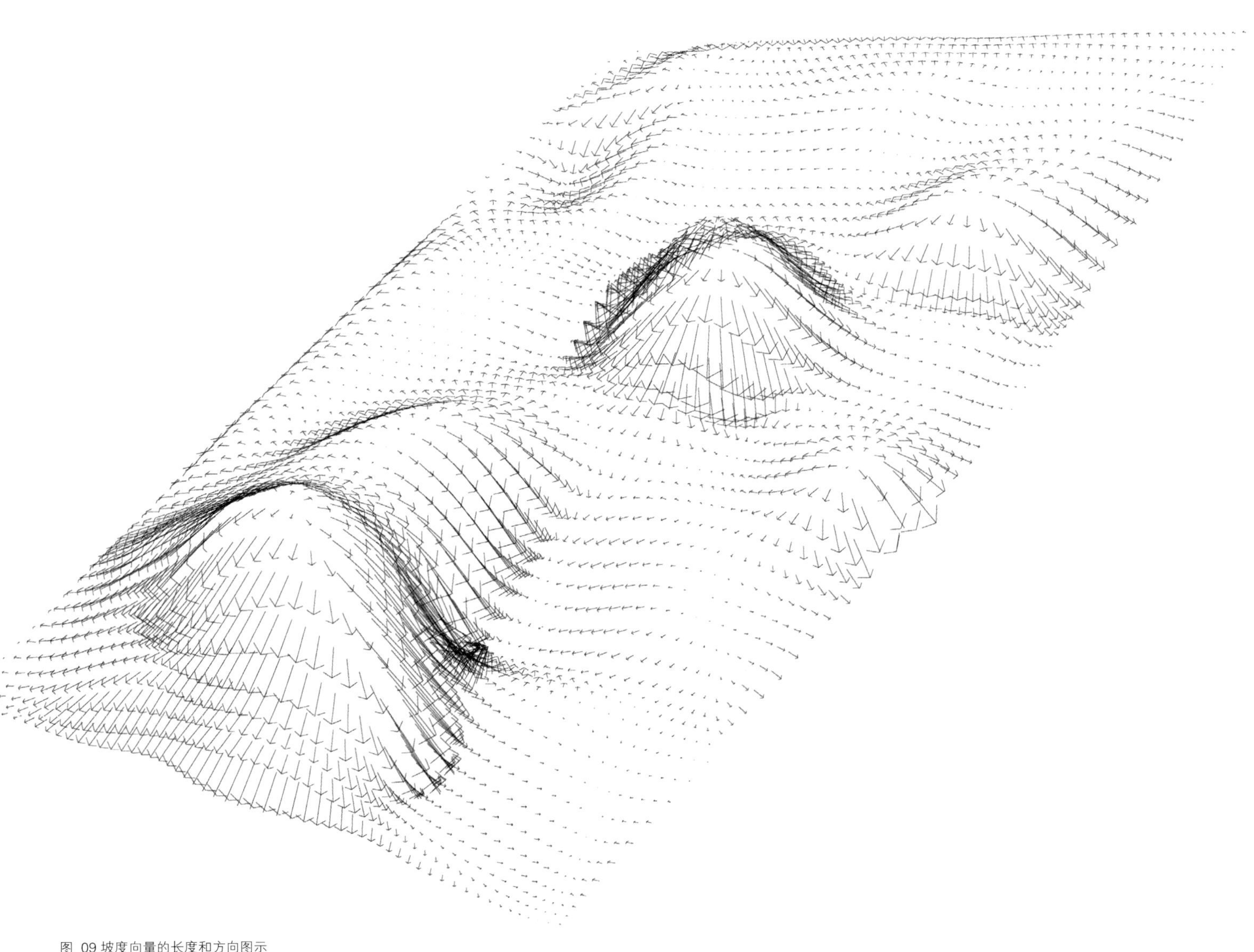

图 09 坡度向量的长度和方向图示
Fig 09 Study of vector lengths and directions of site slope conditions

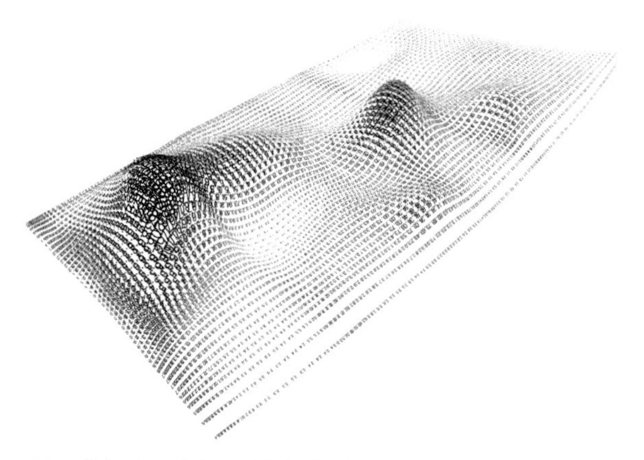

图 10 坡度示意图（数字显示出坡度，文字大小的也显示与坡度的参数相关性）
Fig 10 Slope conditions of site (number display and text size parametric)

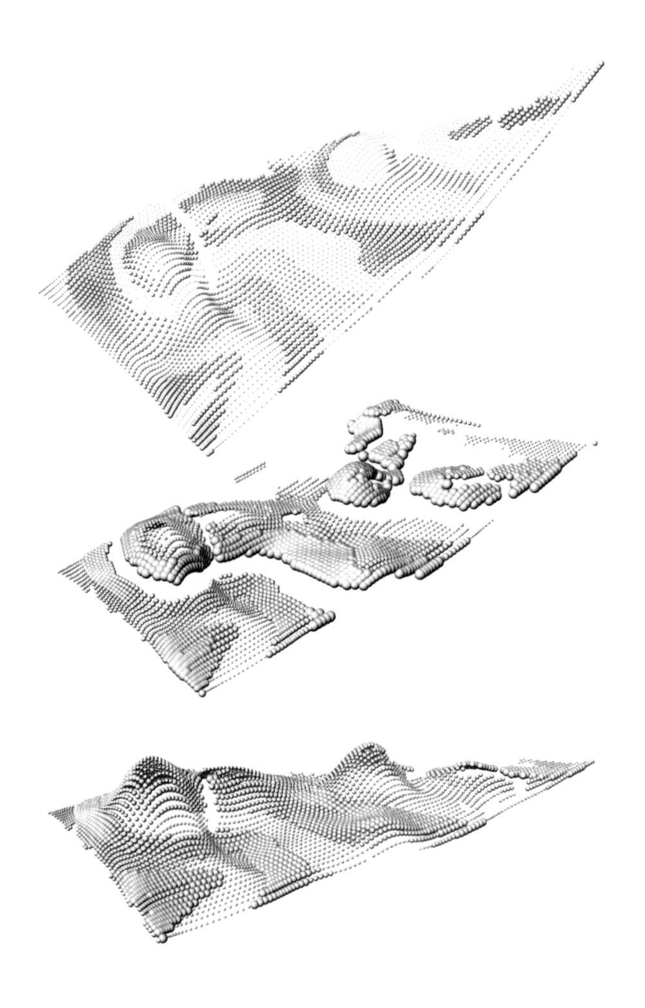

图 11 模拟植物在第一，二年及四年的生长状态
Fig 11 Bands of planting at year 1, 2, and 4.

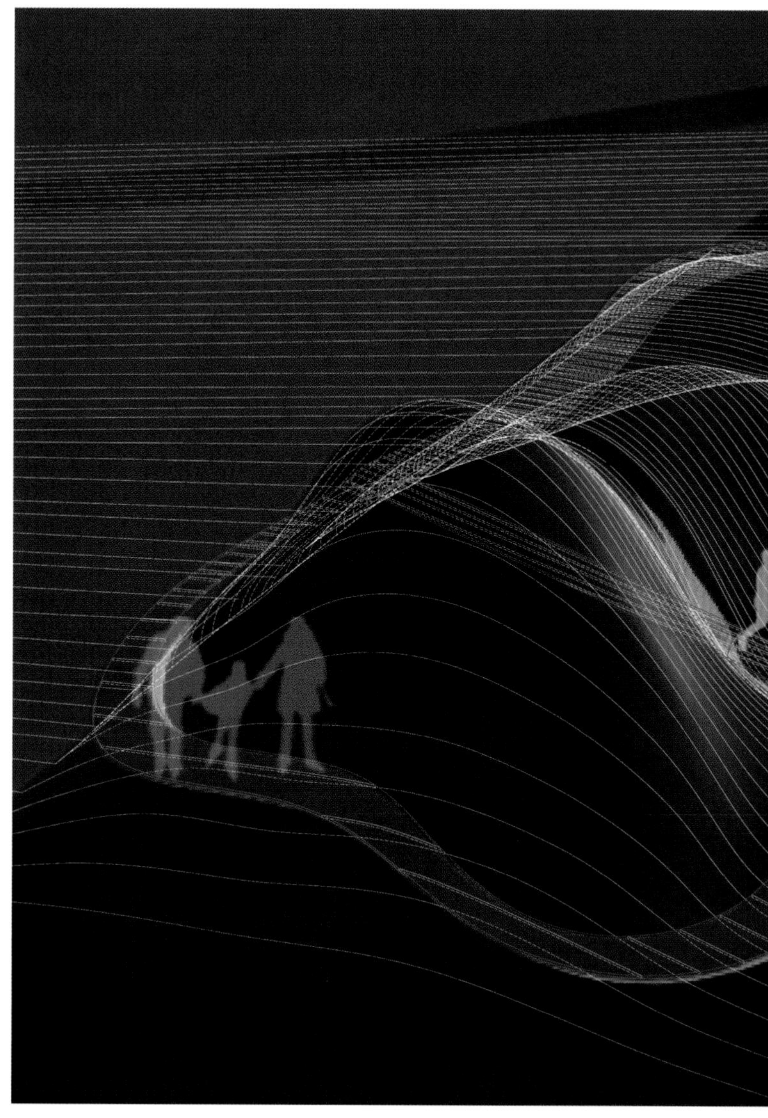

括每日光照时数（图 07–08）、坡度（图 09）、海拔高度（图 10）、风向。计算模型软体将基地细分成许多网格点，然后再逐个格点分析计算这些参数。软体并为各种植物模拟设定标准的指数增长模式，让植物可以根据计算模型缩放比例以反映各个网格点中微气候如何影响植物生长的不同（图 11）。

带状的植栽设计强化了地形的戏剧性，并且帮助稳固地貌。深根系的树木与先进的工程固坡技法结合成有效系统来维护坡形（图 12）。人们在园中体验多样的路径系统，路径在园中有上下坡度；有时盘旋于地表之上，有时又把土地錾开通过，或者就沿着地表而行；提供了观看山坡与低谷等多变的视野（图 13–14）。我们根据博览会的目标来

图 12 声波园在 2013 年 3 月 26 日的建设情况（摄影：凌皓欣）
Fig 12 The garden as of March 26, 2013.（Photo Credit: Hao-Hsin Ling）

To capture over 140 different site conditions (Fig.05-06), Balmori constructed a parametric computational model of the garden that adapts to and aligns with transient information flows. Advanced programming methodologies allow the model to analyze year-round natural conditions of a particular area of the site, including sun hours per day (Fig.07-08), slope conditions (Fig.09), altitude (Fig.10), and wind exposure. The model performs by subdividing the site into a fine grid of points, which are then analyzed individually. A 'standard' index growth pattern for each plant species is identified, situated, adapted and 'scaled' by the

图 13 路径与活动
Fig 13 Occupy the paths

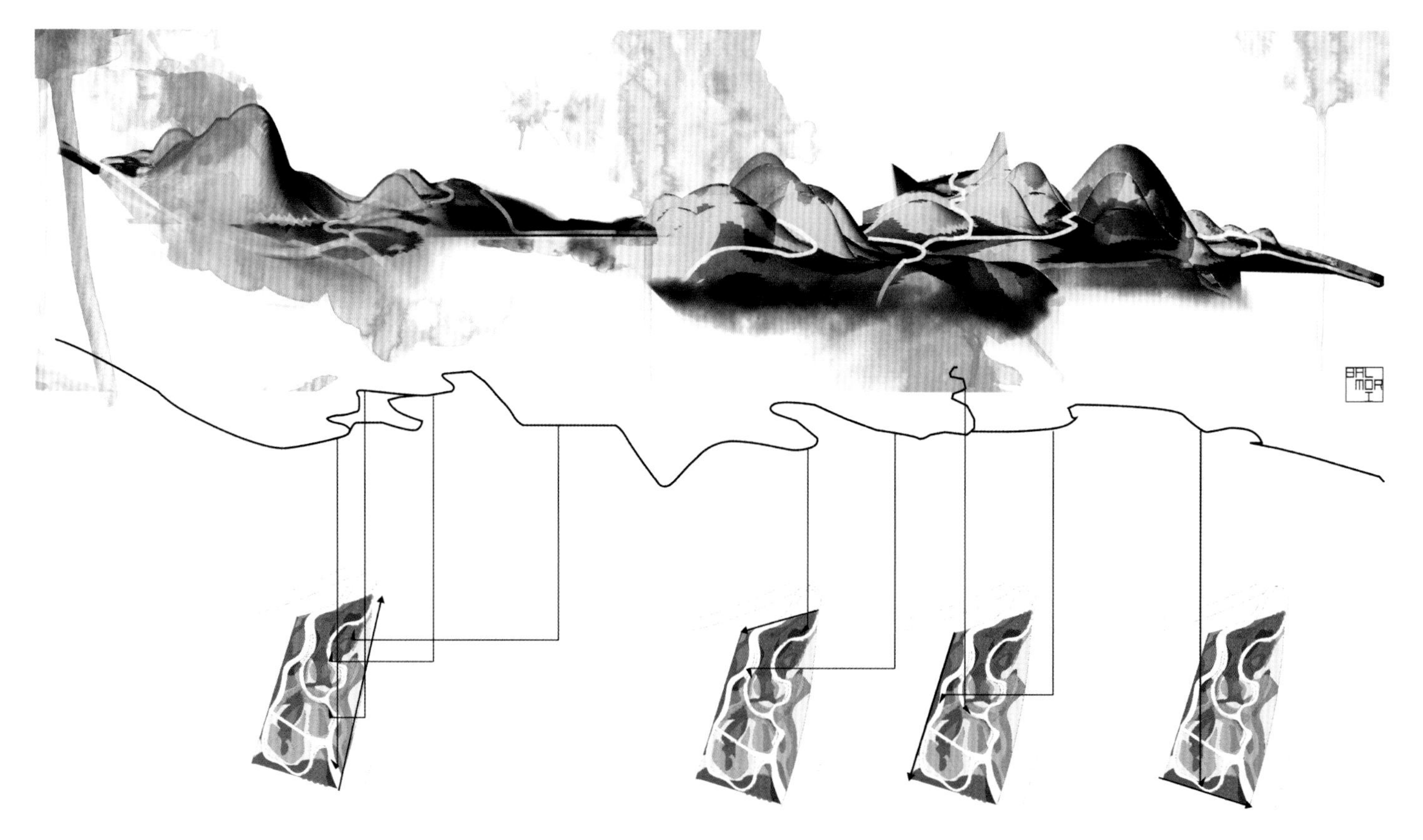

图 14 花园的多视角效果图从动线上开展开来
Fig 14 Perspectives unveil through the circulation in the garden

挑选植物并专注于植物的颜色季节变化（图 15–16）、质地、气味和它们净化空气的能力来选择。所以在植栽设计方面，巴默瑞的花园致力于全年为附近居民与游客提供与自然接触的空间，让人流连其中（图 17）。□

computational model to reflect the effect of the varying local condition the site may have on its growth (Fig.11).

Ribbons of trees reinforce the reading of the dramatic topography. The trees' deep root systems combined with the latest technology in engineered earth stabilization maintains the shape of the landforms (Fig.12). The experience is shaped by various paths that ascend and descend through the garden, hovering above and cutting through the site to offer perspectives to the hills and over the valleys (Fig. 13-14). Balmori's selection of plants builds upon the goals of the Expo with a focus on seasonal colors (Fig.15-16), textures, smells, and capacity to clean the city's polluted air. To this end, Balmori's garden will provide a yearlong retreat for locals and visitors alike, offering a place to connect with nature and linger about (Fig.17).■

作者简介：

黛安娜·巴默瑞，博士，美国景观设计师协会资深会员（FASLA），国际景观建筑师协会会员（IFLA），纽约市巴默瑞联合设计事务所（Balmori Associates）公司主要设计可持续性设施作为景观和建筑之间的纽带，已获得国际间广泛认可。

公司的设计理念是通过景观来探索自然和建筑物之间的交界。通过创造性思维、水文学、生态学和四维空间分析，形成艺术和功能设计结合的美学。这一设计理念涉及建筑、工程、生态和艺术等多种领域，目前我们为欧洲、亚洲和美国的公共和私营业主设计的多个项目囊括了多项大奖。

她自 1993 年起就在耶鲁大学建筑学院以及林学与环境研究学院担任评论员。20 多年来，她通过发表众多出版物和在业内活跃的表现，成为一名理论和艺术设计师。她的最新著作是《基础：景观与建筑之间》（Groundwork: Between Landscape and Architecture，兰登书屋，2011 年，与乔尔·桑德斯 Joel Sanders）合著《景观宣言》（A Landscape Manifesto, 耶鲁大学出版，2010 年。）

Biography:

Diana Balmori, Ph.D., FASLA, IFLA, established Balmori Associates in New York City. The practice is recognized worldwide for designing sustainable infrastructures that serve as the interface between landscape and architecture.

The firm's approach is rooted in the exploration of the boundaries between nature and structure through landscape. Inventive thinking, and analysis of hydrological, ecological and temporal dimension leads to an artistic and functional design aesthetic. This approach involves architects, engineers, ecologists and artists and has led to a portfolio of award winning projects for public and private clients in Europe, Asia and America.

Diana Balmori has been a Critic at Yale University since 1993, both in the School of Architecture and the School of Forestry and Environmental Studies. For more than 20 years, Balmori has established herself as a theoretical and artistic designer through a variety of publications and active role in the community. Her most recent books include Groundwork: Between Landscape and Architecture, written with Joel Sanders (Random House, September 2011) and A Landscape Manifesto (Yale University Press, 2010.)

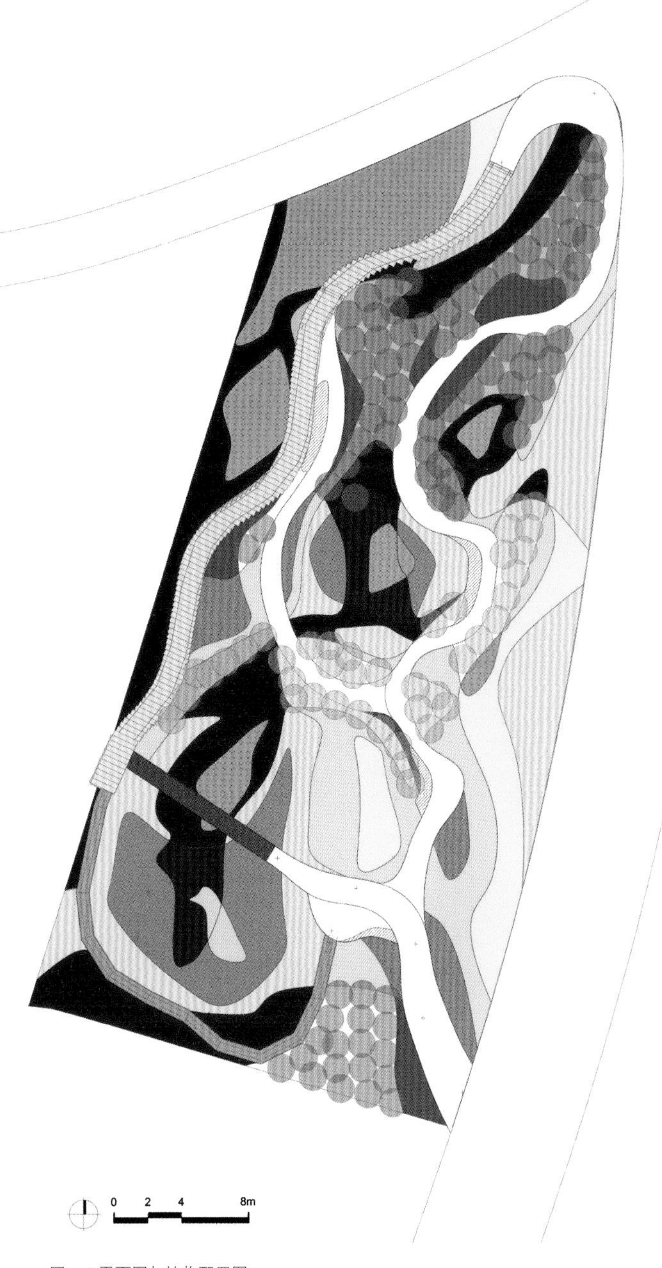

图 15 平面图与植物配置图
Fig 15 Plan and planting palette

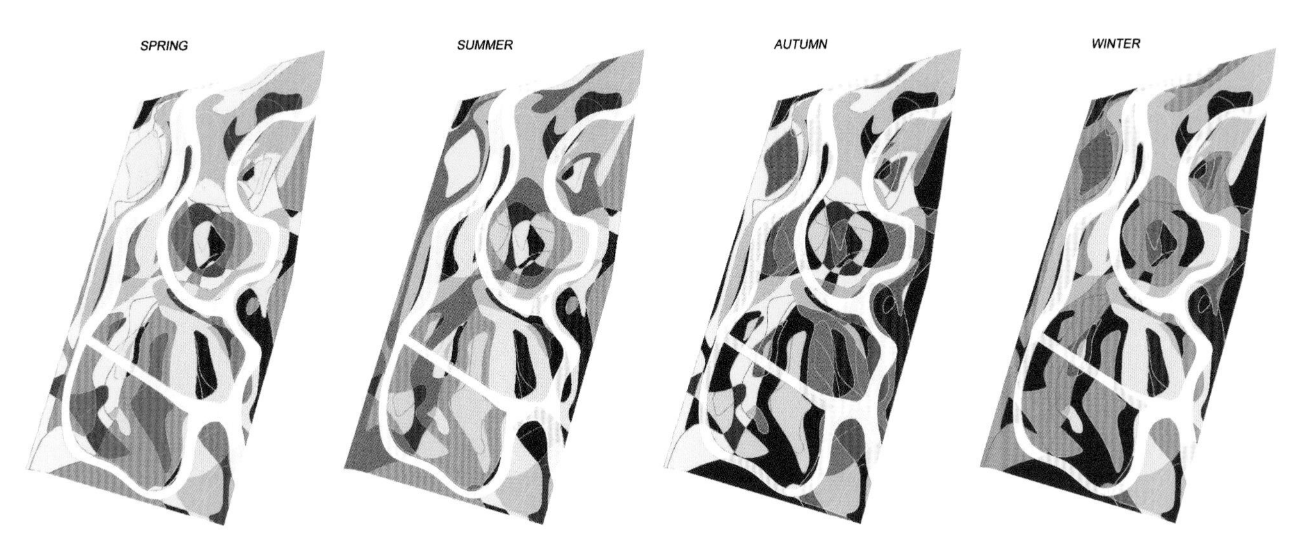

图 16 四季色彩图示
Fig 16 Colored plans for each season

图 17 在声波园中
Fig 17 In the garden

凌皓欣（中译）
Translated by Hao-Hsin Ling

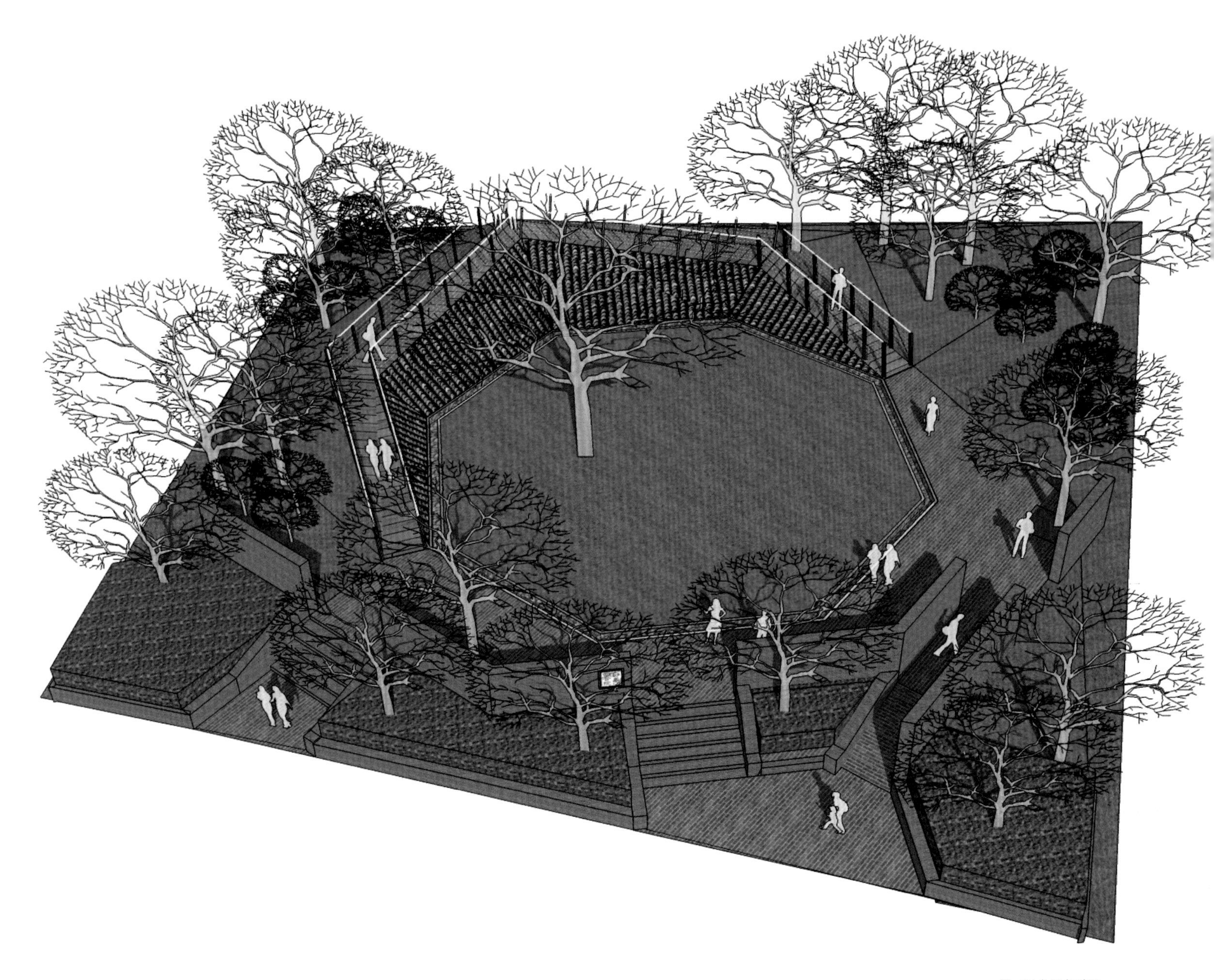

图 02 全园鸟瞰图
Fig 02 bird's eye view

印象四合院 IMPRESSION QUADRANGLE

张新宇　　Xinyu Zhang

项目位置：中国，北京，第九届北京园博会
项目面积：1,000m^2
委托单位：第九届中国（北京）国际园林博览会组委会
设计单位：北京市园林古建设计研究院有限公司
景观设计：张新宇
完成时间：2013 年 5 月

Location: The 9th China (Beijing) International Garden Expo, Beijing, China
Area: 1,000m^2
Client: The 9th China (Beijing) International Garden Expo committee
Designer: Beijing Institute of Landscape and Traditional Architectural Design and Research Ltd. Co.
Landscape Design: Xinyu Zhang
Completion: May, 2013

图 01 总平面
Fig 01 site-plan

❶ 木栈道 wooden trestle
❷ 瓦坡 slop of watt
❸ 八边形水池 octagonal courtyard
❹ 栏杆 handrail
❺ 灰砖铺地 gray brick floor
❻ 斜面草坡 grass slope
❼ 景墙 landscape wall
❽ 台阶 step
❾ 种植池 ramp

图 03 从入口看印象四合院
Fig 03 view from entrance

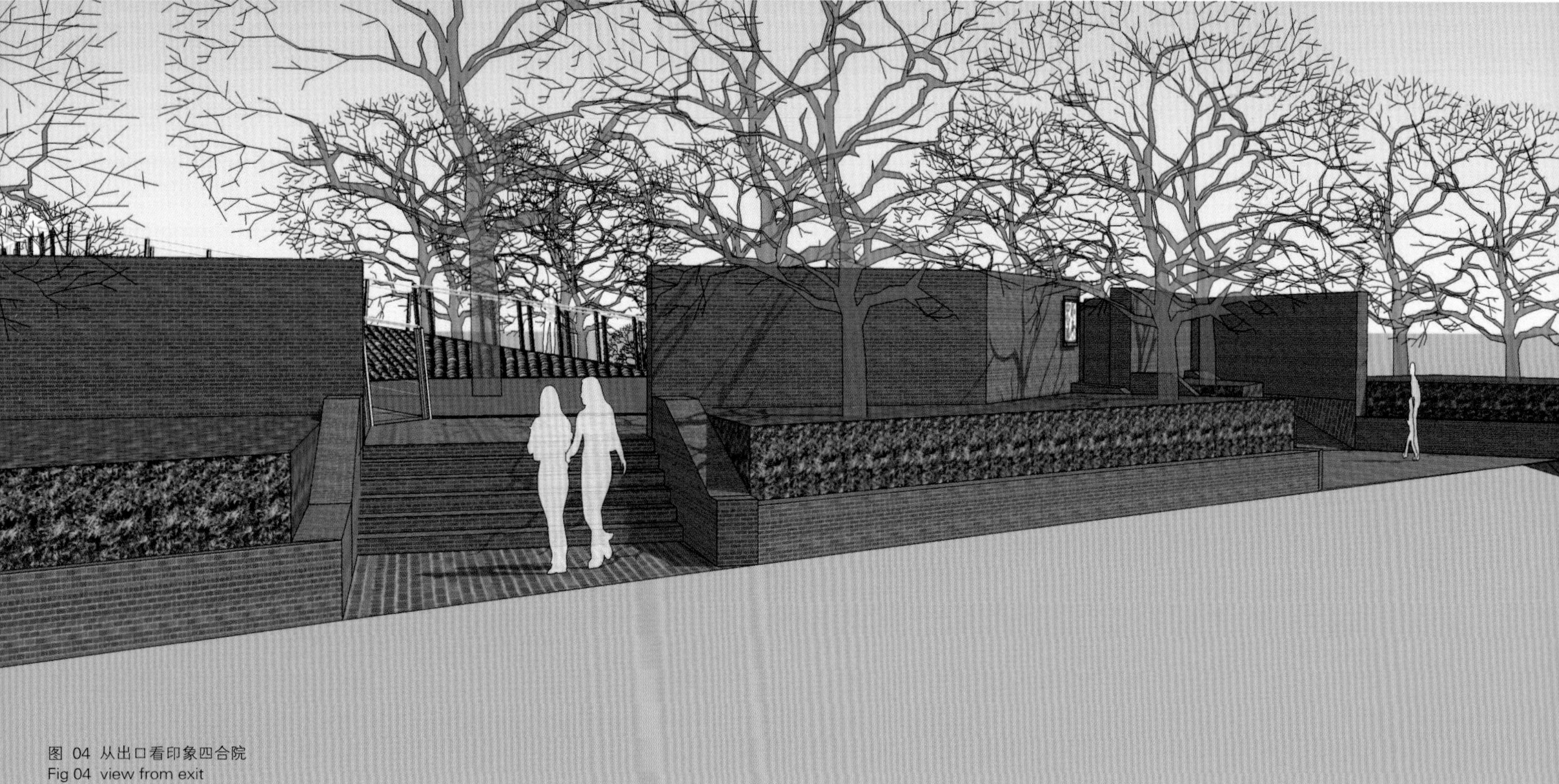
图 04 从出口看印象四合院
Fig 04 view from exit

该项目位于园博园内锦绣谷北侧，永定河南岸设计师广场，为第九届中国（北京）国际园林博览会 6 个设计师园之一，东侧与日本设计师三谷徹的大师园相邻，西侧与英国设计师的“北京花园”相邻。整个设计师广场由 3 个大师园和 6 个设计师园组成，代表了目前国际一流的设计水平。

该园总面积 1,000m^2（图 01–02），由阶梯式入口（图 03–05）、坡道、灰砖院墙、八边形庭院（图 06–09）、瓦坡跌水（图 10）几部分组成。以北京四合院为创作题材，通过现代的设计手法，将传统与现代材料创新应用，艺术地再现了人们记忆中的四合院。

印象四合院 美丽北京城

印象四合院之所以能够获得第九届园博会设计师园评比优秀奖，我想应该与这个题材的选取有很大的关系。四合院是地道的北京元素，而且是贴近大众的选题，同时它也勾起了许多人久远的记忆。作为土

As part of the 9th China (Beijing) International Garden Expo, the project is located on the south bank of Yong Ding River in one of the six designer's gardens in Designer Square. The site is next to Beijing Garden and Master's Garden designed by the famous Japanese designer San Guche. The area includes 3 master's gardens and 6 designer's gardens, which are representative of the highest level of contemporary landscape design.

The project area is 1000m2 (Fig.01-02). Consisting of an entrance(Fig.03-05), a ramp, walls, an octagonal courtyard (Fig.06-09)and a waterfall (Fig.10)etc., the garden is constructed of traditional courtyard elements inspired by historical courtyards in Beijing. Combining traditional and contemporary elements, it arouses the memory.

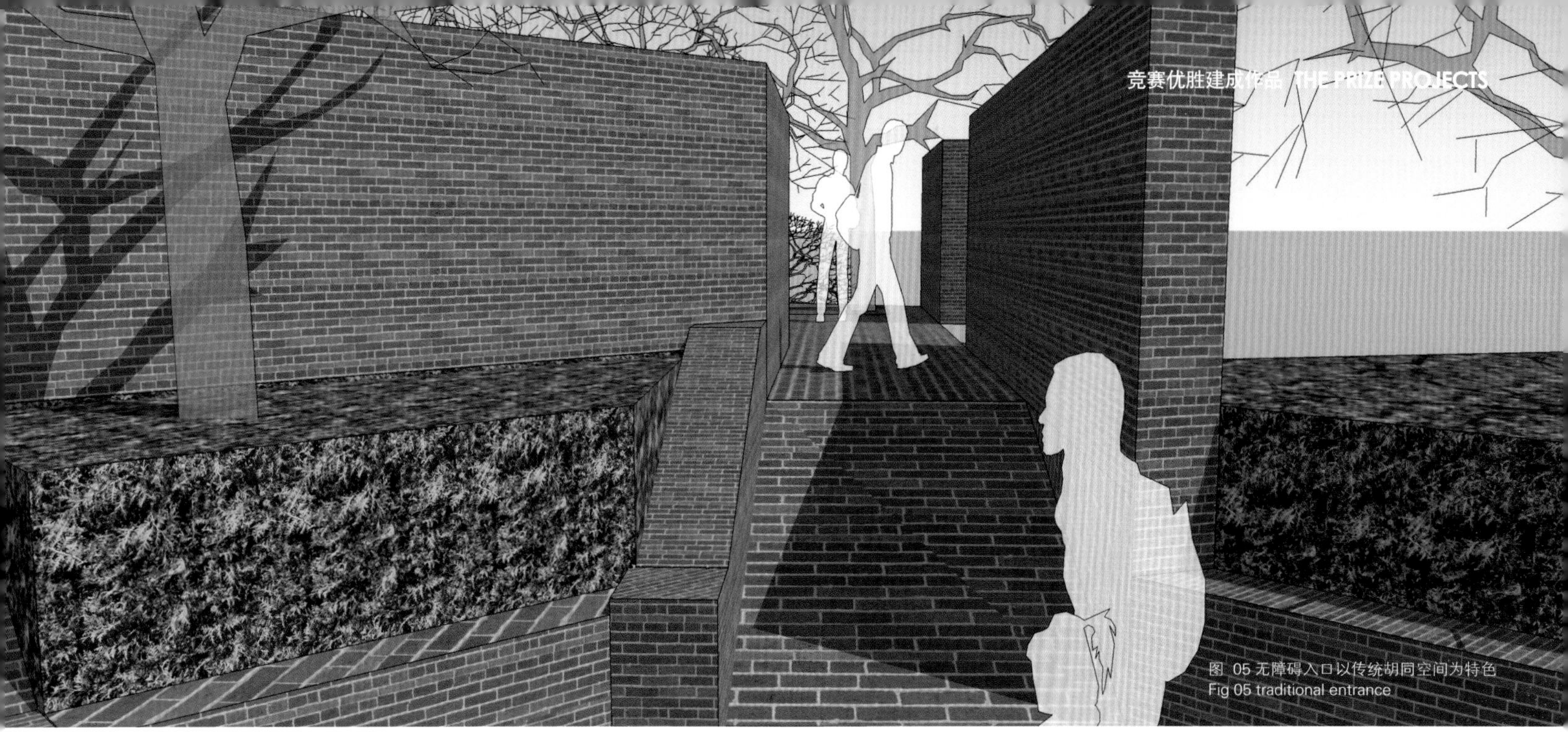

图 05 无障碍入口以传统胡同空间为特色
Fig 05 traditional entrance

图 06 印象四合院内部空间
Fig 06 interior space of Impression Quadrangle

生土长的北京人，参加设计师园的设计方案征集首先想到的就是如何突出北京的特色，展现北京的文化，因此方案的设计主题很快就锁定为四合院。

对于四合院这种传统的居住形式，不同的人对其有不同的认识，而且这种认识上的差异还是大相径庭，截然不同的。有人认为四合院代表了等级与秩序；有人认为它代表了封闭与保守；还有人认为它代表了懦弱与苟且；当然更有人认为它代表了人与自然的和谐。林语堂在《迷人的北平》一文中写到“世界上没有一个城市像北平一样的近于理想，注意自然、文化和生活的方法……这是一个理想的城市，那里有空旷的地方使每个人都得到新鲜的空气”。

在建设世界城市的今天，我们身处钢筋水泥的丛林，拥挤而喧嚣。我们应如何看待四合院在现今城市中的作用与地位，发现并重新认识

Impression Quadrangle

I think the reason Impression Quadrangle (also called Sihe-Courtyard in Chinese) won the "Award for Excellence" in the garden competition for landscape architects at the 9th China (Beijing) International Garden Expo is its theme. A quadrangle is a historical type of residence that was commonly found throughout China, most famously in Beijing. People are familiar with this topic, which can trigger people's long-faded memories. As a Beijing native, when I tried to find a theme for my project, the first thing that came to my mind is how to highlight the features of Beijing and show Beijing culture. Then I settled on the design theme of the courtyard.

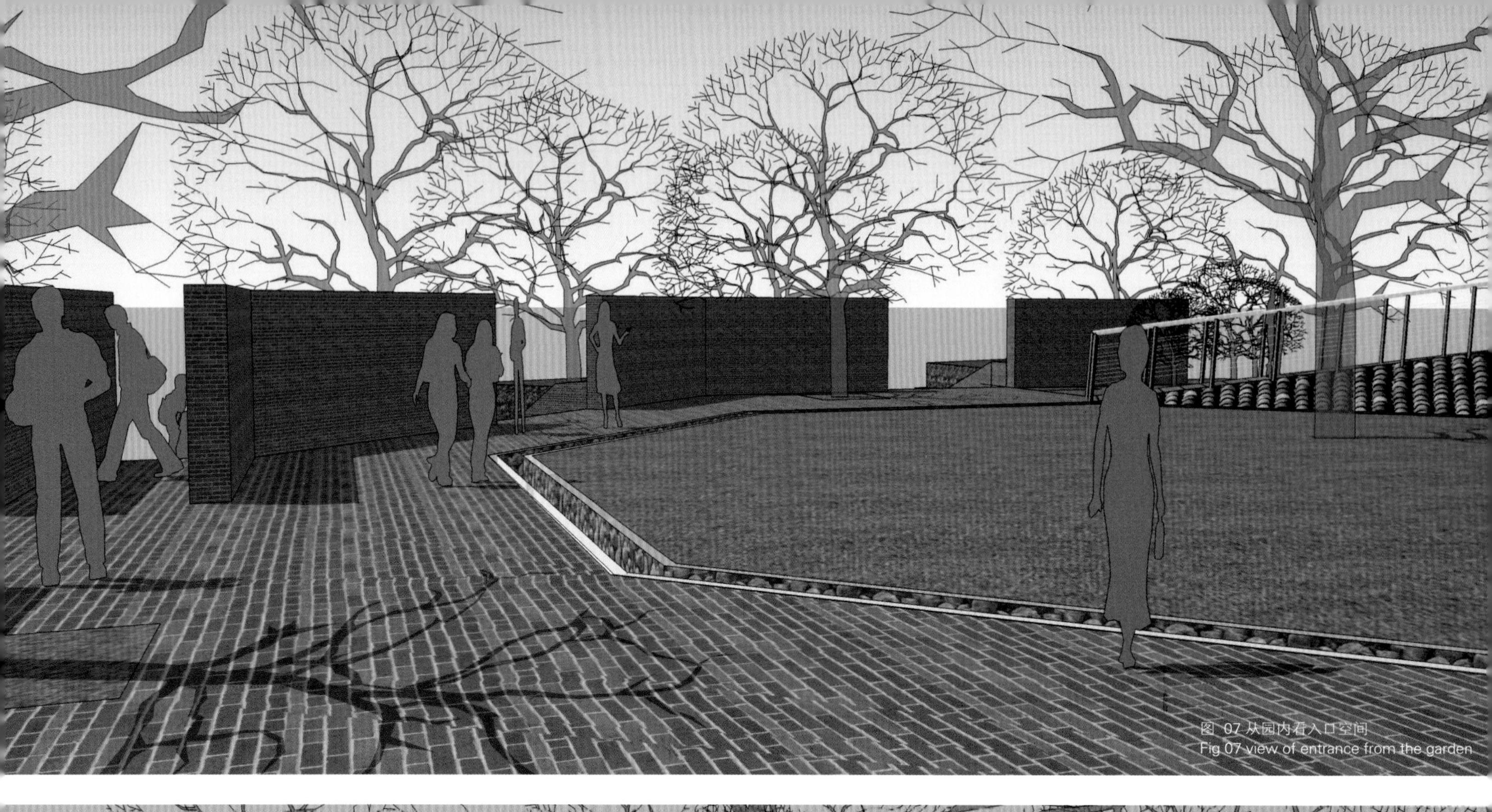

图 07 从园内看入口空间
Fig 07 view of entrance from the garden

图 08 从栈桥看院景
Fig 08 view of garden from little bridge

四合院蕴含的“实用价值、精神价值、文化价值”，这应是我创作《印象四合院》的初衷吧。

“要想领略北平的美，最好是坐飞机来一个鸟瞰，房子隐隐呈现在枝叶下面，粼粼散开，匀整而不单调”。

北京城之所以曾经成为一个只见树木不见屋顶的绿色都市，正是因为整座城市都是由四合院这样的“绿色细胞”构成。如今的北京已特色不在，一个重要的原因就是大片四合院的消失。

2008 年，沿着西二环从复兴门向北行驶，以前还可以俯瞰到大片的浓绿不见了，取而代之的是南起复兴门立交桥北到阜成门立交桥，

For such a historical type of residence, people hold varied opinions, which can be dramatically different. Some think it means hierarchy and order; some believe it stands for privacy and conservatism; some consider it as cowardice and resignation; some others regard it as harmony between humans and nature. "There is no city in the world like Peking which is an ideal city focusing on nature, culture and living... It is an ideal city in which everyone gets plenty of fresh air", Yutang Lin wrote in Peking—A Fascinating City.

In today's world, in the process of transforming Beijing into a

图 09 从花园空间看全院
Fig 09 view of the whole area from the garden

图 10 瓦坡
Fig 10 slop of watt

全长 1700 余 m、宽 600 余 m 的范围内，崛起的一片具有浓郁时代色彩的建筑群——金融街核心区。金融街地区在十余年建设间共消失胡同 53 条，而闹市口大街、太平桥大街、广宁伯街、武定侯街、锦什坊街、成方街等虽然名称依然存在，但早已是旧貌换了新颜。如果将名存而实亡的胡同算上去，金融街地区已消失胡同 64 条，占西城区消失胡同的近三分之一。金融街建成后，虽形成各类绿地近 30 余公顷，得到绿地的同时树木的总量却降低了，绿色消失了。“得到想要的，失去拥有的”一得一失之间是喜抑或是忧呢？

四合院无论是一处院落还是一片院落，也不论是小院子还是大宅门，不管你从什么位置、角度、距离观察它，绿树浓荫都是映入眼帘、

global city, it has become a concrete jungle, crowded and noisy. The original intention of the design was to propose the question—'how should we think of the function and role of courtyard in today's cities', and make people rediscover the practical value, spiritual value and cultural value of courtyard.

"If you want to enjoy the beauty of Peking, the best way is to get a bird's-eye view in a plane. You will see houses indistinctly under branches and leaves, orderly scattered, even and aesthetic".

The reason why Beijing used to be an environmental-friendly

铭刻于心的最深印象。

正如谢冰莹在《北平之恋》中所述“北平市，就像一所大公园，遍地有树，处处有花；每一家院子里，不论贫的富的，总栽得有几棵树，几盆花。房子的排列又是那么整齐，小巧。所以谁都说北平最适宜住家。在胡同里的小院子里，你和孩子们一家过得很清静，很舒服，绝对没有人来打扰你；即使住在闹市附近，也没有那么多的车马声传进你的耳鼓。”。无论如何，以上所述的一切都不是要把人们引向对过往的留恋，更不是要是古非今责难于谁，而是希望大众真正认识到四合院是最绿色生态、和谐宜居的生活模式。它比高楼大厦更接地气，它比花园洋房更加静谧温馨。

北京旧城，也就是二环路凸字形范围内，面积约62.5km^2。此范围内的生态建设，除了要发挥北海、景山、天坛等历史名园的作用外，更要保护好大面积四合院区，保护好四合院内的树木。北京划定的25片历史文化保护区占地10.38km^2，占旧城总面积的17%。其中皇城中有14片，皇城外的内城有7片，还有4片分布在外城。保护好这些四合院，并尽可能扩大保护范围，则有可能使皇城范围变成北京旧城范围内的“绿心”，保护好四合院就是最大程度地保护了北京的绿肺。

坐落在鼓楼与地安门之间的南锣鼓巷，至今已有740多年的历史。近几年，这里已经逐渐取代三里屯和后海，成为北京新的时尚地标。美国《时代》周刊，最近精心挑选出了亚洲25处你不得不去的好玩儿的地方，中国有6处被选中，南锣鼓巷名列其中。徜徉在绿荫浓郁的南锣鼓巷，传统与时尚的碰撞使古老的街区散发出与时俱进的活力，不同地域、不同肤色的人们在此找寻与探究着各自心中的北京。

今天的北京已不再是二环路内皇城根下的旧京，胡同四合院也已不再静谧，在北京城变得越来越多元越来越混搭的当下，四合院与胡同的生态、文化与精神价值愈发彰显。在国家五位一体的建设布局中，生态文明建设与文化建设是关乎城市发展的两个重要层面。实践使我们清晰地认识到四合院正是集文化与生态和谐于一的载体。□

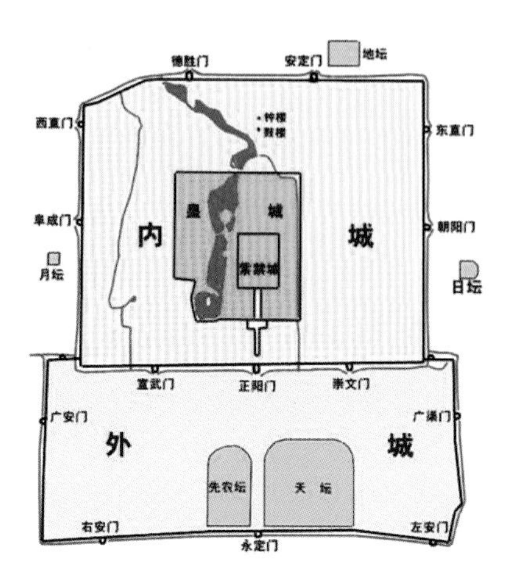

city, is that the whole city was a “green cell” much like the quadrangle. However, these characteristics no longer exist, partly because of the disappearance of the quadrangle.

In 2008, along the road of Fuxingmen to the north I found overpasses from Fuxingmen to Fuchengmen instead of the large number of trees that used to be there. Within an area of 1700X600m a large group of outstanding contemporary buildings appeared at the heart of the financial district. 53 Hutongs (traditional alleys) disappeared during the decade of rapid expansion on old streets like Naoshikou Street, Taipingqiao Street, Guangningbo Street which no longer carry any of their past look. 64 Hutongs disappeared within the financial district, while the total number of Hutongs disappearing in Xicheng district is about 180. After the financial district was rebuilt, there were almost 30 hectare of new green land but fewer trees than before. It is like

a comedy as we get what we want without retaining what we already have.

Whatever Quadrangle or hutong, whatever small yard or big house, whatever location or district, what deeply impresses our hearts is plants.

As Xie Bingying described in Love in Beijing, "Beijing city is like a big park with trees and flowers everywhere. In every yard, no matter poor or rich, there are always a few trees and some flowers. The arrangement of the houses is so neat and compact making Beijing the most suitable place to live. Inside little yards and alleys, people with their kids can enjoy a very quiet and comfortable time with no worry about others disturbing them, even if they live near downtown." Anyway, the purpose of all the above is not to lead people to the past, nor to lay blame, but to help people realize that green ecological courtyards offer a harmonious mode of living. They are more meaningful than high buildings and more warming than standard houses.

Beijing old city, which refers to the '凸' shaped area inside the Second Ring road, covers 62.5 square kilometers. Ecological sites in this area include Beihai park, Jingshan, and other famous historical gardens, but there should also be a focus on protecting traditional quadragles. The 25 historical and cultural protection zones in this metropolis cover a total area of 10.38 square kilometers, accounting for 17% of the total area of the old city. Among them are 14 areas in the imperial city, 7 areas outside the inner city and four slices in the outer urban area. It seems that maintaining the quadrangles and expanding the scope of protection as much as possible could allow the Forbidden City to become the "green heart" of the old district. If we want to protect the 'green lung' of Beijing as much as possible, we need to care for the quadrangles.

Located between the Drum Tower and the Ground Gate is the Nanluoguxiang, which has existed for 740 years. In recent years, it has gradually replaced Sanlitun and Houhai, to become the city's new fashionable area in Beijing. Recently, America's Time magazine, selected 25 recommended places in Asia. Of the 25 places, six were Chinese, including Nanluoguxiang. Roaming in the shade of Nanluoguxiang, people can feel the collision of tradition and fashion. What's more, the ancient block sends out the vigor of the contemporary world. People from different regions, with different colors of skin explore and experience this new Beijing.

Nowadays, Beijing is no longer the old city inside the Second Ring of Beijing. The Hutong and the quadrangle house are now mixed with an increasing number of elements that strengthen the city's ecological, cultural and spiritual value. In the national 'five to one' integrated construction plan, both the construction of ecology and civilization are stressed. We can be sure that the quadrangle house is representative of both nature and culture in harmony.■

作者简介：

张新宇 / 男 / 本科 / 北京市园林古建设计研究院副院长 / 教授级高工 / 北京市园林古建设计研究院有限公司 / 中国北京

Biography:

Xinyu Zhang / Male / Bachelor of landscape architecture, Beijing Forestry University / professor, senior engineer / Beijing Institute of Landscape and Traditional Architectural Design and Research Ltd. / Beijing, China

查尔斯 · 沙（校订）
English reviewed by Charles Sands

从左到右依次为：
孙虎、王受之、梅卫平
聘
景观设计师
Landscape
Design
山水比德设计集团
王受之设计工作室

创意经典·落地运营

第九届中国（北京）国际园林博览会规划设计

THE 9TH CHINA (BEIJING) INTERNATIONAL GARDEN EXPO PLANNING AND DESIGN

张　果、孙志敏、陈　健、李明媚、吕　露、汪可微、陈星竹、吴　琦、孟范嵩、忻　欣、宋亚男、郭　雪、李祎龙

Guo Zhang, Zhimin Sun, Jian Chen, Mingmei Li, Lu Lv, Kewei Wang, Xingzhu Chen, Qi Wu, Fangsong Meng, Xin Xin, Yanan Song, Xue Guo, Yilong Li

合 作 单 位：北京山水心源景观设计院
Cooperation: Beijing XY Landscape Architecture Design co.Ltd

图 01 总平面图
Fig 01 Master Plan

图 02 从鹰山鸟瞰园区
Fig 02 Bird' s-eye view from Yin Mountain

图 03 原场地的永定河及高铁
Fig 03 Yongding River and high-speed rail in its previous condition

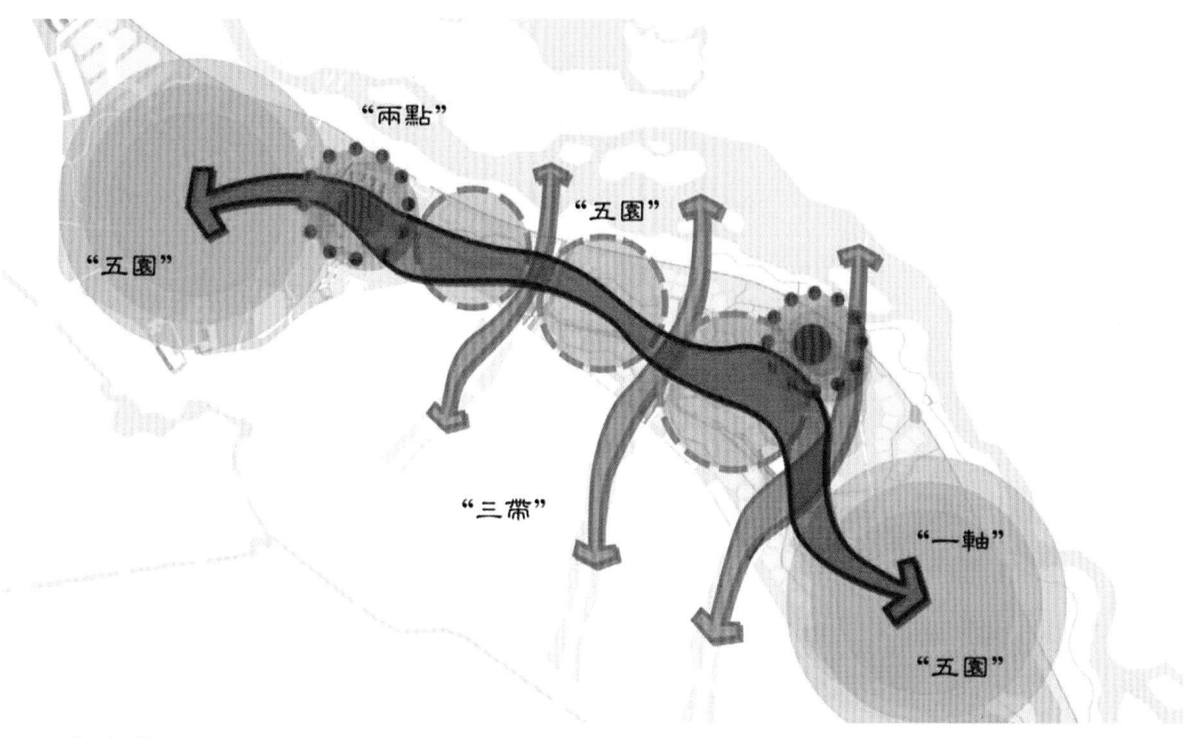

图 04 一轴两点三带五园
Fig 04 one axis, two point, three belts and five areas

项目概述

第九届中国（北京）国际园林博览会园区规划面积为 267km²，位于北京市丰台区永定河以西地区，北至莲石西路，西至鹰山公园西墙，东临永定河新右堤，南到规划梅市口路，西南接射击场路（图 01）。

场地分析

园区场地狭长，现状条件较为恶劣，是建筑垃圾填埋场，并伴有少量生活垃圾，场地西北是鹰山森林公园主山，山体相对高差 60m 左右，山上植被良好，场地东侧有规划高架京石高铁横穿用地（图 02）。场地东部、北部为永定河主河道，已常年无水，河道干涸，生态环境较差；隔河相望为南大荒公园和永定河东堤，现为苗圃地和河道绿化地（图 03）；场地西部射击场路西侧地块现为首钢料场用地，规划为中关村科技园区丰台园西区用地。

规划理念及原则

在规划初期确定了"文化传承、生态优先；服务民生、永续发展"的规划理念，并在此理念上确定了五大规划原则：①功能齐全，设施完善，满足会展需求；②主题突出，特色鲜明，表达展会理念；③文化建园，贴近百姓，体现地域文化；④生态环保，注重科技，提倡创新应用；⑤延展理念，拓展功能，确保展后利用。

Project Overview

The planning area of 9th China (Beijing) International Garden Expo is 267 hectare, it located in west side of Yongding River, Fengtai, Beijing. The area of project is to the north of Lianshi West Road, and to the west of Yinshan Park, stretching to Yongding River's new dike to the east, and reaches Meishikou Road to the south and Shejichang Road on the south-west. (Fig.01)

Site Analysis

The site of the garden is long and narrow and in bad condition. It used to be a construction waste landfill. In the northwest of the garden lies the main mountain of Yingshan Park which has a 60 meters relative height and a substantial plant cover. On the east side, there is a planned Jingshi high-speed rail which will traverse the site (Fig.02). To the north and east is the bend of Yongding River which is dried up and has a poor ecological environment. Across the river is Nandahuang Park and the east dike of Yongding River which has now become a nursery and green space (Fig.03). On the west side across Shejichang Road is

图 05 主轴局部效果图
Fig 05 Sketch of main axis

总体布局及结构

根据场地的特色以及规划的理念及原则，在规划设计总体布局上呈现“一轴，两点，三带，五区”结构（图 04）。

一轴：园林博物馆至功能性湿地区的东西向景观轴线，串联起不同展园，同时也是贯穿全园的重要快速交通流线。轴线的铺装上线型种植冠幅较大的落叶乔木，在明确流线的同时也有遮荫的效果，增加游览时的舒适性（图 05）。采用轴线转折形式，避免轴线一览无余，轴线采用人车分流，不同铺装材质区别道路的不同功能，根据预测人流量对轴线铺装进行仿真人流测算，故将轴线铺装（含电瓶车道）平均宽度确定在 26m。轴线铺装功能决定形式，轴线边界一直一曲，宜宽则宽，宜窄则窄。“直”方便电瓶车的通行，“曲”则是绿地与铺装相互渗透，完成了空间上的交流与沟通，能更好地使主轴景观绿化和展园的绿化相互渗透，使人游览时能体会园林美好意境。并模仿河道水流，人流汇集处（大门入口、园区入口及服务区周围）轴线局部放大，形成港湾，设林荫广场；人流迅速通过处收紧尺寸，把空间让给绿化（图 06–07）。

在电瓶车道外侧的轴线铺装上留出视线通廊，利用规则种植，使朝向塔的空间适度开敞，把永定塔借景到园中，并在空间许可的范围内设计种植灌木的绿岛。轴上的树阵与服务区内的树阵相互呼应，形成引导轴上面对服务区的树阵，树阵留出适当退让的空间，并设计有足够厚度弧线种植隔离。

两点：位于鹰山脚下的园林博物馆和由建筑垃圾填埋坑改造的锦绣谷。中国园林博物馆是第九届中国（北京）国际园林博览会的一个重要组成部分，是国内第一座以园林为主题的国家级博物馆。园林博物馆坐落于鹰山脚下，鹰山现状为鹰山公园，稍加改造可成为园区及园林博物馆良好的背景与依托。此园坐落西北，面向东南，依山就势沿鹰山东山麓展开，与场地内各展园形成互为借景的格局。园林博物园在园博会期间可结合作为主展馆满足室内布展要求。锦绣谷结合现状占地 10hm^2 的建筑垃圾填埋坑的改造修复，利用其 20m 高差将其调整为逐层下落的台地形式，并利用这些台地布置展园，形成“园林多宝格”。同时利用其小气候良好的特点，大量布置园林花卉，形成“锦

Shougang Factory and the site of a planned high-tech district, as well as Zhongguancun Fengtai west zoo.

Planning concept and principle

The design determined “inherit culture and ecology first; serve livelihood and sustain development” as the design concept and set five principles based on this foundation: 1. Provide functions and facilities that satisfy social demands. 2. Highlight the theme and characteristics that display the exhibition’s concept. 3. Include elements that highlight the local culture. 4. Protect the environment, emphasize technology and promoting innovation. 5. Allow for expanding functions to guarantee further utilization.

The overall layout structure

On the basis of the site’s features and planned concept and principles, the overall layout is based on “one axis, two point, three belts and five areas”. (Fig.04)

One axis: the line formed by the landscape museum and the functional wetland acts as a landscape axis which links the various Expo Gardens and also acts as a convenient boulevard through the garden. On this axis are large deciduous canopy trees, which enhance the pedestrian experience by providing shade and comfort (Fig.05). The surface treatment is varied, providing visual interest and defining the different flows of people and vehicles. By forecasting pedestrian flow rate, the width of the axis was set at 26 meters. The form of the boulevard, straight in one border and curved in another with variations in width, was decided by its function. The “straight” line provides convenience for small electrical vehicles, while the “curved” line negotiated the relationship between the green areas and the pavement. This curved line also recalls the meandering flow of a stream. At the gathering areas the boulevard opens into tree-lined squares. In

图 06-07 主轴及雨水花园局部
Fig 06-07 Details of axis and rainwater garden

绣谷”景观。（图 08-09）锦绣谷是生态改造重要展示，谷内设计了原始土壤断层展示，见证了园区整体改造的过程，园林绿化的手段让宗地重新焕发生机，昔日的建筑垃圾填埋场，可以变成今日的锦绣花谷。

三带：三条景观绿廊联系园博园和永定河。三条景观绿廊设置是根据规划的中关村科技园区丰台园西区道路规划，由于场地平面呈条带形状，为加强园区与永定河道的景观联系和划分空间的要求，结合园区主出入口布置三条功能性绿色景观带，满足会期间的快速集散要求。三带所在区域为园博园主展区。同时在会后也可以成为城市与自

areas for quick passage the boulevard narrows to allow for more green space. (Fig.06-07)

The vehicular side of the axis is provided with viewsheds and uses strategic plantings to visually include Yongding Tower in Expo Garden site. There is also a green island, planting with shrubs where space permits. The array of trees corresponds with the axis of the fultional areas.

Two points: the Landscape Museum is located at the foot of Ying Mountain and the Jinxiu Valley which is located of the reclaimed landfill. The China landscape Museum is an important part of the 9th China International Garden Expo. It is the first

图 08-09 锦绣谷
Fig 08-09 Jinxiu Valley

然的重要联系纽带。

五园：由园林博物馆和三条景观廊道划分出的五大区域，其中一个区域为功能性湿地区，其他四个为园博会展区。

各展园分布是第九届中国（北京）国际园林博览会的最重要的规划，展园分区以园博馆为起点，结合总体布局，以从古到今的时间序列和由国内到国外的空间序列进行展园布置（图 10）。国内展园面积分为 800 到 3000m^2 不同地块类型共 77 个，国内展园又细分为传统园林和现代园林两个类型，各个城市可根据各自的需求选择一个或多个地块，

national museum with the theme of landscape architecture. The Landscape Museum is located on the site of a park and can be a good background for the museum with little need of renovation. The museum lies on a northwest and to southeast axis, unfolding itself along Yin Mountain and borrowing the scenery from other parts of the garden. It also serves as the main exhibition hall offering indoor displays during the expo. The Jinxiu Valley is a reclamation project currently in construction to the west of the landfill with an area of 10 hectares. The designers made use of a 20 meters height differential to create a gradually descending, multi-tiered series of terraces. These terraces also act as exhibition

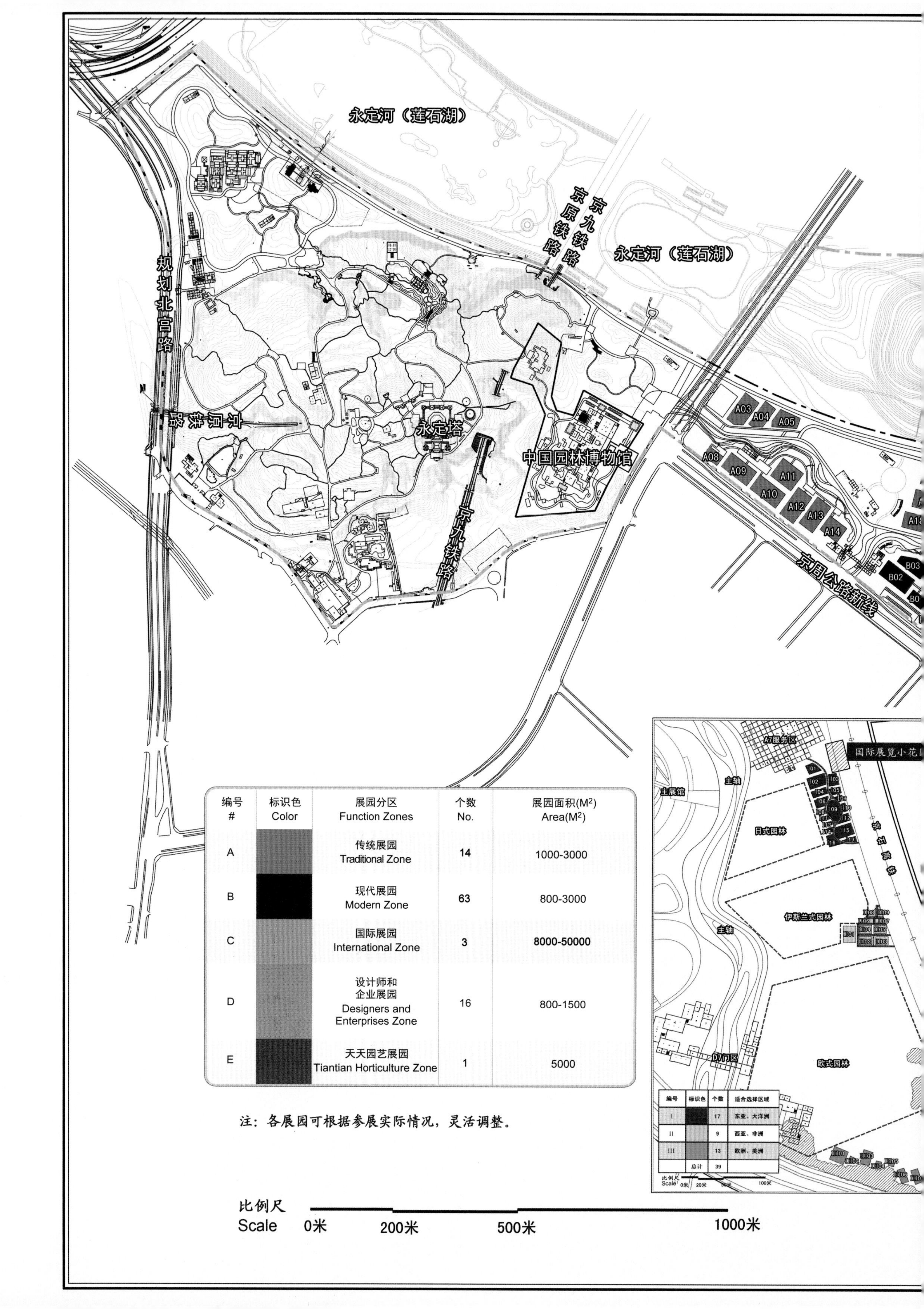

编号 #	标识色 Color	展园分区 Function Zones	个数 No.	展园面积(M²) Area(M²)
A		传统展园 Traditional Zone	14	1000-3000
B		现代展园 Modern Zone	63	800-3000
C		国际展园 International Zone	3	8000-50000
D		设计师和企业展园 Designers and Enterprises Zone	16	800-1500
E		天天园艺展园 Tiantian Horticulture Zone	1	5000

编号	标识色	个数	适合选择区域
I		17	东亚、大洋洲
II		9	西亚、非洲
III		13	欧洲、美洲
	总计	39	

图 10 展园规划图
Fig 10 Planning of Expo Garden

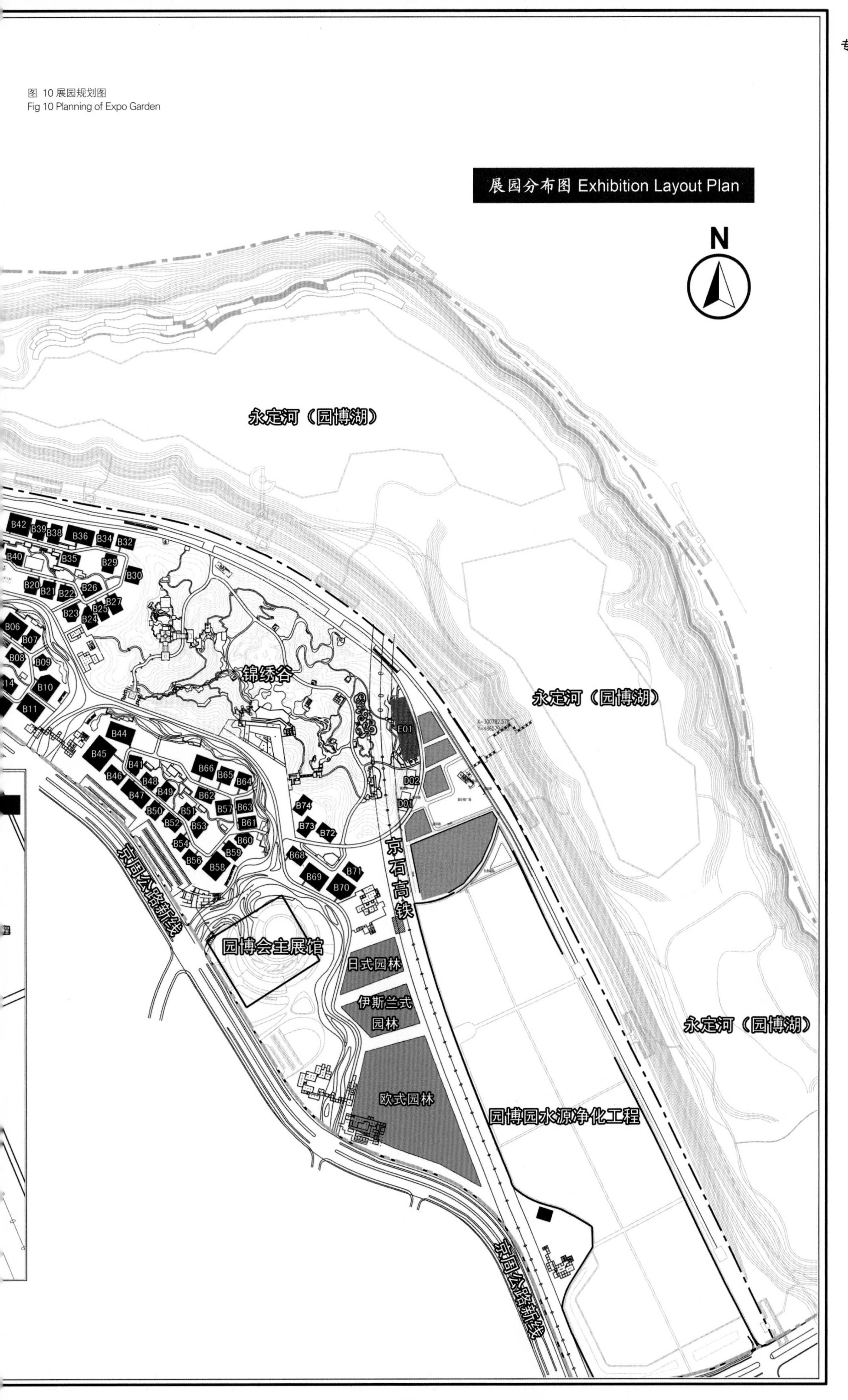

图 11 绿岛局部效果图
Fig 11 Green island district rendering

并且集中建园、以省建园，可以让各个城市最大限度的展示城市的园林特色。国际展园根据世界园林体系设置 8000 到 50000 平米三大块，有利于系统展示国外园林的特色，并在国际展园周边设置国际展览小花园，国际展览小花园面积在 50~200 平方米之间。企业设计师展园是历届园博会的重头戏，不仅设计师可以自由的表达，而且还能在一定程度上影响着园林设计的发展趋势，企业设计师园分为大师园和设计师园，大师园面积 2500 平方米，设计师园面积 1000 平方米。

竖向规划

园区竖向规划可用一句话来概括：延山，调谷，整治河道。延山：修补西侧地块山体东西两侧破损处，使山体向东西延伸与平地自然相接，形成东西走向山脉形势，使之与东侧地块自然过渡。调谷：结合建筑垃圾填埋，在东侧地块适当调整现有大坑，结合本地块南北部分地形处理，形成东西走向的谷地景观。一方面适当遮挡南北方向不佳远景，另一方面丰富整个场地地形变化，为各省展园的塑造提供良好的环境条件。整治河道：永定河此段河道长 4.2km，面积为 246hm^2。因此处防洪等级很高，河道内不允许搞任何永久性建构筑物。因此，规划中对河道地形进行适当整理整治，营造九曲河流穿草甸的景象，隐喻永定河被称为“小黄河”的历史典故。并在近锦绣谷段增加水面面积到 90hm^2，弥补园区内没有较大水面的遗憾。

水系规划

北京是一个极度缺水的城市，所以在园区的水系规划中未设计规模较大的景观水面，园区水系规划强调源流、东西贯穿；强化主轴、丰富景观；汇集雨水、体现生态。主轴边缘的公共绿地设置雨水花园，雨季时，雨水流入雨水花园，形成植物层次丰富的湿地景观，并可以进行雨水回收，旱季时雨水花园内的植物与卵石形成旱溪景观，丰富了轴线的四季景观（图 11–13）。

种植规划

坚持适地适树原则，大量选用适合北京地区地理及气候特点和城市绿化中表现相对稳定的绿色植物作为园区绿化的基础，以期快速形成相对稳定的绿色背景。因地制宜，结合现状进行山体绿化改造，适当增加色叶植物充实山体绿化，形成色彩丰富的山林景观。结合地形地貌的改造，努力创造适宜的小气候环境，适当引种部分边缘地带性植物，丰富区域植物种类。注意乔灌草的复层种植，认真研究植物间的相互作用，防止外来植物品种对本地的不利影响，营造抗性较强的

sites creating a "landscape Curio Box" effect. The micro-climate there benefits a variety of garden flowers forming a beautiful image of Jinxiu Valley (Fig.08-09). As a key work of ecological reconstruction, there is a visible section of the underlying soil which verifies the entire process of the reconstruction. With the landscape greening, the site comes back to life, demonstrating that a waste landfill can be transformed into a valley of flowers.

Three belts: there are three greenways connecting the Expo Garden with the Yongding River, which were designed based on the road planning in Zhongguanyuan west area. These functional landscape belts are combined with the garden's main entrance allowing for a rapid flow of people during the expo period. All the belts are in the main area of the Expo Garden and will become important connections between the city and nature after the Expo.

Five areas: the landscape museum and three belts are divided into five areas, which include a functional wetland and four exhibition areas.

The distribution of the exhibition areas is the most important aspect of the planning process fro the 9th International Garden Expo. It starts from the landscape museum and is organized based on the geographical relationship between China and world(Fig.10). The China exhibition area contains 77 sites based on geographical areas ranging from 800 to 3000 kilometers. It is also subdivided into classical and modern gardens and each city can choose one or more sites to construct gardens based on their own characteristics and distinctive culture. The international exhibition is separated into three areas ranging from 8000 to 50000 square meters according to the world landscape system, which promotes the unique features of each foreign garden. There are also small gardens of 50 to 200 square meters surrounding the international exhibition area. The corporate designer gardens have always been the most attractive part of the Expo, the allow designers to fully express themselves and may lead the trend in international garden design. The corporate designer's garden

植物组团，减轻后期养护的压力。

服务设施规划

根据园区总体布局，在三带上布置相对集中的服务设施。在东侧布置服务于园区的管理及配套设施以及湿地展示馆，在沙坑边布置建筑垃圾处理展示馆。根据《公园设计规范》要求，在全园合理配置服务于游客的功能性设施

防灾避险

园区现状被铁路、公路、高压线分隔严重，因此在园区内划定两块相对安全的防灾避险用地，满足区域内住民防灾避险要求。同时在这两块用地上布置适宜的防灾避险设施。园区南侧预留和城市联通的出入口，结合区域规划中的道路绿带形成快速进入园区的绿色安全通道，方便灾时区域住民的快速疏散。结合园区喷灌设施营造完善的山林防火体系，提高园区本身的防灾抗灾能力。

生态技术运用

园区内建构筑物采用节能环保材料进行建造，在建筑设计中注意低能耗设计技术的使用，同时尽量利用建筑物墙面和屋面布置太阳能板，转化为电能供建筑物使用。注重中水的应用。园区喷灌和景观用水主要以中水为主，建筑物冲厕用水也尽可能使用中水。在园区中设置环保示范主题园，运用园林造景的手法将环保材料和环保技术展示出来，向广大游客进行宣传教育，强化环境保护意识。□

consists of a master garden of 2500 square meters and a designer garden of 1000 square meters.

Vertical planning

The vertical planning focuses on "extending the mountain, adjusting the valley and renovating the river bend". Extending the mountain means fixing the western mountain's damaged areas on the east and west sides and connecting it with its surroundings. This will also block undesirable scenery to the north and south and enrich the form of the land. Renovate the river bend: the 4.2 kilometer long, 246 hectare Yongding River provides important flood control, which precludes any kind of permanent construction on the river bend. The form of the river has been adjusted to produce an image of a meandering streams running through meadows and referring to historical allusion which refer to the Yongding River as the small Yellow River. In Jinxiu Valley, the river expands into a big lake making up for the deficiency of a large waterscape.

Water system planning

Beijing is a city short of water resources. Because of this, the Expo Garden has no large waterscape. The water system in the garden is based on existing sources and runs from east to west.

图 12 种植设计
Fig 12 Planting design

图 13 主轴及雨水花园局部 2
Fig13 Details of axis and rainwater garden 2

It enhances the axis, enriches the scenery, gathers the rainwater and demonstrates ecological functions. The public greenland between the axes is designed with a rainwater garden. On rainy days, the rainwater goes into the garden, creating a vivid wetland when gathering water. During dry seasons, the plants and stones there form an arid creek, enriching the seasonal landscape of the axis(Fig.11-13).

Planting planning

The planting strategy is based on the existing environment conditions. It is largely comprised of plants which adjust to the local climate and perform well in the urban landscape as basic materials of fundamental greening with the goal of constructing a green foundation in a short period. Renovating the mountain with colorful plants on the existing landform can produce a vibrant scenery. To introduce marginal plants and create a diverse plant system, the design rebuilds the landform on the basis of existing conditions to create microclimates suitable for these plants. Attention is also paid to the interaction between plant species to prevent damage for imported exotic plants, and create hardy plant groupings which can survive into the future with little maintenance.

Service facilities planning

Based on the overall layout, the service facilities are mainly set on the three belts. The garden management facilities lie on the east side of Expo Garden. There is also a construction waste disposal exhibition near the sand pit. The public service facilities are set around the gardens based on the principle of "Garden Design Standards".

Emergency and disaster-prevention

The site is currently separated from railway, highway and high voltage lines. For this reason, two emergency areas have been provided on the Expo site satisfying local residents' demand for disaster-prevention. These two areas will have enough

北京市建筑设计研究院有限公司简介：

北京市建筑设计研究院有限公司（简称 BIAD），是与共和国同龄的大型国有建筑设计咨询机构。业务范围包括：城市规划、大型公共建筑设计、园林景观设计、建筑智能化系统设计、工程概预算编制等领域。伴随着新中国的社会发展，BIAD 不仅在首都北京完成了许多重大项目的设计作品，而且在全国范围内完成了一大批城市标志性建筑。BIAD 始终将设计精品工程和保持技术领先作为自己的责任，在多年的工作过程中，BIAD 逐步形成了"建筑服务社会"的企业核心理念。开放的 BIAD 愿与社会各界进行广泛合作，为创造人类美好的生活环境而不懈努力。

Beijing Institute of Architectural Design profile:

The Beijing Institute of Architectural Design (BIAD) is a large-scale state-owned architectural design and consulting institute established in October,1949,following the founding of the People's Republic of China.BIAD's business scope includes:City Planning、Large-scale public building design,civil engineering、Landscape Design、Design of intelligent systems for buildings、etc. BIAD has been an integral of China's development designing a large number of major projects in the capital city of Beijing. BIAD has established goodcooperative relations with many well-known foreign design offices.BIAD hass also graduually developed its key corporate concept-"Architecture serves society". BIAD is willing to join hands with all circles of society, making sustained efforts to create a wondderful living environment for all human beings.

facilities for emergencies. On the south part of the garden is an entrance which links to the urban space allowing for a fast rapid evacuating during a disaster. The irrigation system is combined with a forestry fire protection system enhancing the garden's disaster prevention capailities.

Ecological technology application

The constructions in the garden are all built with environmental friendly materials. An emphasis is placed on applying low-cost technology and installing solar panels on walls and roofs of structures to generate electricity. Attention is also paid to recycling water. The irrigation and landscape water sources are predominantly recycled water. A demonstration park is also designed to educate the public, raising their awareness of environment protection through a series of landscaping methods to display environmental-friendly materials and technology. ■

查尔斯·沙（校订）
English reviewed by Charles Sands

国学造园与国学赏园
——感悟于第九届中国（北京）国际园林博览会

北京市园林绿化局第九届园博会组委会办公室协调处处长、首都经贸大学旅游研究中心研究员　　崔勇

图 01 江苏园
Fig.01 Garden of Jiangsu

GARDEN-CONSTRUCTION AND GARDEN-APPRECIATION WITH STUDIES OF CHINESE ANCIENT CIVILIZATION — INSPIRATIONS FROM THE 9TH GARDEN EXPO IN BEIJING, CHINA

CHIEF OF OFFICE OF COORDINATION OF THE 9TH GARDEN EXPO ORGANIZING COMMITTEE OF BEIJING MUNICIPAL BUREAU OF LANDSCAPE AND FORESTRY RESEARCHER OF THE TOURISM RESEARCH CENTER OF CAPITAL UNIVERSITY OF ECONOMICS AND BUSINESS Yong Cui

图 02 福建园
Fig 02 Garden of Fujian

“自古天府多胜境，从来日下汇名园”，这是第九届中国（北京）国际园林博览会北京园万象朝晖阁上的一幅楹联。正如此联所言，公元 2013 年 5 月 18 日，一座集中国乃至世界园林流派于一体的园博园在北纬 39 度 52 分 11.05 秒，东经 116 度 11 分 09.9 秒周边 513hm^2 的土地上正式落成了。自此，历史文化名城北京城西南角的永定河畔，以“园林城市，美好家园”为主题，彰显人类生态理念、生活智慧、哲理演绎、文化内涵、艺术传承、造园技巧、科技应用的靓丽画卷渐渐呈现出来。

“国学”，一国所固有之学术也。指一个国家的传统历史文化与学术。“国学造园”是指以中国优秀传统文化为核心、以传统造园思想为指导、以传统造园手法和技巧为手段、结合现代景观设计理念与现代高科技技术，建成以“国学”为基础的充满中国优秀传统文化内涵的园林。

在“国学造园”的同时，还要提倡和推广“国学赏园”。即发挥中国园林艺术的教育和引导功能，提高全民对中国优秀传统文化的了解与理解，加深对生态文明理念和素质的修养，并自觉转变成对中国园林艺术鉴赏的兴趣和能力。第九届中国（北京）国际园林博览会的盛大开幕，是“国学造园”的一次重要实践，也为“国学赏园”提供了现实的教科书。

中国园林艺术是中国传统优秀文化有形的综合传承平台，既是传承形式又是被传承的内容，主要体现在：

中国古典园林艺术是精神文化产品，历史上曾经是帝王的奢侈品、士大夫的精神隐园、佛与道的解惑境界。时代发展了，园林艺术已经成为全民的文化休闲精神消费品。

本届园博会中，移天缩地、恢宏大气的皇家园林（北京园），曲径幽深、清新淡雅的江南园林（忆江南）（图 01），通透精致、中西合璧的岭南园林（广东园），色彩明快、自然奔放的闽南园林（福建园）（图

A couplet on the Wanxiangchaohui Pavilion of Beijing Garden states: ‘Our nation has abounded in beautiful scenery since ancient times, and no sites are more replete with beauty than our gardens.’ The Garden Expo, which was inaugurated along the Yongding River in Beijing on May 18th, 2013, emulates this couplet with the theme of ‘Garden City, Beautiful home’ .

In the Expo one can see grand imperial style gardens such as the Garden of Beijing; fresh and elegant gardens like the Garden of Remembering of the South of the Yangtze River(Fig. 01); Southern gardens with both Chinese and Western characteristics like the Garden of Gaungdong; colorful and unrestrained gardens like the Garden of South Fujian (Fig. 02); and quiet and natural gardens like the Garden of Chongqing (Fig.03) . All of these genres are available to be experienced and enjoyed. The Yinchuan Garden (Fig.04), Hohhot Garden (Fig.05), Lhasa Garden (Fig.06), and Urumqi Garden (Fig.07) with their unique ethnic features, combine landscape design with national history and culture, providing visitors an array of enlightening sights.

The Chinese Garden Museum (Fig.09), the only professional national museum focusing on landscape design, is an outstanding achievement of the Expo and guides one through the development of garden design in China and worldwide. With both charm and intelligence, the museum guides viewers through garden history, explaining and demonstrating the value of one of the most important of Chinese cultural traditions. This is most apparent with indoor gardens such as Chang Yuan Garden, Pianshishanfang Garden, Yuyinshanfang Garden, and outdoor

图 03 重庆园
Fig 03 Garden of Chongqing

图 04 银川园
Fig 04 Garden of Yinchuan

图 06 拉萨园
Fig 06 Garden of Lhasa

图 05 呼和浩特园
Fig 05 Garden of Hohhot

图 07 乌鲁木齐园
Fig 07 Garden of Urumchi

图 08 天津园
Fig 08 Garden of Tianjin

图 09 园博馆
Fig 09 Museum of the Expo

图 10 永定塔
Fig 10 Yongding Tower

02），飘逸宁静、拙朴自然的巴蜀园林（重庆园）（图 03）等各个园林流派淋漓尽致地得到完整体现。充满民族情调的银川园（图 04）、呼和浩特园（图 05）、拉萨园（图 06）、乌鲁木齐园（图 07）等展园则完美地把园林艺术和民族独特的历史文化紧密结合起来，呈现给游人一个个目不暇接的美丽景观。

传承不是守旧，"国学造园"需要在继承中创新。"梦之园"上海园、"沽"文化园天津园（图 08）、以无边框水景表现"楚风"精髓的武汉园等展园，在传承优秀传统文化理念的基础上，以现代化的景观理念和手法表现了人类对美好家园的向往。

世界上唯一以园林艺术为主题的国家级专业博物馆——中国园林博物馆（图 09），是本届园博会的一个重要杰作，具有展示、收藏、科研、教育、服务等基本功能，是收藏园林历史见证物、弘扬中国优秀传统文化、展示中国园林艺术魅力、研究中国园林重要价值的国际园林文化中心，承担着世界园林艺术大众科学普及和高端学术研究双重使命，在中国乃至世界园林艺术发展史上有着重要的意义。特别是室内的畅园、片石山房、余荫山房，室外的塔影别苑、染霞山房、半亩一章，集中国园林艺术经典作品于一身，使游人陶醉在与园林、与自然相融相亲的意境，流连忘返。与园博馆并称园博园三大标志性建筑的还有辽金风格的永定塔（图 10）和展示园林与生活的主展馆（图 11）。

徜徉园林之间，细细品味文化的差异和特色，是精神境界的一种享受，亦是园林艺术的精神附加值。

gardens such as Tayingbieyuan Garden, Liangxiashanfang Garden and Banmuyizhang Garden; all of which combine contemporary landscape design with classical garden design to enchant visitors with nature and art.

The International Garden Expo also provides a platform for international exchange. Masterpieces created by the world's leading landscape architects such as Peter Walker (Fig.13) from America, Peter Latz from Germany and Toru Mitani from Japan display the latest contemporary design strategies and techniques, making the Expo stand out from the preceding ones in terms of the participation of these international masters. □

中国古典园林的造园理念、效果、设计、手法、技巧、特性等诸多方面充满了东方的智慧、哲学、文化、和谐，是中国传统优秀文化有形的综合传承平台。

本届园博会"化腐朽为神奇"的理念与中国古典园林"天人合一"的造园理念一脉相承。对 140hm^2 的建筑垃圾填埋场进行生态修复，使城市的废弃地、污染源变成了一处精品园林，体现了人与自然相亲、相融的生态文明理念。特别是利用原来未填满的一处面积 20hm^2 深 30m 的大坑，因地就势、巧妙构思，规划建设了一个花团锦簇的亮点——"锦绣谷"（图 12），体现了高超的设计理念和造园手法。

图 11 主展场夜景
Fig 11 Night view of the main exhibition park

中国古典园林集建筑、山石、水系、植物、动物、家具、饰件、匾额、楹联、雕塑、彩画、诗词、书画、碑文、戏曲、音乐、餐饮等诸多有形的元素于一体，是物化、有形的传承平台和具有可视性的文化传承载体和传承内容。每一种元素都有着上百年甚至数千年的发展历史和文化血脉。本届园博会 128 个展园中，不同地区的园林艺术作品，呈现着这些元素不同的表现形式，集中、综合地表现着这些元素在不同历史时期、不同地理区域的动态“写真”。游人们在悠悠品味、细细欣赏中可以感悟到：

“虽由人作，宛如天工”的造园效果。追求尊重自然、顺应自然，回归自然的思想境界。

“缩天移地，曲径通幽”的造园设计。取自然之精华，以小见大的气概和含蓄、婉转、自然流露的心态和审美观。

“师法自然，巧于因借”的造园方法。点明了园林艺术和园林工程原理和技艺手法，总结了民间匠师的造园经验，反映了中国古典园林建筑的艺术成就。“应时而借”、“应景而借”，增加了园林时空艺术的感染力。

“精工营造，匠心独具”的造园技巧。是设计者、施工者千百年来智慧和艺术的结晶。在园博园中，中国古典园林的亭、台、楼、阁、轩、榭、斋、馆、坞、塔等各种建筑形式一应俱全。各展园在配置植物时，注重对植物赋予人格化的寓意，使之成为既是观赏对象，又是情景交融、人与自然相亲相融的重要元素。公共园区有以樱花、紫薇、月季、牡丹、丁香等植物造景为主的 5 个植物花园，

“寓情于景，内涵外丽”的造园特性。中国古典园林独有的艺术表现形式匾额和楹联，是中国传统文学最好的传承载体和教科书。匾额和楹联构思于园林，指导意境造景；寓意于园林，引导意境感受；增彩于园林，融合意境元素。匾额和楹联既有文字意境，其造型的样式、材料、彩画本身也是一种美的元素。

园林艺术是中国传统文化中的代表性元素，是国际交流中重要的文化符号。园博会为园林艺术国际交流与展示搭建了重要平台。

被称为“世界园林之母”的中国园林艺术，与瓷器、丝绸、茶叶一样，早已经成为国际交流中重要的中国文化符号。本届园博会是一次园林艺术国际交流与展示的平台。以世界著名园林景观设计大师美国的皮特·沃克（图 13）、德国的彼得·拉茨和日本的三谷徹领衔设计、建造的 9 个大师园和设计师园，显示了前沿的理念、独特的设计，也因此使本届园博会国际级大师参与程度为历届园博会之最。

为了表现世界三大园林体系的风格，园博园以“创造加摹仿”的方式，精心设计、建造了完整的欧式园林和伊斯兰式园林两个展园。而梦唐园（图 14），则是中国古典园林艺术“出口转内销”的作品，一定程度上记载了亚洲国家园林艺术在汉唐时代中国古典园林影响下的传承与发展，比如像“枯山水”等表现形式，展示的波斯地毯等艺术品也像在回味着丝绸之路的辉煌。34 个国际城市和机构的展园，精彩纷呈，在展示园林艺术的同时，促进着多元文化的交流。

文化的传播归根结底是生活方式的传播，中国园林在新形势下，以其平和、自然、生态、和谐的特性有助于在国际上树立“爱好和平、和谐相处”良好的中国形象。

当代社会对生态文明建设和良好公共环境的需求，大众对提高生活品质与品位的要求，决定了园林艺术有着更为广阔的发展空间。中国园林艺术产业要乘文化大繁荣、大发展的东风，乘势而上。从事园林艺术事业的人，不能“孤芳自赏”，要充分发挥园林艺术的教育和引导功能，争取更多的社会各界人士认识园林、理解园林、支持园林、“消费”园林，使人们对“不出城郭而获山水之怡，身居闹市而有林泉之志”理想生活的向往尽早成为现实。■

图 12 锦绣谷
Fig12 Jinxiu Valley

图 13 皮特・沃克园
Fig13 Garden of Peter Walker

图 14 梦唐园
Fig 14 Garden of Dreaming of the Tang Dynasty

查尔斯・沙(校订)
English reviewed by Charles Sands

“虽由人作，宛自天开”是中国古代造园大师计成在《园冶》中的名句，道出了造园的最高境界。以质量立身，以景效立命的天开集团恪守这份造园理念，抱着“为有限的城市空间营造无限的自然理想”的初衷，励志将公司打造成为“人居环境景观营造”的标杆式企业。天开园林首家公司由现任总裁陈友祥和副总裁谭勇于2003年在重庆联手创立，从此开启了天开时代的奋斗历程。2004年北京天开公司成立，天开集团进一步加快前进步伐，业务量逐年稳步增长，公司知名度也在大幅提升。2012年，天开完成了全国布局和业务升级，分公司覆盖中华各大区，北京、天津、上海、重庆、成都、长沙、哈尔滨等地。源于对景观工程的质量和景效的高要求，天开在行业内获得众多如万科、龙湖、纳帕、中旭、泰禾等高端合作伙伴的信赖，2012年3月，蓝光和骏地产也正式与天开签约结为战略联盟。时至今日，天开已发展成为一家集园林景观设计、园林工程施工、家庭园艺营造、苗木资源供应及石材加工为一体的领军式园林公司。我们的实力、品质和战略布局，能为客户有效降低沟通管理成本，更是全国地产企业所有项目卓越品质的保障。

China has a long history in Gardens and Landscape. “Materpiece of Nature, although Artificial Gardens”, as the first and the most acknowledged philosophy about garden building in the world, was initiated from the book Yuanye which was written by Ancient Chinese gardening master Jicheng, and Tiankai was named after it. Tiankai company was established in Chongqing in 2003 by President Chen youxiang and Vice-president Tan yong. We come from nature and back to natue, that’s the reason why we love nature gardens. Tiankai hopes to create infinite natural idea for the limited city. Today, Tiankai has become to comprehensive group corporation including garden construction, design, plants maintenance, seedling resources and so on. Tiankai has the second class qualification of garden construction, and B class qualification of landscape design. There are many international and outstanding designers in Tiankai. Tiankai is the best company in China in garden design and construction field. Tiankai has established 10 branches respectively in Beijing , Shanghai, Tianjin, Chongqing, Chengdu, Changsha and Haerbin. Tiankai's contruction business covers all over the country. The partners of Tiankai include Vanke, LongFor, Napa and Blue Ray, which are all very famous enterprises in China. We can effectively save communication and management cost for our clients and keep all the projects in consistent and excellent quality.

镜 园
THE MIRROR GARDEN

张建林 Jianlin Zhang

项目位置：中国，北京，第九届北京园博会
项目面积：1,000m²
委托单位：第九届中国（北京）国际园林博览会组委会
设计单位：西南大学
景观设计：张建林
完成时间：2012年3月

Location: The 9th China (Beijing) International Garden Expo, Beijing, China
Area: 1,000m²
Client: The 9th China (Beijing) International Garden Expo committee
Designer: Southwest University
Landscape Design: Jianlin Zhang
Completion: March, 2012

我见过新疆的胡杨林，她静静地伫立在大漠戈壁
我见过沙漠中的绿洲，是高山的雪水滋润着那一片绿意
我见过江南水乡，小桥流水和沃野千里
我见过南国的山川，消失的山林和石漠的土地
曾几何时，大地干裂，生命的消失……
又曾何时，沙漠变绿洲，让人感受春的气息
人啊，是你的足迹改变了我们赖以生存的环境格局

I have seen the forest in Xinjiang,
Standing quietly in the wild desert;
I have seen the oasis hid in dunes,
Slush from the mountain watering the green;
I have seen the water villages of southern China,
Small bridges over the flowing streams
And fertile land spreading over thousands of miles;
I have seen mountains and rivers in the south,
Forests vanished and land barren.
Once upon a time the earth dried up and life disappeared...
And when is it the desert changed into an oasis,
And spring came back to the land?
O humans, it is your footprint
that causes these changes.
that causes these changes.

图 01　总体鸟瞰图
Fig 01　Overall Aerial View

2012年3月，我有幸被邀请参加北京园博园设计师园方案设计竞标，并选择北京园博园设计师园6号地块做方案；6号地块呈不规则四边形，用地平坦，占地约1,000m²（图01）。

主题创意

古人云“以铜为鉴可正衣冠，以古为鉴可知兴衰，以人为鉴可以明得失，以史为鉴可以知兴替”。以园林为镜，用景观来诠释自然景观演替的内在规律，唤醒人们对自身生存环境的关爱作为6号设计师园表达主题。

人类进入21世纪以来，其生存环境不容乐观。直视人类自身行为与环境的发展轨迹，自然界的枯与荣、生与死、绿色与荒漠的演替给予我们无限的思考。而世间万物都处在对立和矛盾之中，并在特定的条件下对立双方可以互相转化。通过对自然景观演替和传统文化的解读与思考（图02-03），选取自然环境中的荒漠与绿洲两种对立面景观形态，提取干裂的大地、树枝和水纹图样作为花园的设计符号；以树、土地为创作线索，将水与“绿洲”紧密相连，干裂的大地纹样与“荒漠”叠加，无水成荒漠，有水变绿洲，演绎生命的枯荣和自然景观的演替规律。

规划布局

镜园由四段平行、等距、错位的镜面景墙将用地分为东西面积相当的两部分，西高东低的地形处理，隐意中国自然地形的分布和水流方向，在东西两部分高差分界处布置镜面景墙和架空景观平台；紧邻园博园主干西侧部分布置沙漠景观，东侧远离园博园主干道布置绿洲景观（图04-07）。西侧运用“枯山水”的形式来增强场地氛围，展示生态环境遭受破坏，大地的荒漠与枯树景观，表达世界的寂静和环境的冷漠（图08）；东侧以下沉式花园、镜面水展示亲切宜人的景观，体现水、绿洲的自然活力和温馨（图09）。东西两部分相似的布局，截然不同的景观效果，通过上下和虚实的变化处理，将二者联系在一起。在强烈的景观对比和视觉反差下，让参观者直观地感受到因环境失衡所带来的恶果，警示人们应该爱惜水资源、爱护自然环境（图10）。

空间构成与组织

1. 实空间： 中部由四块镜面玻璃景墙分割成东、西两大部分，荒漠和绿洲两个实空间组成，西高东低，配合台阶和架空观景平台，在三个不同高程面上形成不同空间效果，实现不同空间的戏剧性转换（图11）。

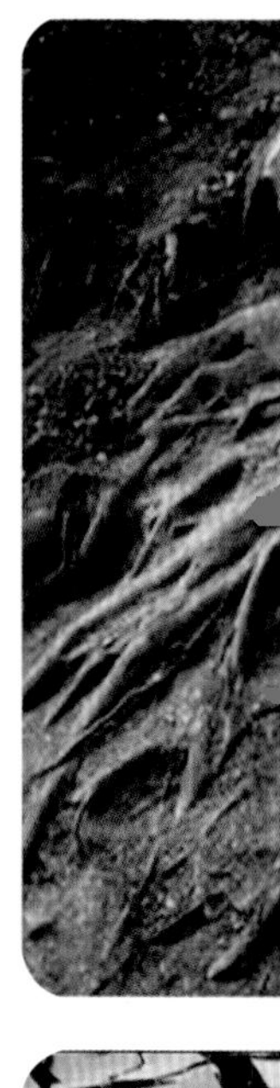

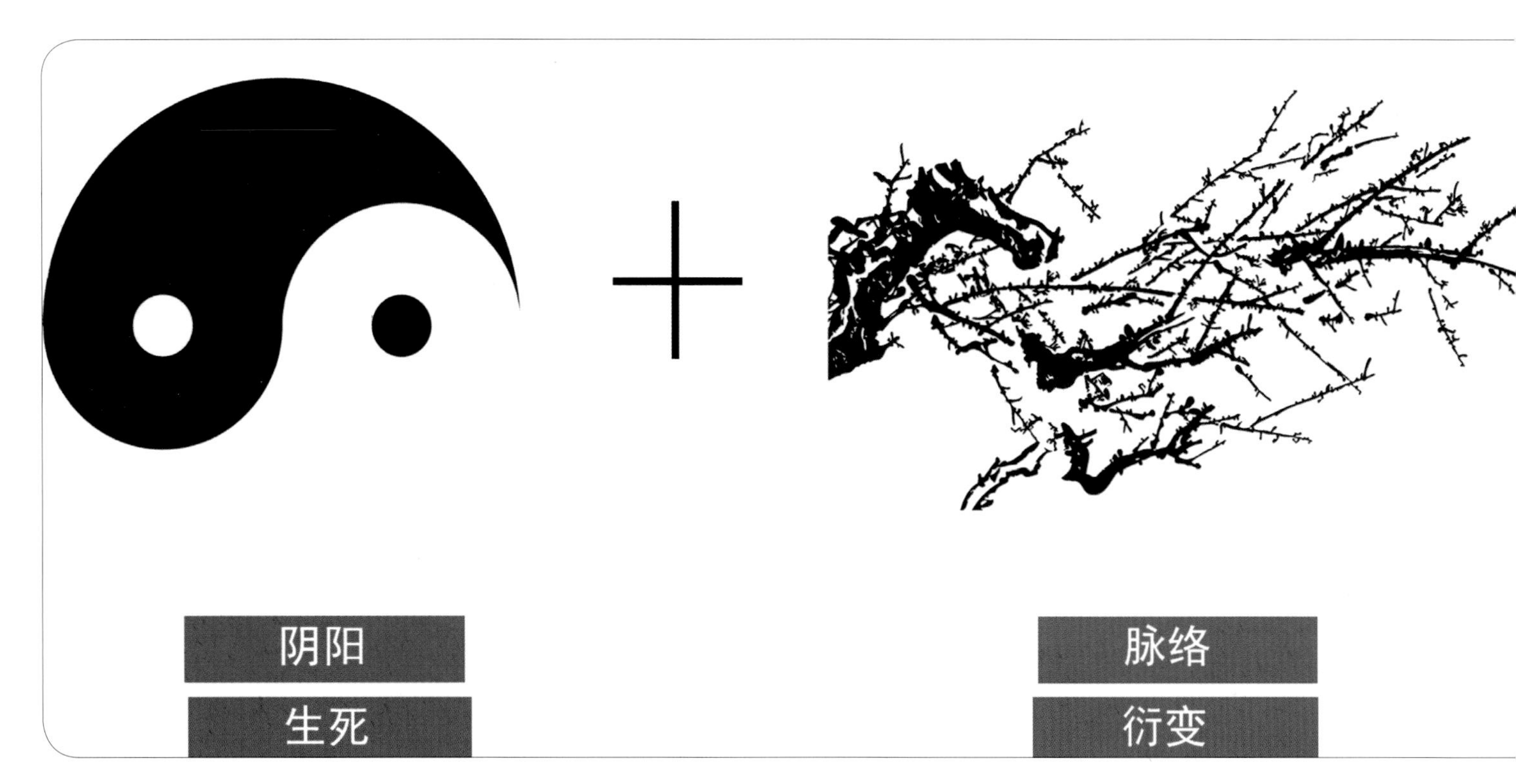

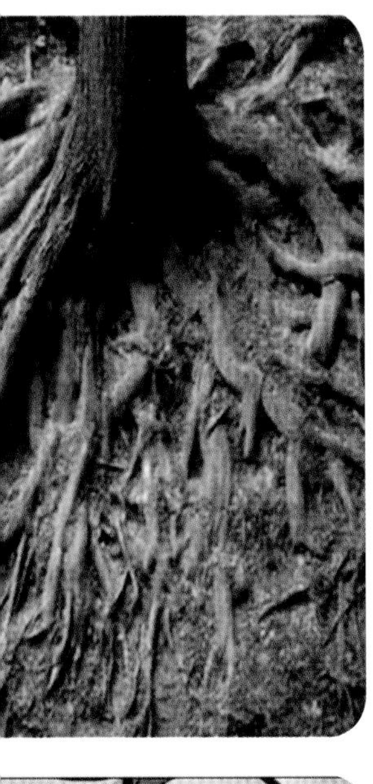

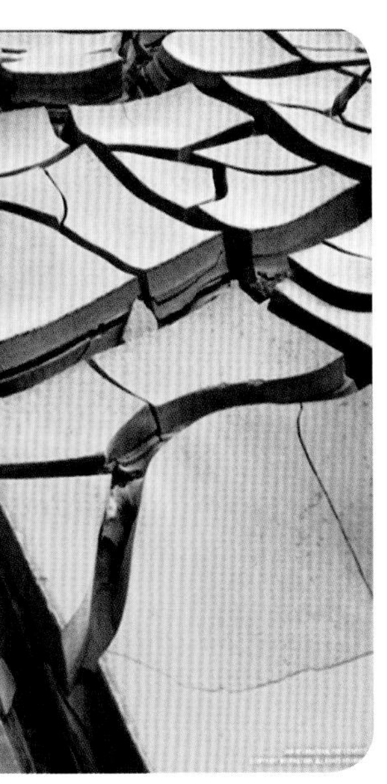

图 02 自然景观解读

干裂的大地：土地因水的枯竭而开裂，植物的死亡，大地变成荒漠与死寂。

树枝：生命的代表。“绿洲”因水而生机蓬勃，“沙洲”因缺水而枯寂。树在有水无水之间所呈现出的两种截然不同的生命状态。

水：生命之源。

Fig 02 Interpretation of Natural Landscape .

Chapped land: Earth cracking for the lack of water, the death of plants, and the earth turned into desert and deathly stillness.

Branches: Representatives of life. The oases grow because of water while the dunes perish for lack of water. Trees appear two entirely different life state with the existence of water.

Water: Origin of life.

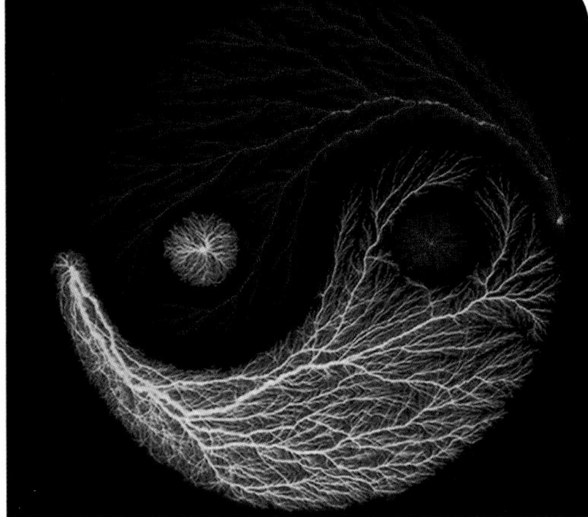

生与死

生命之道

图 03 对传统文化的思考

Fig 03 Cogitation of Chinese Traditional Culture

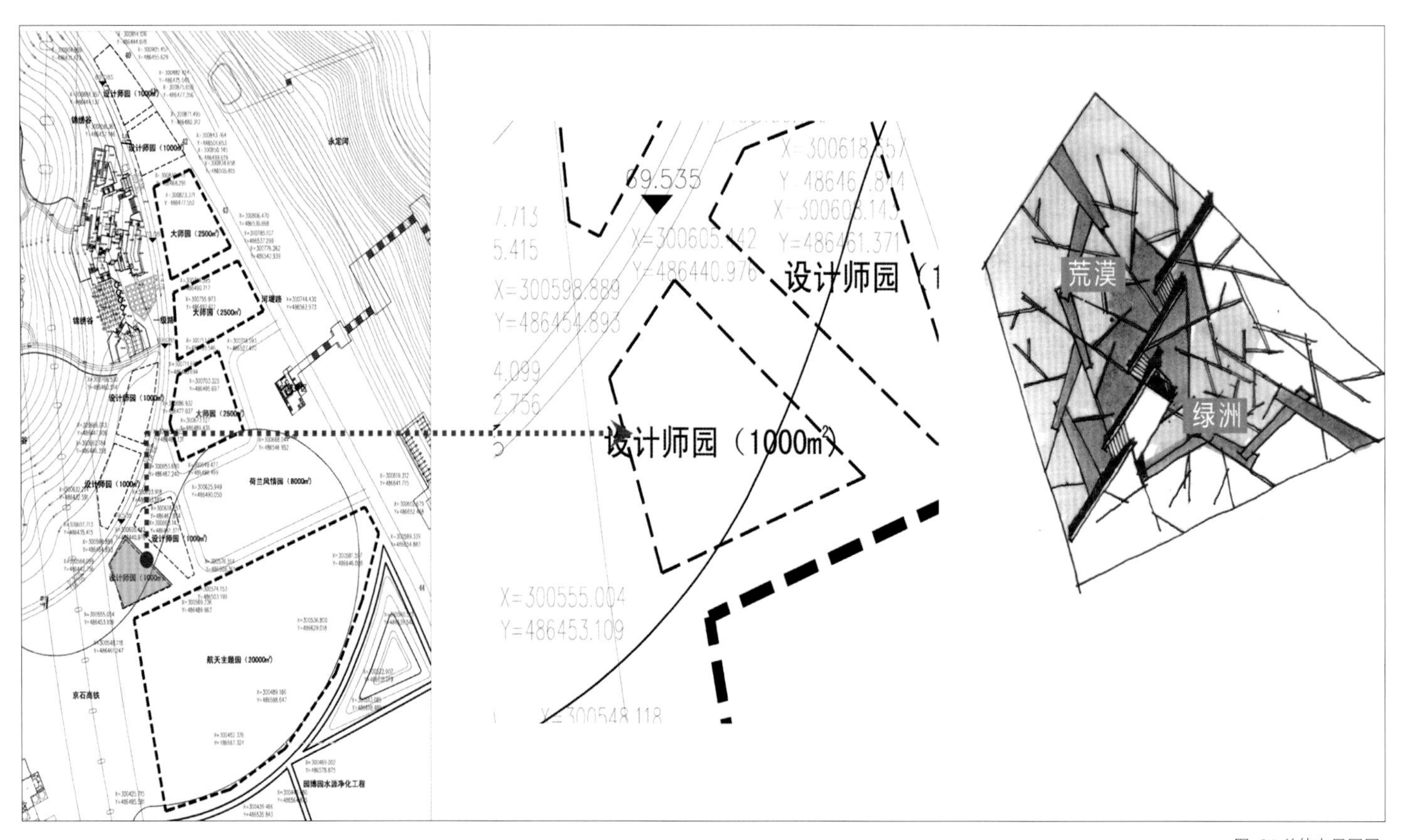

图 04 总体布局图图
Fig 04 Site Overview

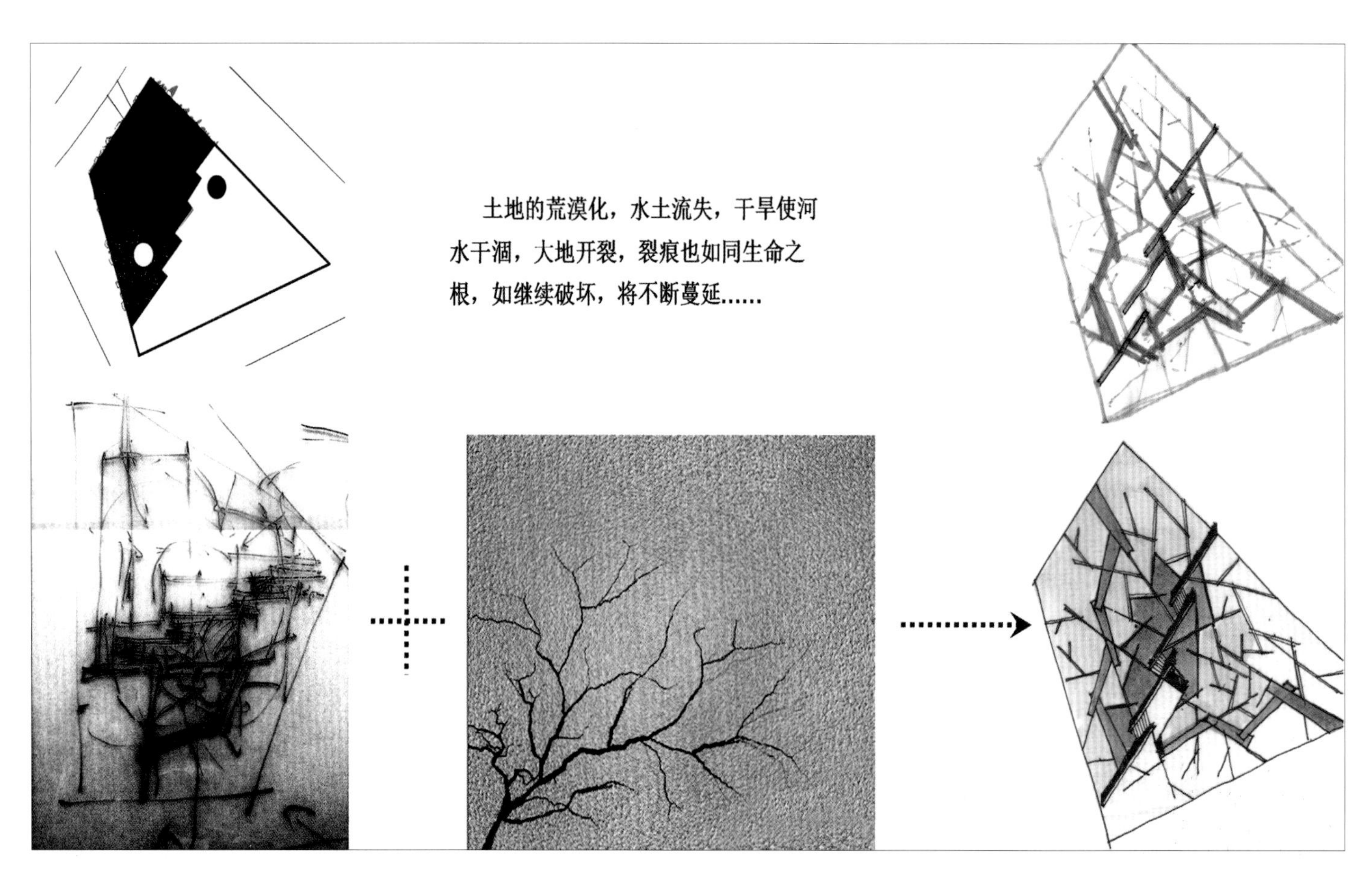

土地的荒漠化，水土流失，干旱使河水干涸，大地开裂，裂痕也如同生命之根，如继续破坏，将不断蔓延……

图 05 构思草图
Fig 05 Idea Sketching

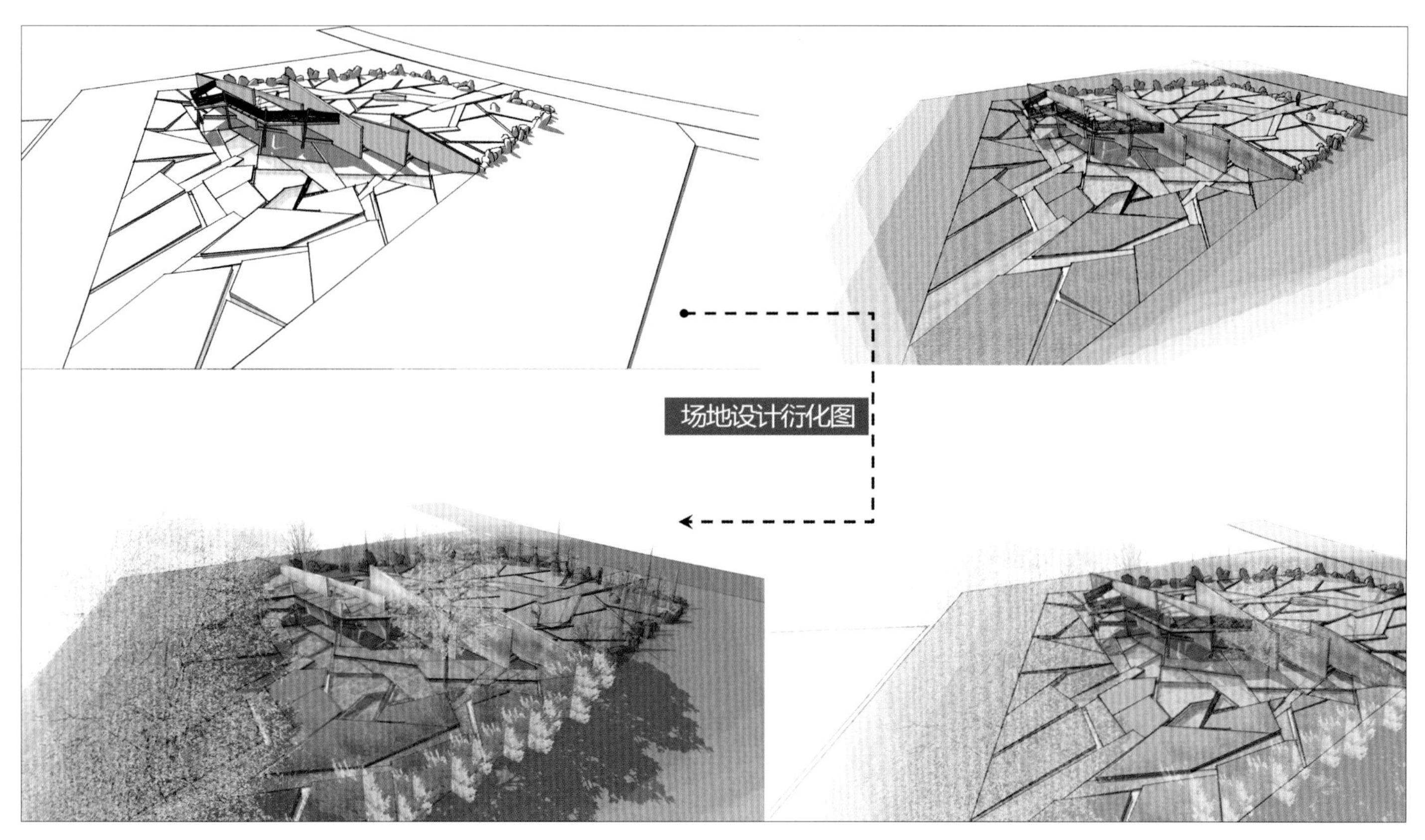

图 06 场地设计衍化图
Fig 06 Derivative Site Design

第一层面空间：利用原有平坦的地形静静地向游人展示荒漠的死寂（图 12），布置 1.8m 高的镜面景墙作为临界面，将沙漠的寂寥与广阔感扩大（图 13）。

第二个层面空间：低于第一个层面空间 1.05m 的下沉式水景花园（图 14-15），三面绿色植物与镜面景墙围合成内向、封闭的空间，架空平台与镜面景墙的组合，形成休息廊。分隔出一个安静、平和、惬意的下沉空间。

第三个层面空间：高于第一个层面空间 1.50m 的景观平台（图 16）。在平台上俯瞰“荒漠”和“绿洲”，亦可借园外之景。

三个层面的空间之间两两互通，形成“寂静迷惑——惊然矛盾——惬意反思”的流线，形成游览感受的趣味性（图 17）。

虚空间——镜面营造出来的虚幻空间

全园被镜面景墙分成两部分，营造了两个不同的实体“半空间”，利用镜面景墙，将各自的“半空间”反射，形成虚幻的“半空间”，对应的虚实空间共同构成完整的景观空间形态，扩大了视觉效果和环境的感染力（图 18）。

空间围合与游线

绿洲采用植物围合，荒漠采用粗糙的石墙围合，截然不同的围合方式，进一步强化所展示的景观特性，突出“绿洲”的生机、活力，烘托出“沙洲”的粗犷、死寂，达到加强游人不同的心理体验的目的。由干裂土壤主纹路、树枝主要枝干以及三个层面的空间相互串通，形成镜园的游线，游线突出主游览线的便捷，园内整体形成游览环线（图 19）。

生态与节水

1. 建立全园雨水收集系统

北京属资源型重度缺水地区，地下水位不断下降。为此，充分利用竖向规划设计进行地表雨水的收集，进行滞水设计、节水设计；在“绿洲”部分做小面积的水池，内铺白色鹅卵石，兼顾有水和无水时的景观效果，同时，“绿洲”部分的地表水通过地表肌理收集，最后汇集到水池，“荒漠”部分则通过相应的地下渗水工程措施，将地下水收集到水池（图 20）。

In March 2012, I was honored to make a bid for the Designer Garden of The 9th China (Beijing) International Garden Expo. A 1000m^2 plot was selected for my design(Fig.01).

Theme and Idea

An ancient Chinese saying goes "You may trim your garments if you take brass as mirror. You may know rise and fall if you take history as warning. You will understand gain and loss if you take people as examples ." Taking the mirror as the theme of the garden, the landscape is designed and built reflecting the inherent laws of natural succession, which evokes the public's concern for our living environment.

The condition of our living environment today, Does not reflect the optimism once felt by our ancestors. Looking into human activity and the environment's developing path, the successions of wilt and flourish, oasis and desert, life and death, presents unlimited thoughts. While everything in the world exists in a duality, under certain circumstances the two sides can be mutually transformed. After pondering the natural succession of the landscape and traditional culture(Fig.02-03), we were inspired to select the two opposite forms of desert and oasis, with cracked earth, branches and ripples as the design motifs. Trees and water create the oases while the cracked earth pattern references the desert, the life magic plays its effective role and tells the story of nature's vicissitudes and the natural law.

Planning and Design Strategies

The Mirror Garden is divided into two equal areas from east to west by four skewed, parallel and same-length mirror walls. The high western terrain and low eastern terrain reminds visitors of the topography of China and the direction of water flow. The mirror walls and a platform built on stilts have been placed at

the boundary where the east to west heights differs. While

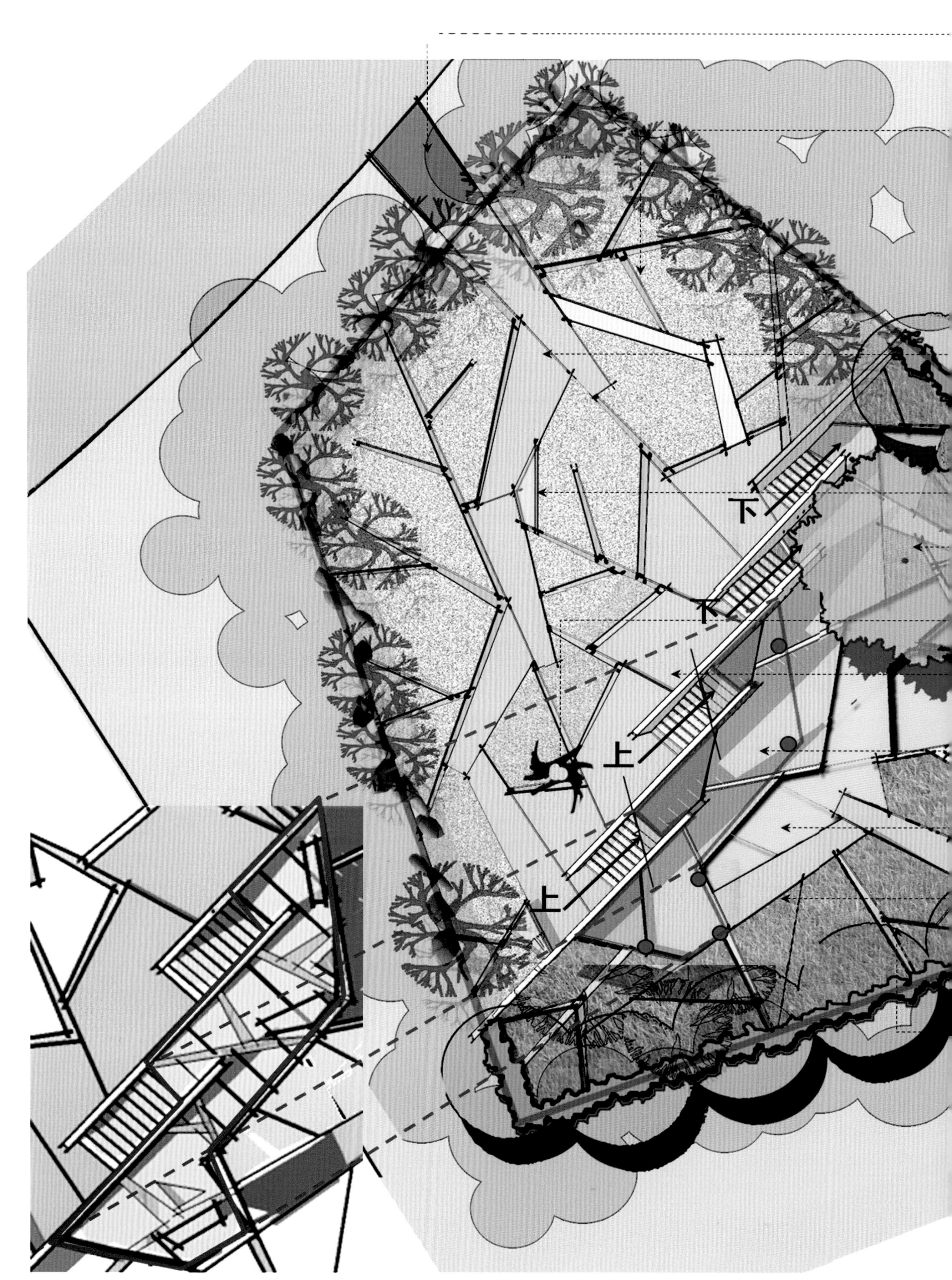
下
下
上
上
上

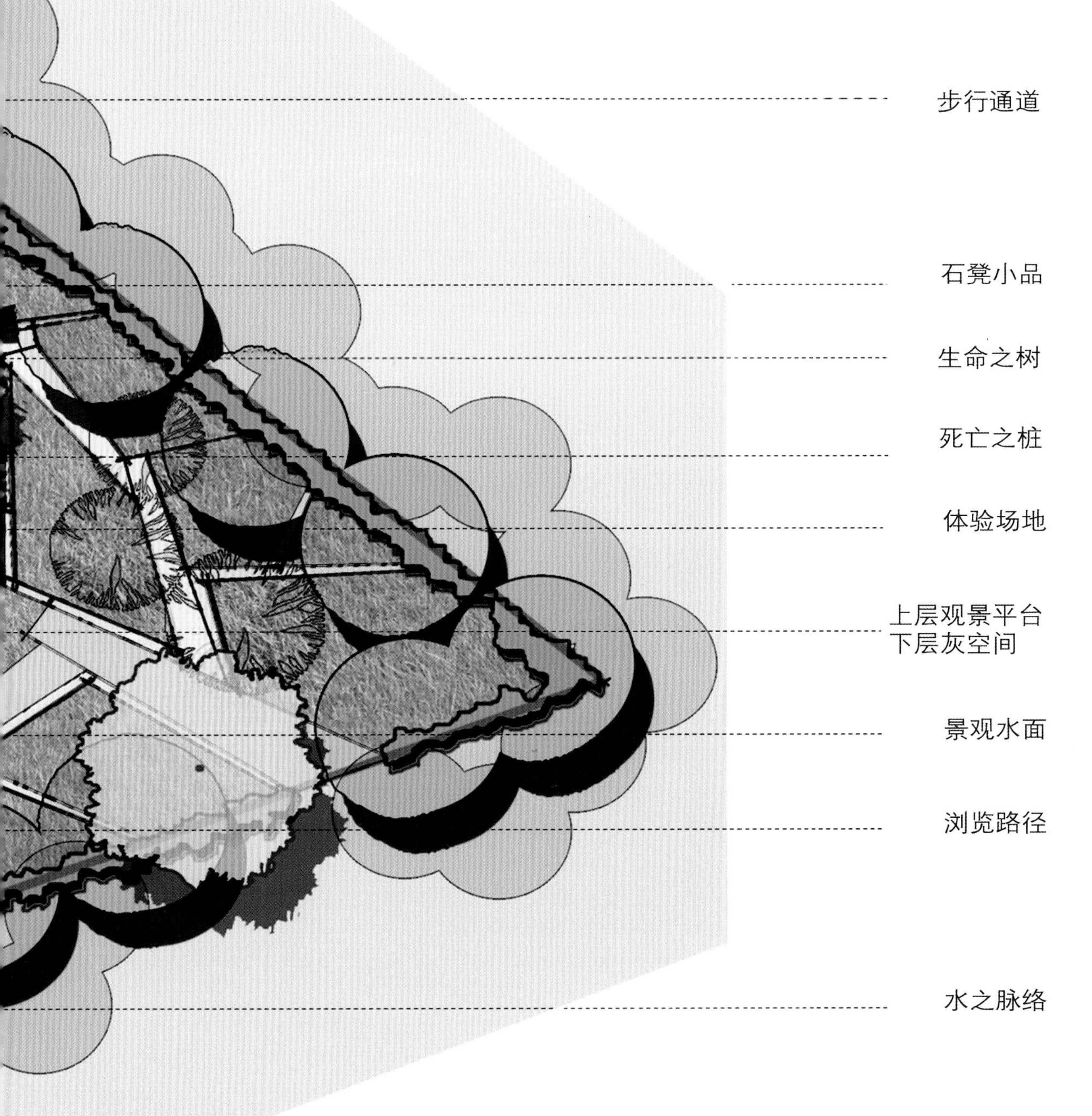

土地的荒漠化，水土流失，干旱使河水干涸，大地开裂，裂痕也如同生命之根，如继续破坏，将不断蔓延……

图 07 总平面图
Fig 07 Master Plan

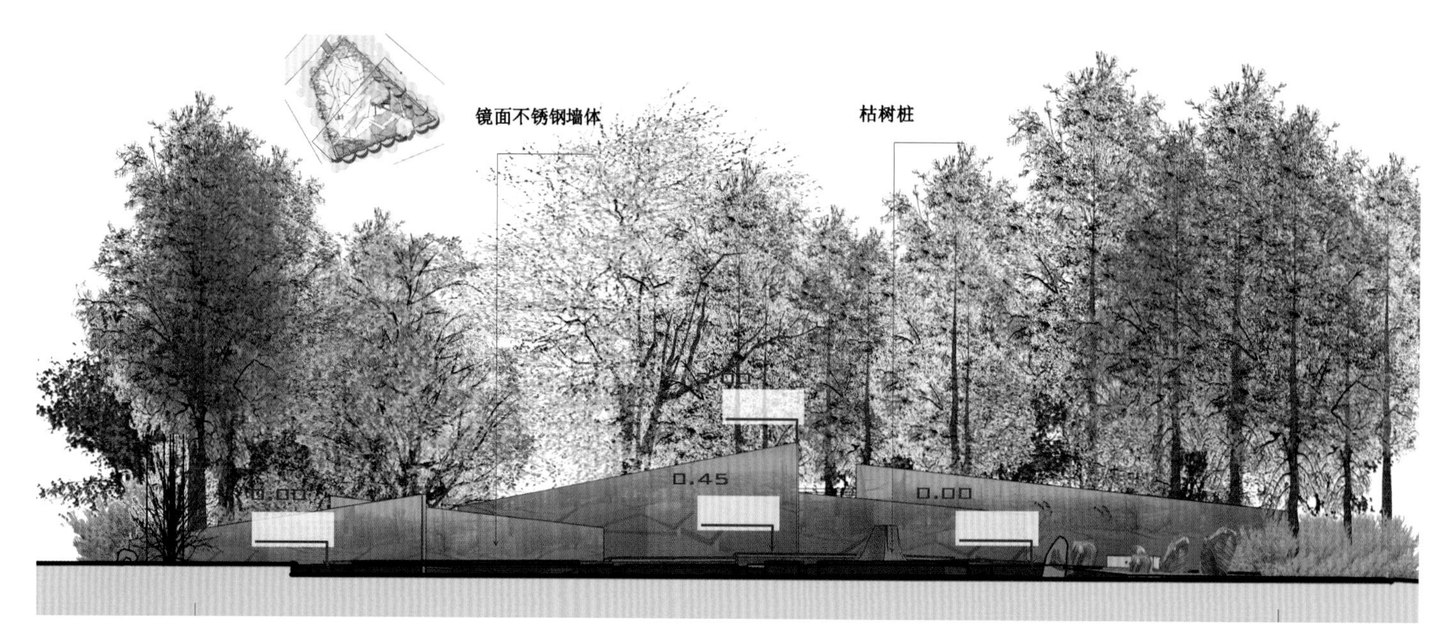

图 08 荒漠区剖面图
Fig 08 Section Plan of the Desert Part

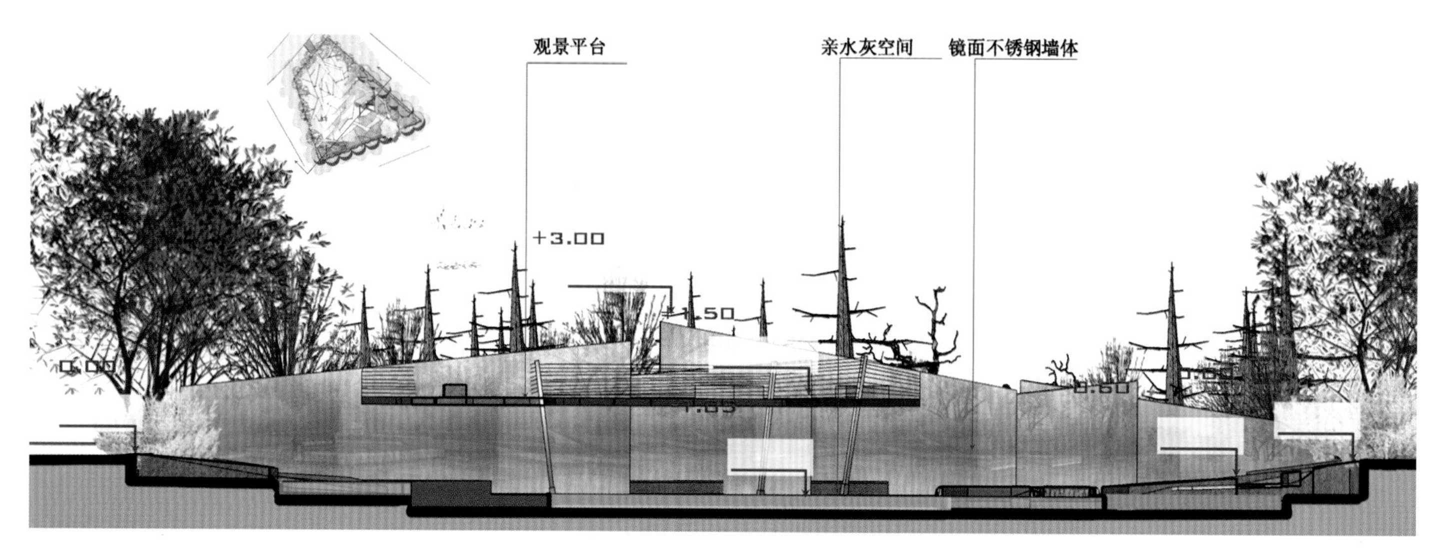

图 09 绿洲区剖面图
Fig 09 Section Plan of the Oasis Part

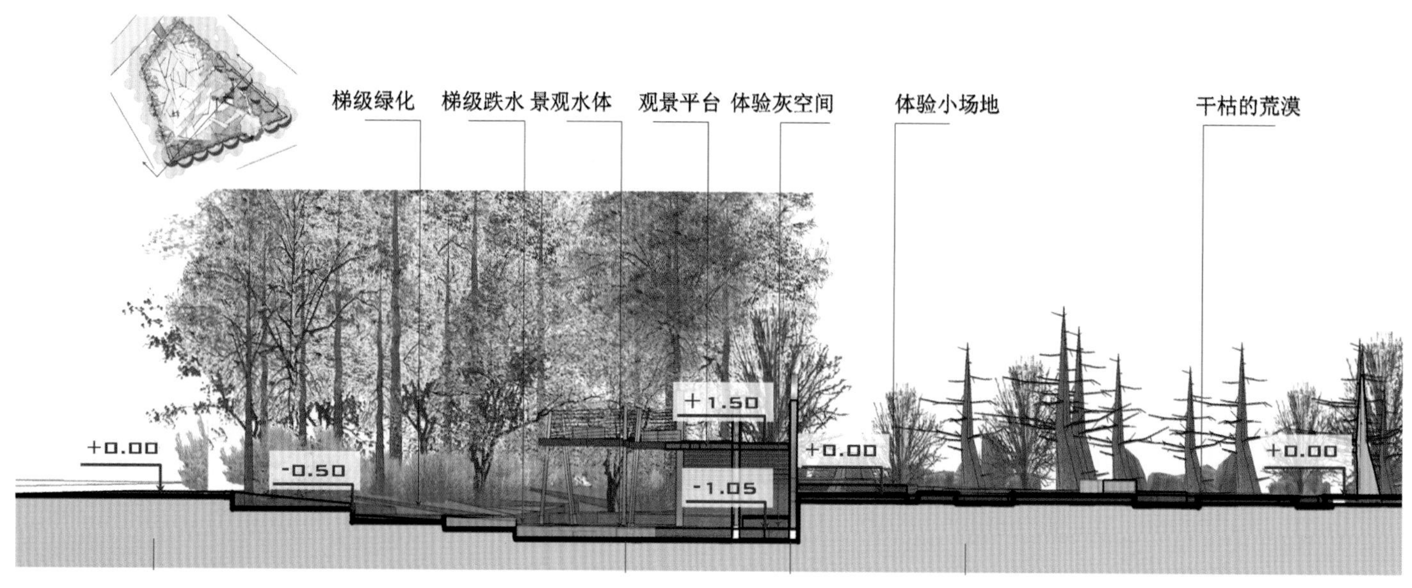

图 10 总体剖面图
Fig 10 Overall Section Plan

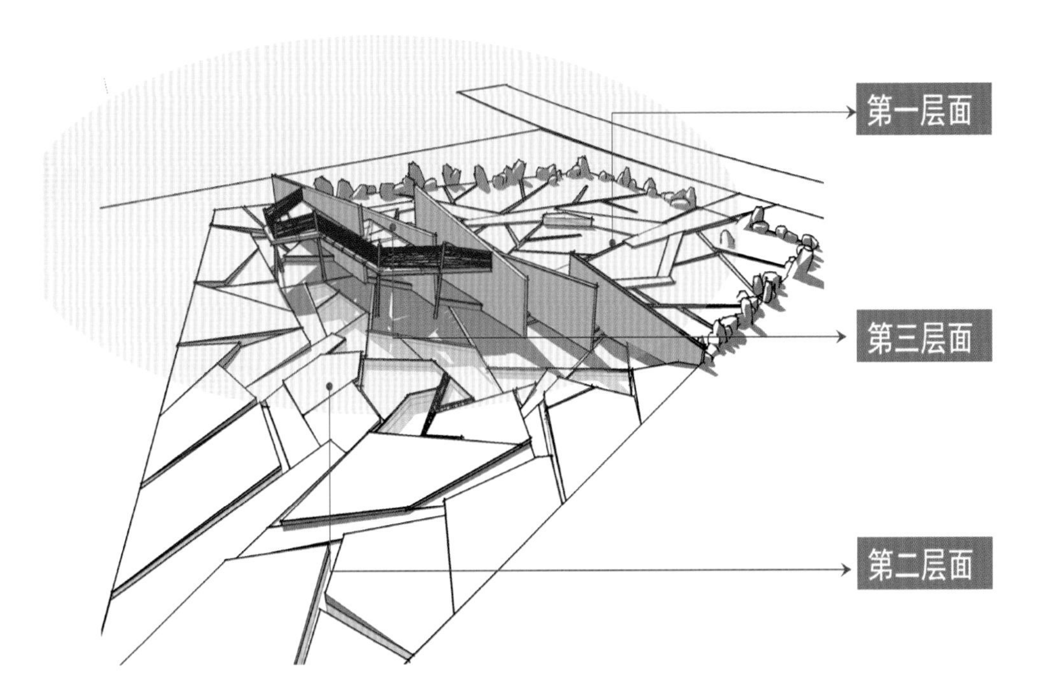

图 11 实空间转换图
Fig 11 Transformation of Real Spaces

图 12 第一层面空间透视一
Fig 12 Scenograph Ⅰ of the First Level Space

图 13 第一层面空间透视二
Fig 13 Scenograph Ⅱ of the First Level Space

图 14 第二层面空间透视一
Fig 14 Scenograph Ⅰ of the Second Level Space

图 15 第二层面空间透视二
Fig 15 Scenograph Ⅱ of the Second Level Space

图 16 第三层面空间透视
Fig 16 Scenograph of the Second Level Space

图 17 空间组织转换图
Fig 17 Space Transition Diagram

12

13

14

15

16

17

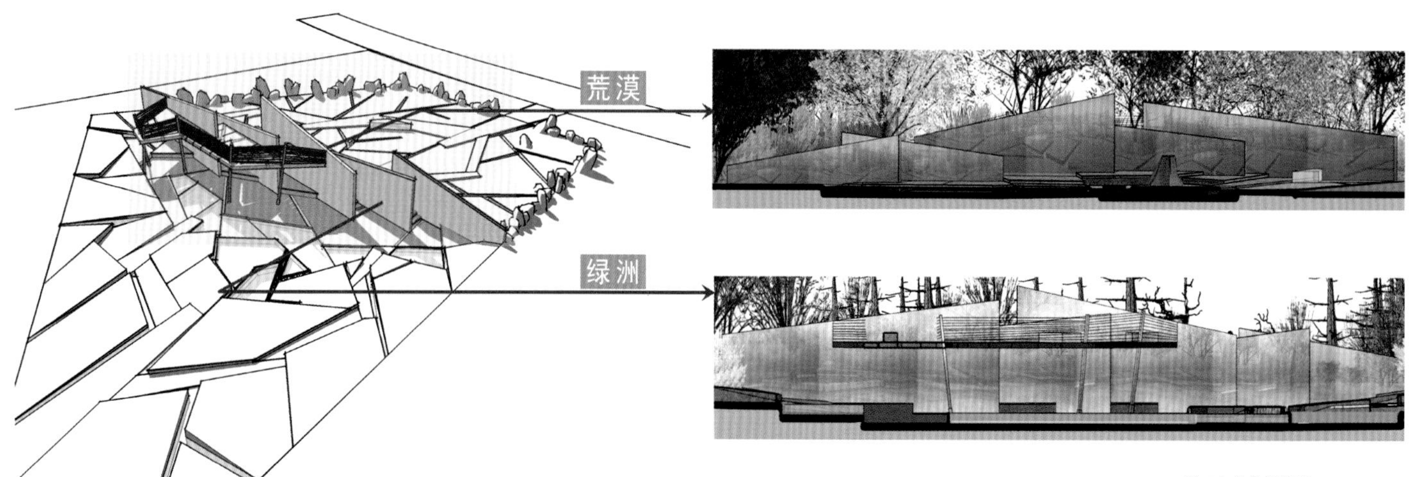

图 18 虚空间演绎
Fig 18 Virtual Space Diagram

2. 营建不同的立地空间条件，满足植物生长

下沉式地形空间，为植物生长提供小的气候环境，减少景观水池内水的蒸发。

3. 自然材料和透水工艺

全园铺装地面和道路，采用天然石材、卵石和沙；垫层采取碎石垫层，整体具有透水性。利用小块的水面及枯寂的山石，反映对水的保护与设计师的思考。

植物景观

以北京地区的常用园林植物为材料；"荒漠"部分选用在园林建设过程中死亡的、树形优美的植物或死树桩作为软景材料；"绿洲"部分突出郁郁葱葱，以杨树为基调树种同时结合空间效果局部栽植挪威槭、榉树及樱花等（图 21）。

干枯的树木、冰裂的土地；清澈溪流、葱绿的山林……自然的演绎给人类诸多启示，她是一面镜子，映射出自然发展的规律。园林是集社会经济、文化艺术和科学技术为一体的综合体，她是一面镜子，映射出人类对自然环境的态度和行为，科学技术的发展水平。□

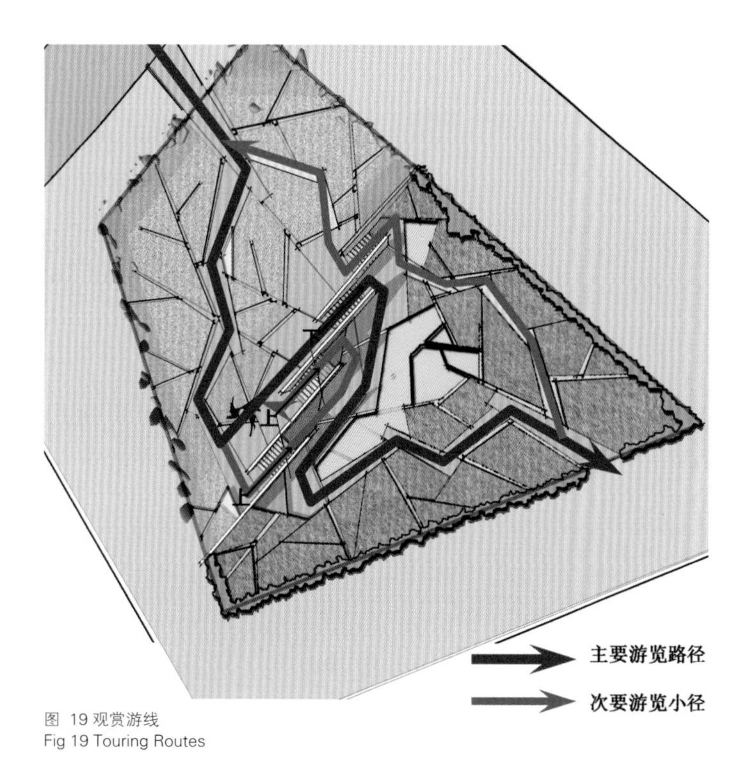

图 19 观赏游线
Fig 19 Touring Routes

the western area close to the EXPO main road has the layout of a desert landscape, the eastern area takes the form of an oasis (Fig.04-07). A Zen garden on the western side adds a peaceful atmosphere, suggesting that

the environment is being destroyed, it expresses the silence and barrenness of the world (Fig.08). Opposite this, the sunken garden and the mirror pool on the east side appear balmy and pleasant, and convey the life giving vitality of the oasis(Fig.09). Similar layouts are used for both the east and west sides but to opposite effect. Visitors will be shocked and informed of the catastrophic consequences of environmental imbalance and then encouraged to learn about water conservation and to develop a love for the natural environment. (Fig.10).

Spatial Composition

1.Real Space

The central area is divided into two parts by the four mirrored walls, and consists of the desert and oasis, two distinct spaces. The western high and eastern low terrains along with the steps and the platform, together create three different spaces at heights with dramatic transformations connecting these spaces(Fig.11).

The first level space shows the dead desert by making use of the existing ground (Fig.12). A 1.8 meter tall mirrored wall is used as the interface boundary, thus it enhances the vastness and solitude of the desert (Fig.13).

The Second Level Space is a sunken garden1.05m lower than the first level (Fig.14-15). Green trees enclose an inward and sealed-in space with the mirrored walls, while the platform combines with the mirrored walls systematically forming a rest corridor. All these elements help to define a quiet and pleasing sunken space.

The Third Level Space is a platform 1.50m higher than the first level space (Fig.16). Visitors can overlook the desert and the oasis on this platform, and also take in the outside view.

The three spatial levels are connected to each other, therefore visitors will experience a series of scenes: first, seeing stillness and becoming confused; then being surprised at the contradiction; finally, feeling pleasure and starting to reflect. Thus, an interesting and dramatic touring route has been formed (Fig.17).

2. Virtual Space—illusive scenes built by mirrors

The entire garden has been separated into two parts, which are two distinct pseudo-spaces .The two pseudo-spaces reflect

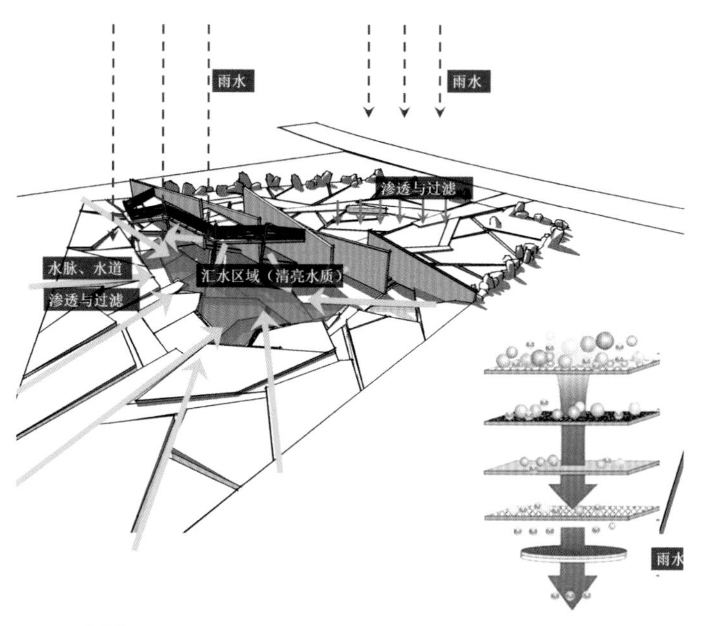

图 20 雨水收集系统
Fig 20 Rainwater Collection System

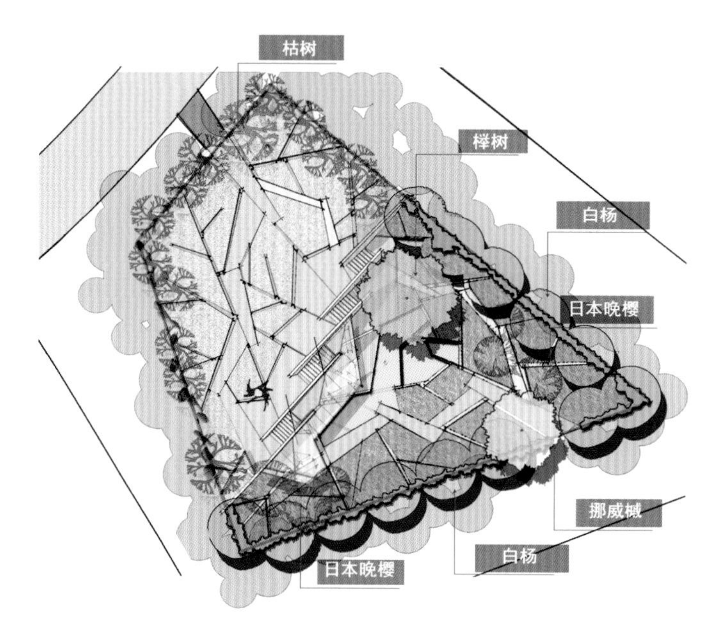

图 21 植物概念设计
Fig 21 The Concept Design of Plants

each other, thus creating an illusory virtual space. The real spaces and the corresponding virtual spaces together comprise the integral landscape spaces, infusing a charming visual effect into the environment (Fig.18).

3. Enclosing Space and Touring Routes

The oasis is enclosed by plants while the desert is enclosed by stone walls, the two entirely different enclosure methods reinforce the landscape characteristics. By making the oasis more vital and the dunes more deathly still, visitors' experiences are enhanced. The touring routes consist of the parched earth paths, and the connections between the three spatial levels, thus they are linked to each other to form a touring loop. These routes are designed to provide a series of orchestrated views (Fig.19).

Ecology and Water Conservation

1.Establish a rainwater collection system for the whole garden.

Beijing has a severe shortage of water resources with its groundwater levels going down year by year. In this case, methods of sectional design are specially used for collecting rainwater. Small pools were built in the oasis area with white cobblestones, which present pleasant scenes in both the wet and dry seasons. Meanwhile, the surface water of the oasis is collected through the surface treatment and eventually flowing into the pools. The underground water seepage project also performs a water conservation role in the desert area by collecting groundwater for the pools (Fig.20).

2. Create different living environment for the plants

The sunken space provides microclimate for plants and reduces evaporation from the pools.

3. Use natural and permeable materials

The road and pavement of the garden are entirely comprised of natural stones , pebbles and sand, even the sub-surface layer is filled with broken stones, which means the entire surface of the garden is permeable .The use of small pools and stones reflects the thinking of the designer and the issue of water conservation.

4. Plantings

The mirrored garden contains commonly used garden plants from the Beijing region. The desert area mainly contains dead trees or stumps, chosen for their beautiful shapes. The oasis area emphasizes luxuriant green, with the main species of tree being poplar, and norway maple, zelkova and sakura trees also planted in the tertiary spaces (Fig.21).

The shriveled trees and parched earth, the crystal stream and the verdurous forests all tell an enlightening story about nature. This story is a reflection of the laws that govern its workings. Gardens and landscape integrate science and technology with society, economics, culture and the arts. They are mirrors that reflect human attitudes and activities towards nature as well as the development of technology. ■

作者简介：

张建林 / 男 / 博士 / 副教授 / 西南大学园艺园林学院副院长 / 风景园林规划设计理论与实践 / 中国重庆

Biography:

Jianlin Zhang / Male / Ph.D / Associate Professor/ Assistant Dean of Horticulture and Landscape Architecture, Southwest University / Chongqing, China

查尔斯 · 沙（校订）
English reviewed by charles sands

林中池塘·平安扣
—园博会设计师园6号地块方案设计
POND IN THE GROVE, PEACE BUTTON
— GARDEN EXPO DESIGNER SQUARE LOT 6 DESIGN

严 伟 Wei Yan

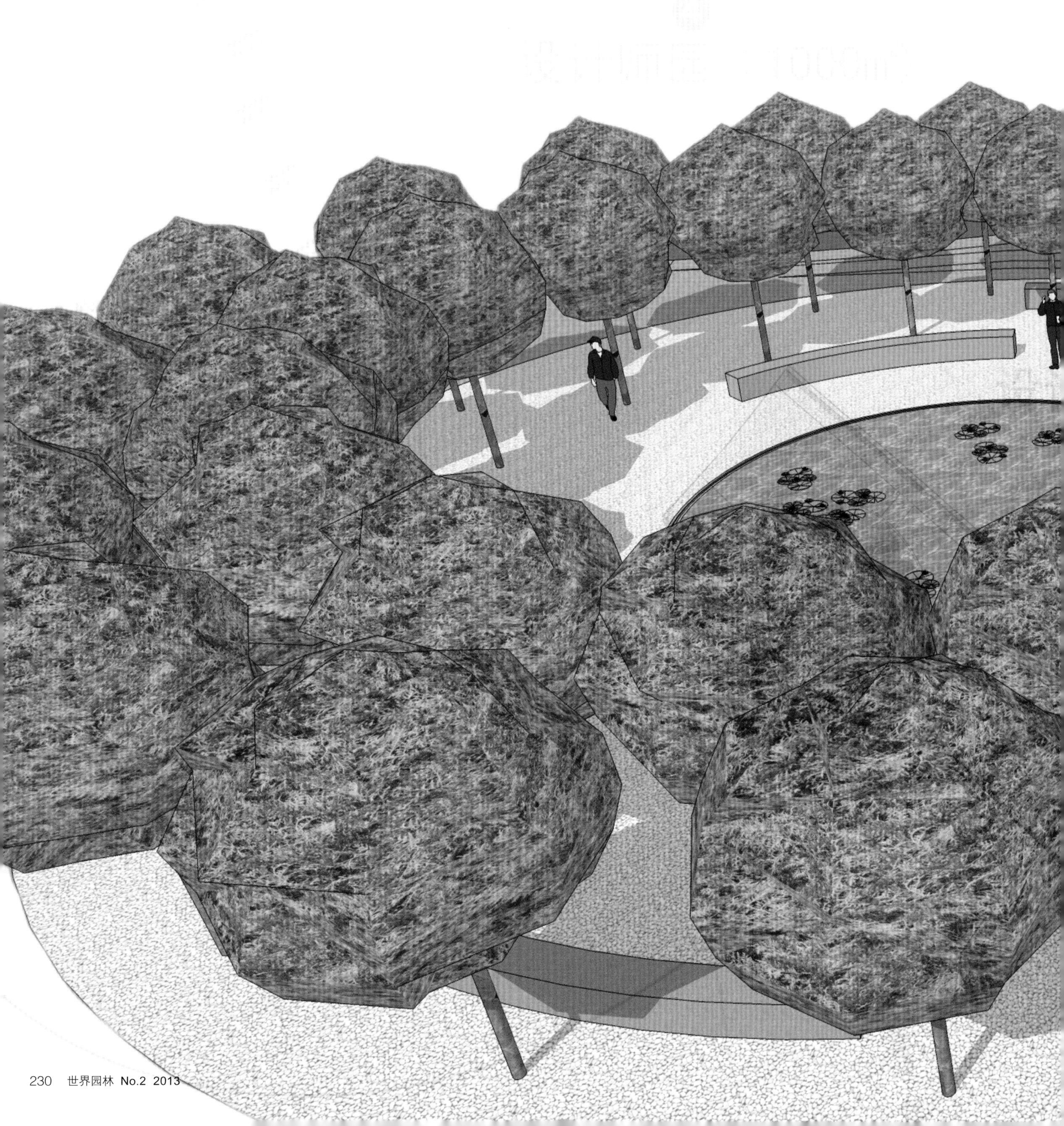

项目位置：中国，北京，第九届北京园博会
项目面积：1,000m²
委托单位：第九届中国（北京）国际园林博览会组委会
设计单位：北京市园林古建设计研究院有限公司
景观设计：严 伟
完成时间：2013 年

Location：The 9th China (Beijing) International Garden Expo, Beijing, China
Area：1,000m²
Client: The 9th China (Beijing) International Garden Expo committee
Designer：Beijing Institute of Landscape and Traditional Architecture Design and Research, China
Landscape Design：Wei Yan
Completion：2013

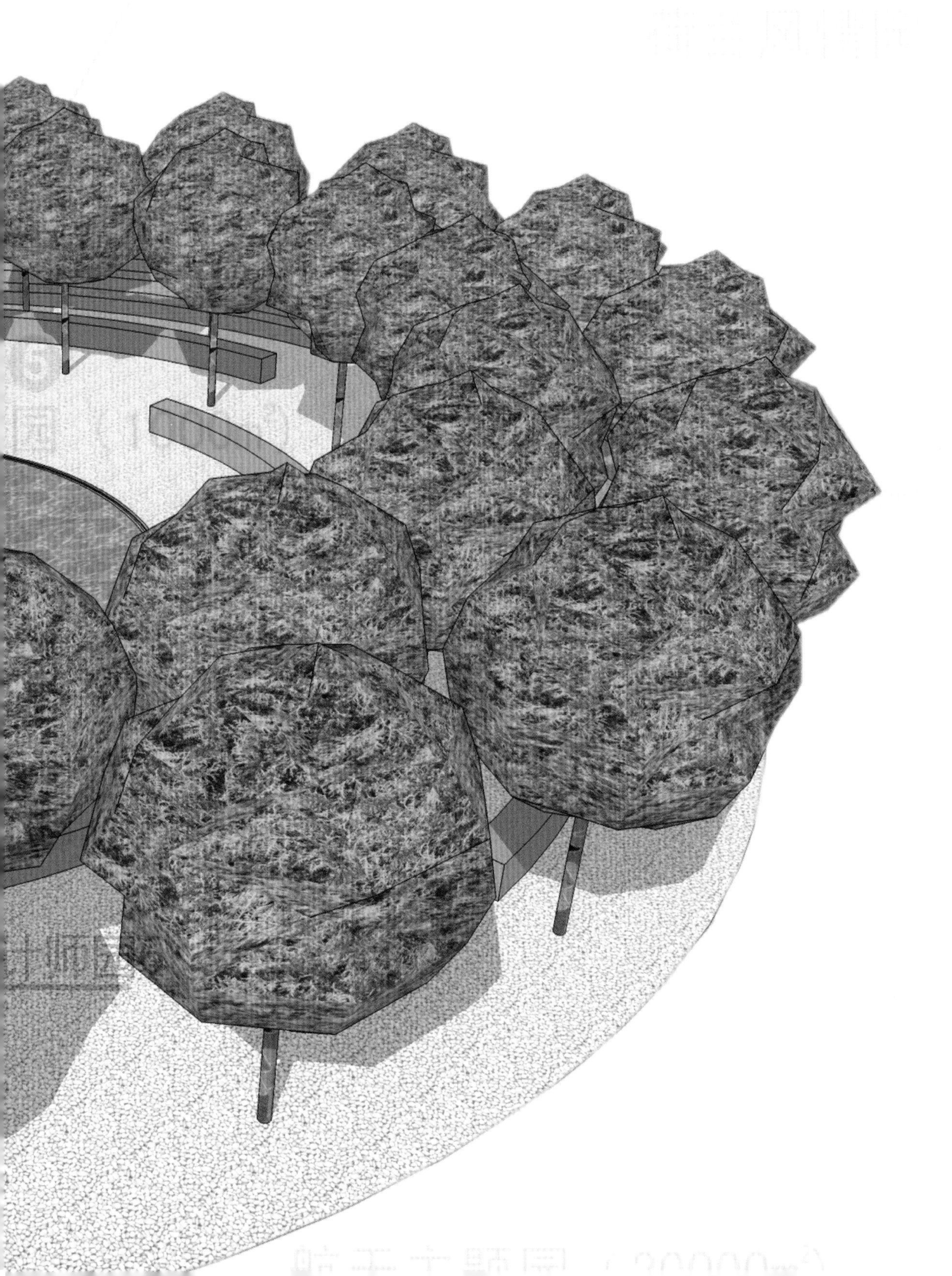

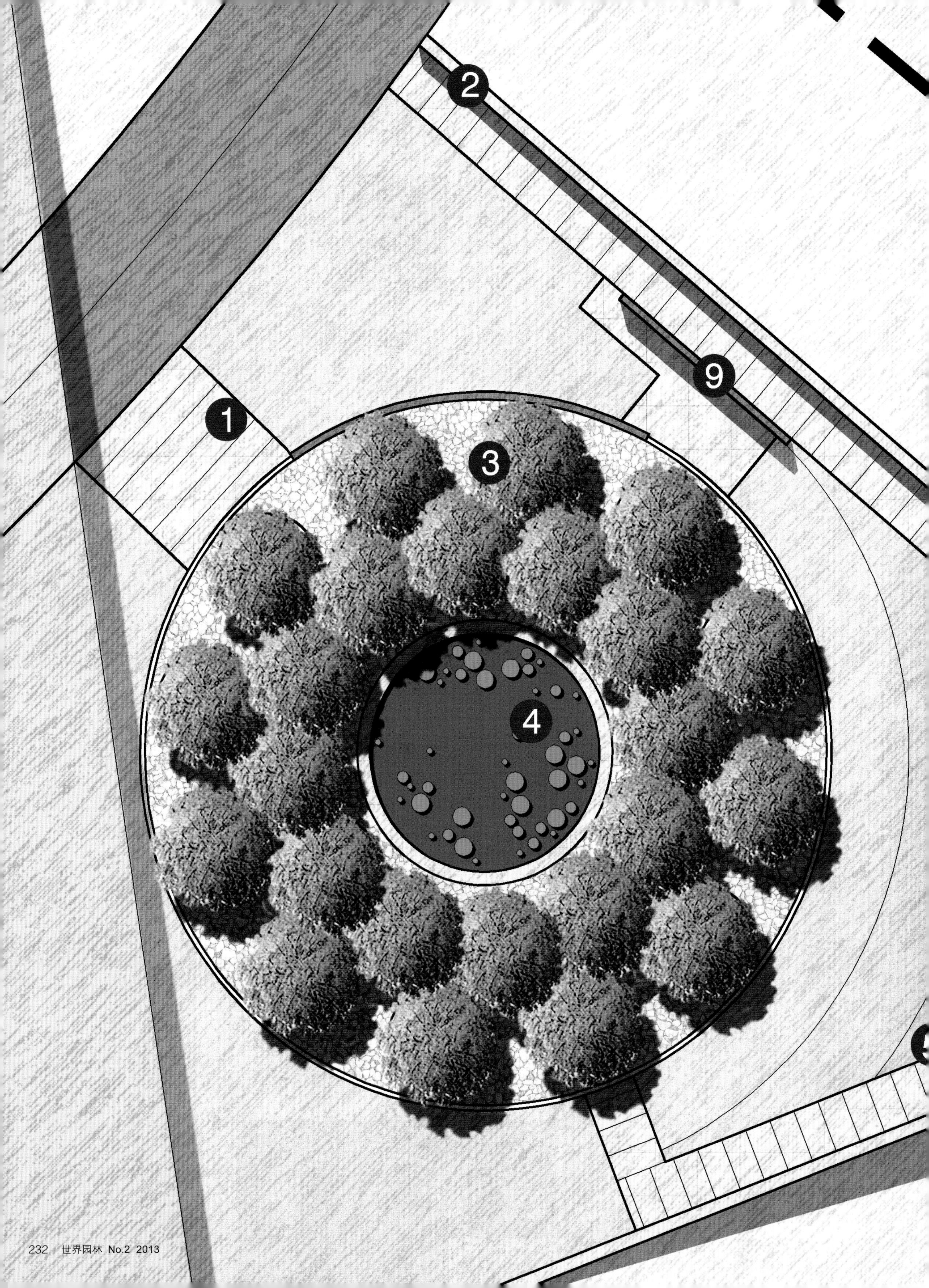
2
9
1
3
4

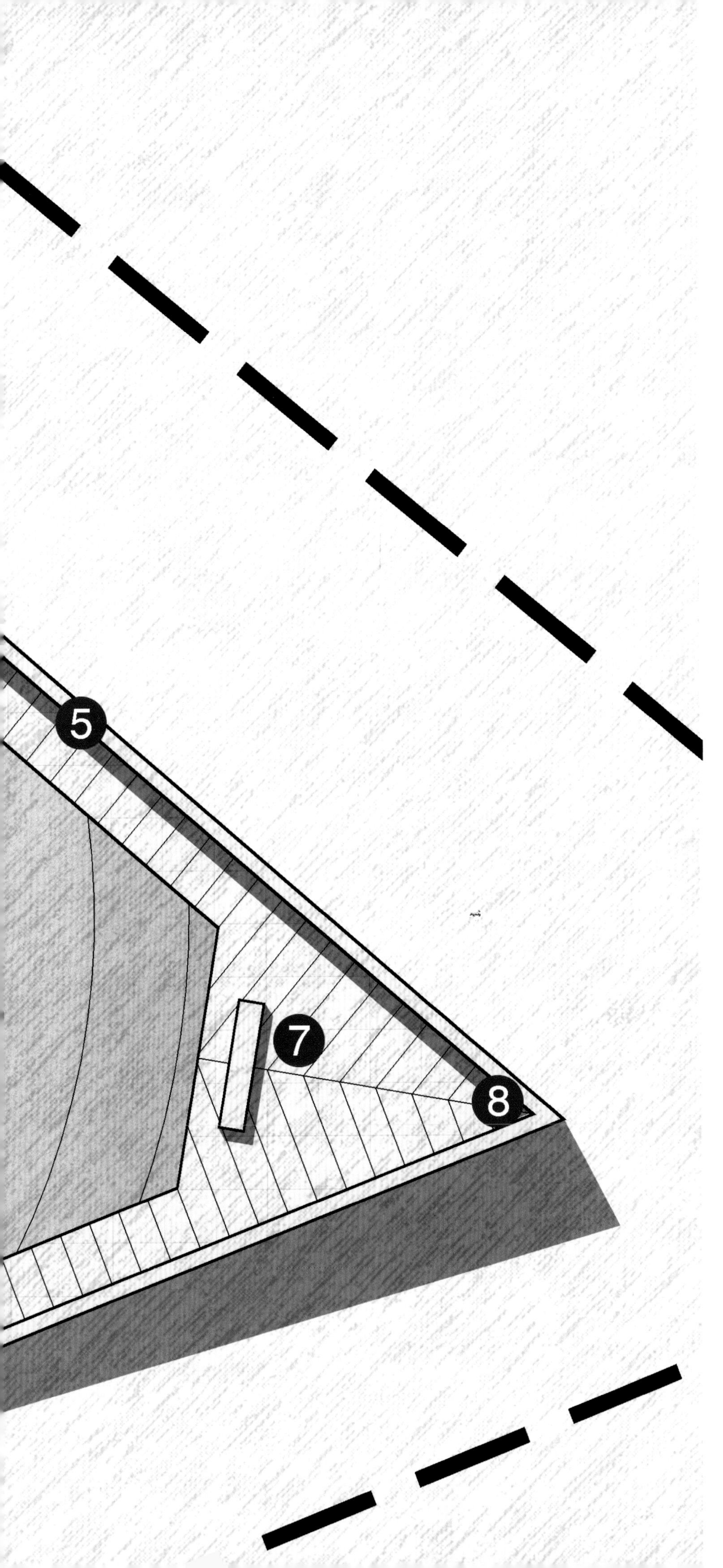

1 入口
Entrance
2 出口
Exit
3 林下坐凳广场
Understory Square
4 睡莲池
Water Lily Pond
5 坡道
Ramp
6 台地花坡
Grass Slope
7 观景平台
Viewing Platform
8 景墙
Scene Wall
9 光影镜面墙
Mirror Wall

图 01 平面图
Fig 01 Plan

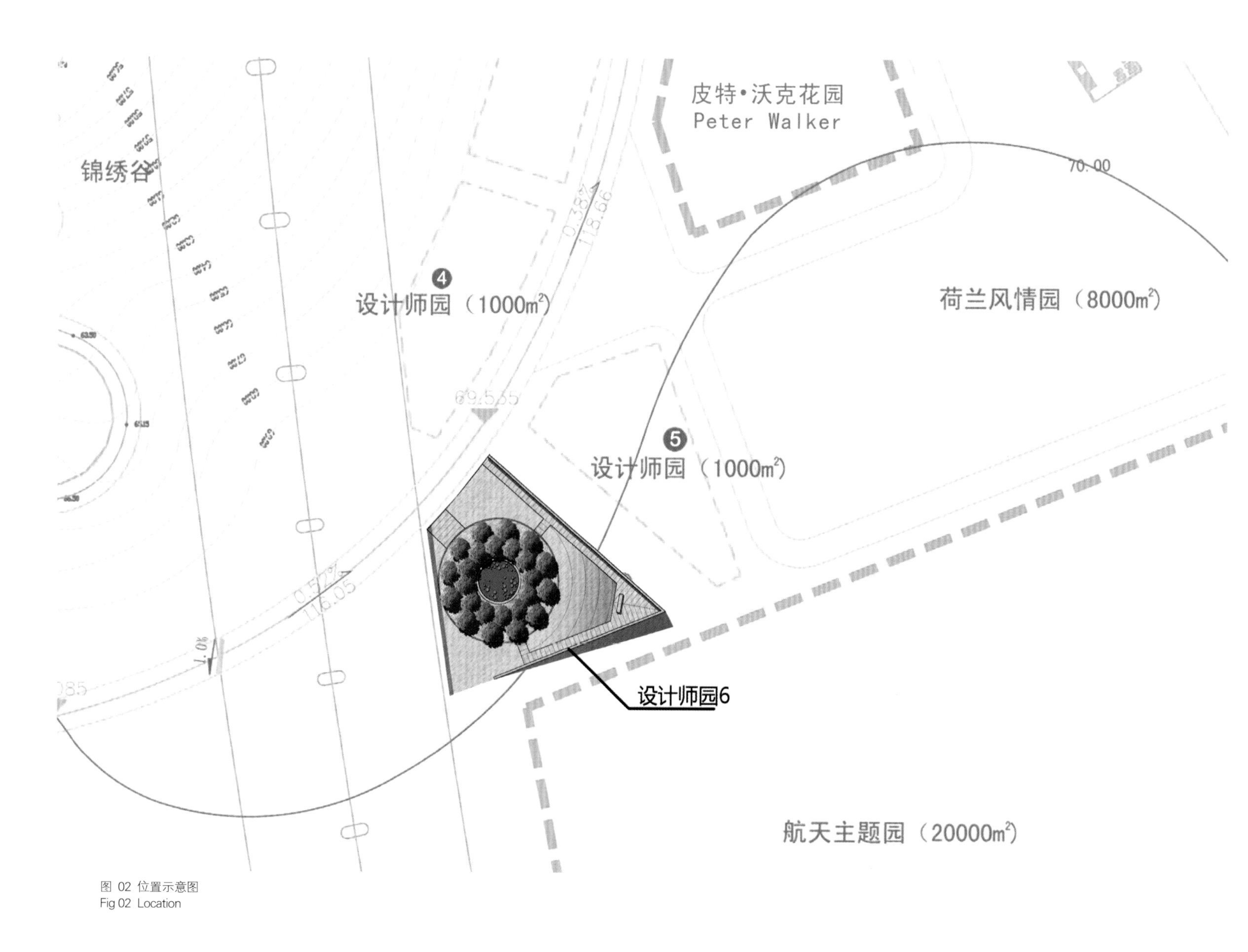

图 02 位置示意图
Fig 02 Location

本届园博会强调游人的参与性，本方案强调人与景观之间的融合、穿插。方案利用场地的特点，形成一个圆形平面与一个三角坡面的交插（图 01–02）。前面圆形树阵强化园林与外部环境的开放性融合。后面的坡地给前面景观营造了很好的背景，同时又提供了不同高程的观景角度（图 03–04）。

设计师园应当引导一个未来的园林设计方向。中国园林讲究师法自然。本方案强调运用自然元素，利用地形植物营造艺术氛围（图 05）。用简约手法表达了林间池塘的宁静氛围：有山、有树、有塘、有花（图 06）。利用自然元素设计也是节约型园林的重要设计方法。

融入人的园林才会出现最美的风景，本方案注重游人的参与体验，希望能给参观者留下美好的记忆。登高远眺的风景（图 07）、内向型的睡莲池、镜面光影墙、背景梯田花坡都能给游人留下美好的记忆（图 08–09）。同时高低两个视点的互望也是一种乐趣（图 10）。

动静的对比是 6 号地想表达的内涵：静静的树林，（图 11）清澈的莲池（图 12），平和的花坡（图 13），包容着流动的人群。面对纷杂变化的社会，园林应能给人提供一个宁静的环境去思考、放松、自我更新。

画好方案，发现图形结构很像一块碧玉。中国人讲玉能镇宅辟邪，因此方案的名字就叫“平安扣”吧。□

One of the priorities of the garden expo is visitors' participation, so this plan considers people as parts of the landscape and integrates all the elements into a whole. (Fig 01)The plan makes use of the site's characteristics, inserting one circular plane into one triangular slope. (Fig 02)The circular array of trees at the front of the site strengthens the connection between inside and outside. (Fig 03)The back slope provides the background for the main attraction as well as different elevations and viewing angles. (Fig 04)

Designer Square guides the direction of the landscape design. Chinese landscape architecture advocates learning from nature. This plan emphasizes making full use of natural elements, like plants and the terrain, to create interesting landscape spaces. (Fig05)Mountains, trees, ponds and flowers are utilized to form a peaceful atmosphere, (Fig 06)at the same time this method is economical.

The landscape design must also be accommodating to visitors. The plan tries to impress visitors with unique experiences. From the distance one overlooks (Fig 07)the peaceful water lily pond, the spectacular lighting wall and the memorable flower slope. (Fig 08-09)In addition, one can experience different aspects of the landscape from either end of the site. (Fig10)

The plan expresses the contrast between dynamic and static elements: the quiet grove, (Fig11)the clear water lilies pond (Fig12) and the colorful flower slope(Fig13). The landscape helps visitors to relax, think and renew themselves in this complex society. (Fig14)

The completed plan resembles a piece of jade. There is a belief in Chinese culture that pieces of jade can protect people from disasters, so they are referred to as "Peace Buttons".■

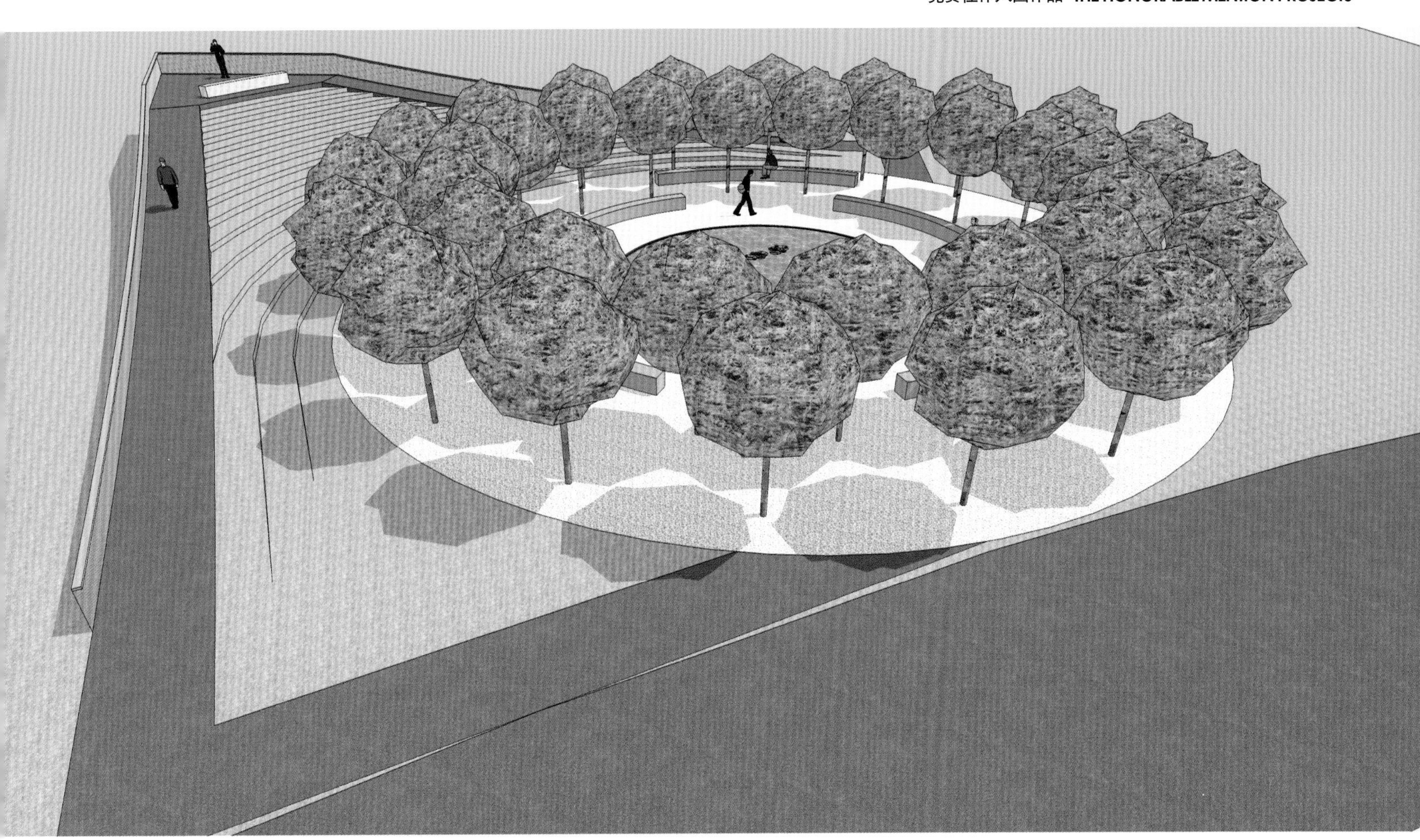

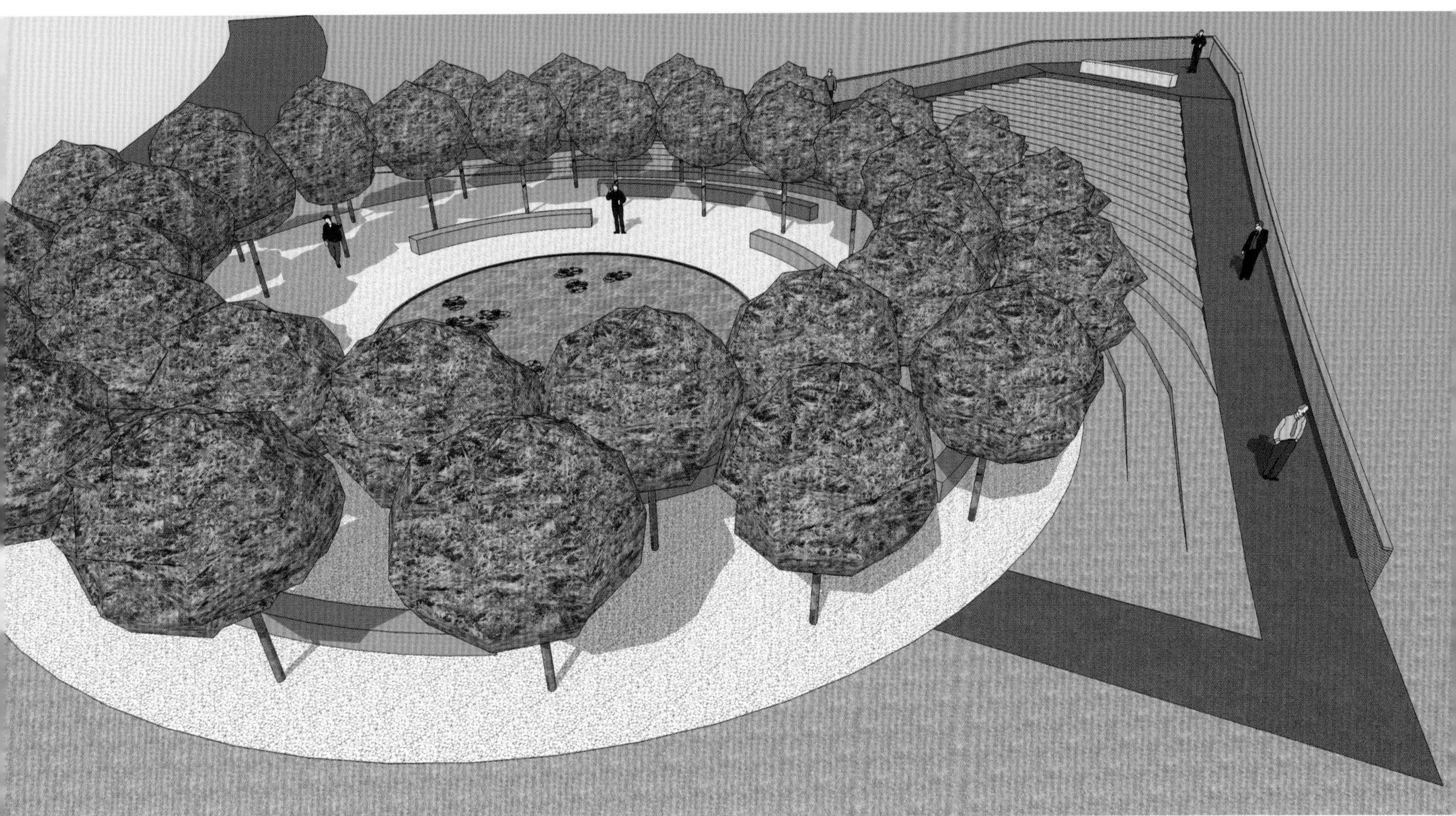

图 03–04 鸟瞰图
Fig 03–04 Bird’s eye view

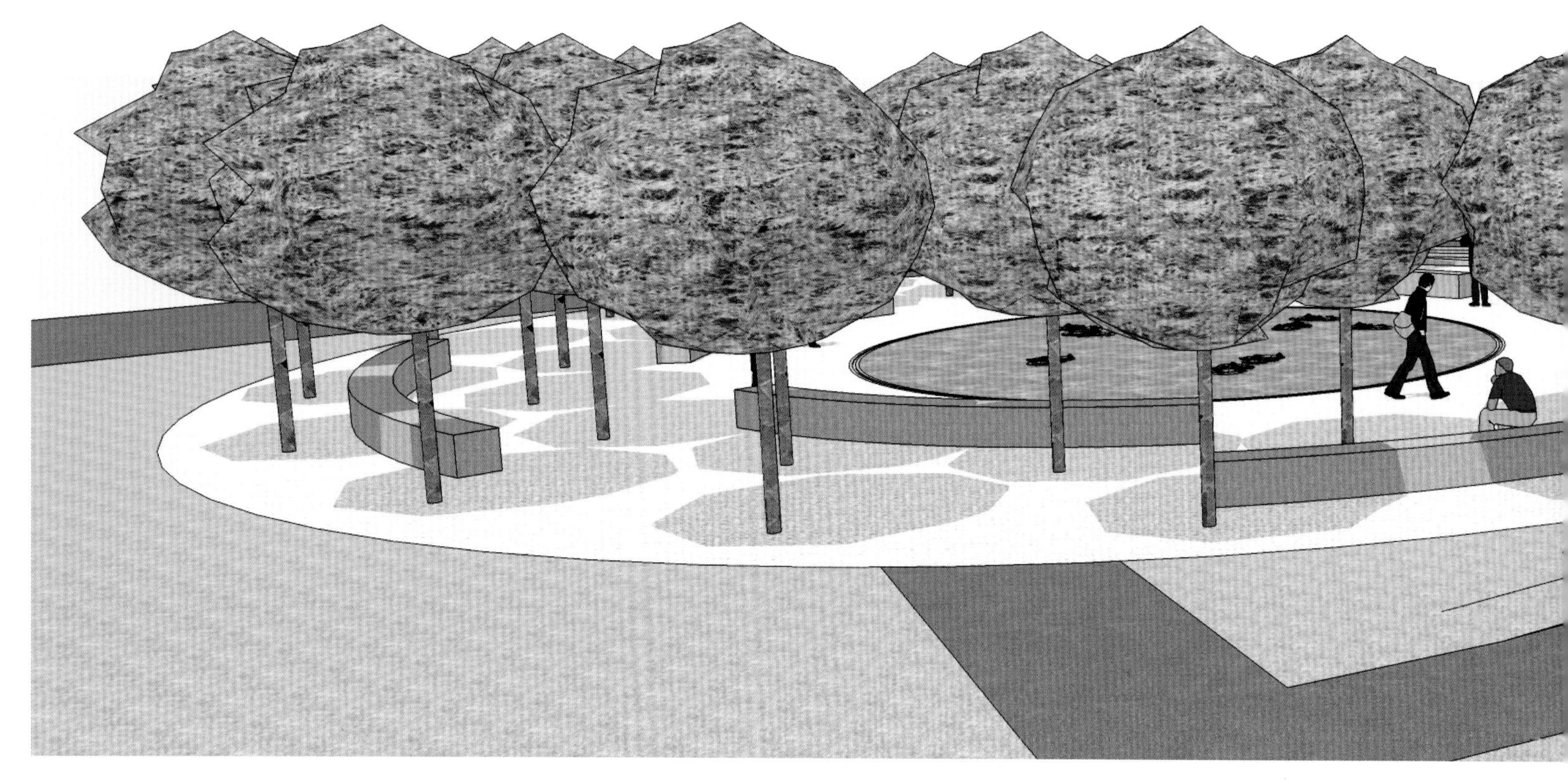

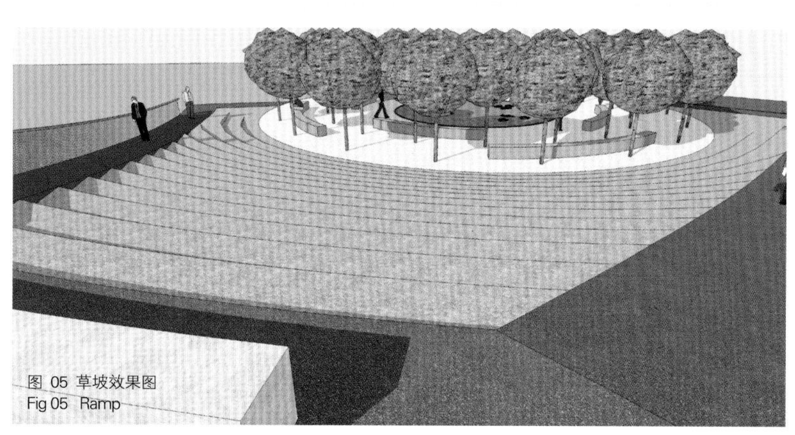

图 05 草坡效果图
Fig 05 Ramp

图 06 池塘效果图
Fig 06 Ramp

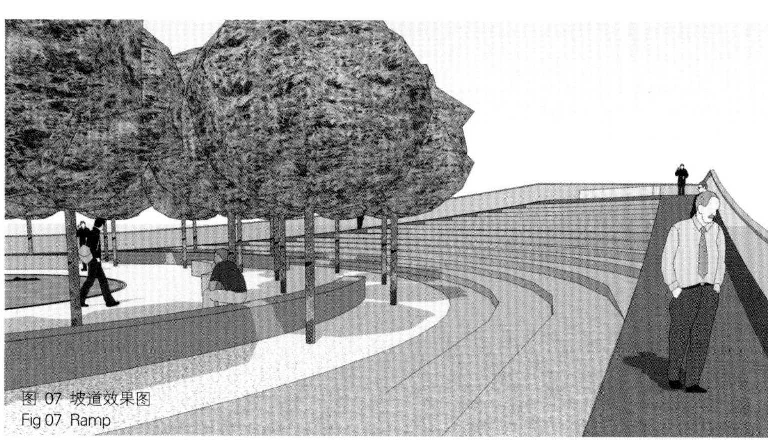

图 07 坡道效果图
Fig 07 Ramp

图 08 池塘效果图
Fig.08 The three l

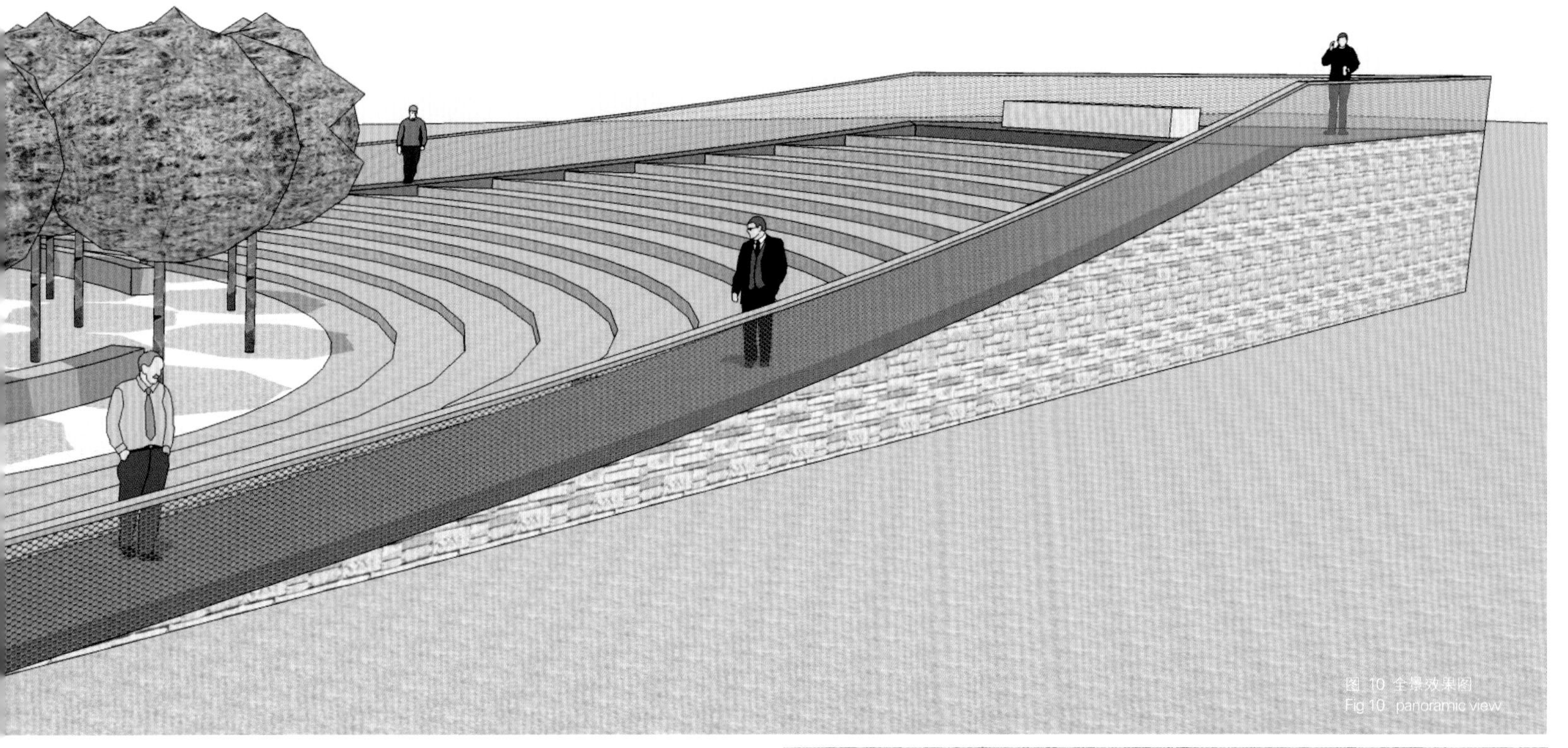
图 10 全景效果图
Fig 10 panoramic view

图 09 林下坐凳广场效果图
Fig 09 Understory Square

图 11 林下坐凳广场效果图
Fig 11 Understory Square

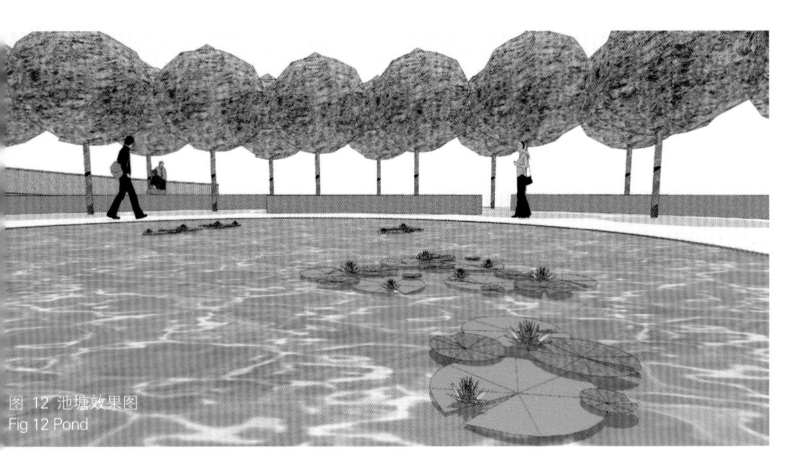
图 12 池塘效果图
Fig 12 Pond

图 13 坡道效果图
Fig 13 Ramp

作者简介：

严伟 / 男 / 高级风景园林师 / 北京市园林古建设计研究院有限公司 / 中国北京

Biography:

Wei Yan/Male/Senior landscape architect/Beijing Institute of Landscape and Traditional Architecture Design and Research/Beijing, China

查尔斯・沙（校订）
English reviewed by Charles Sands

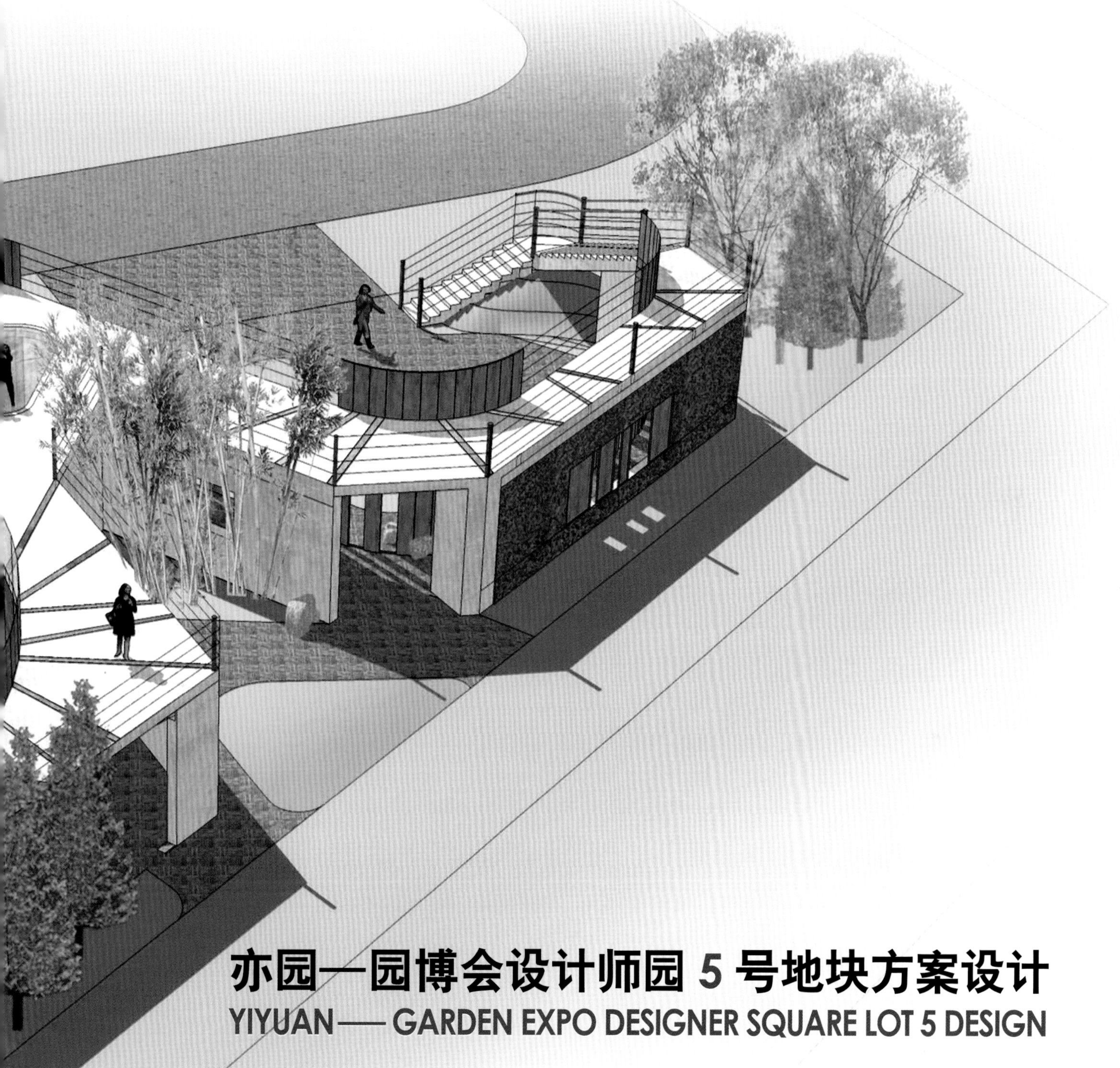

亦园—园博会设计师园 5 号地块方案设计
YIYUAN— GARDEN EXPO DESIGNER SQUARE LOT 5 DESIGN

毛子强 潘子亮

Ziqiang Mao Ziliang Pan

项目位置：中国，北京，第九届北京园博会
项目面积：1,100 m^2
委托单位：第九届中国（北京）国际园林博览会组委会
设计单位：北京市园林古建设计研究院
景观设计：毛子强、潘子亮
完成时间：2012 年 3 月

Location: The 9th China (Beijing) International Garden Expo, Beijing, China
Area: 1,100 m^2
Client: The 9th China (Beijing) International Garden Expo committee
Designer: Beijing Institute of landscape and Traditional Architechtural Design and Research
Landscape Design: Ziqiang Mao, Ziliang Pan
Completion: March, 2012

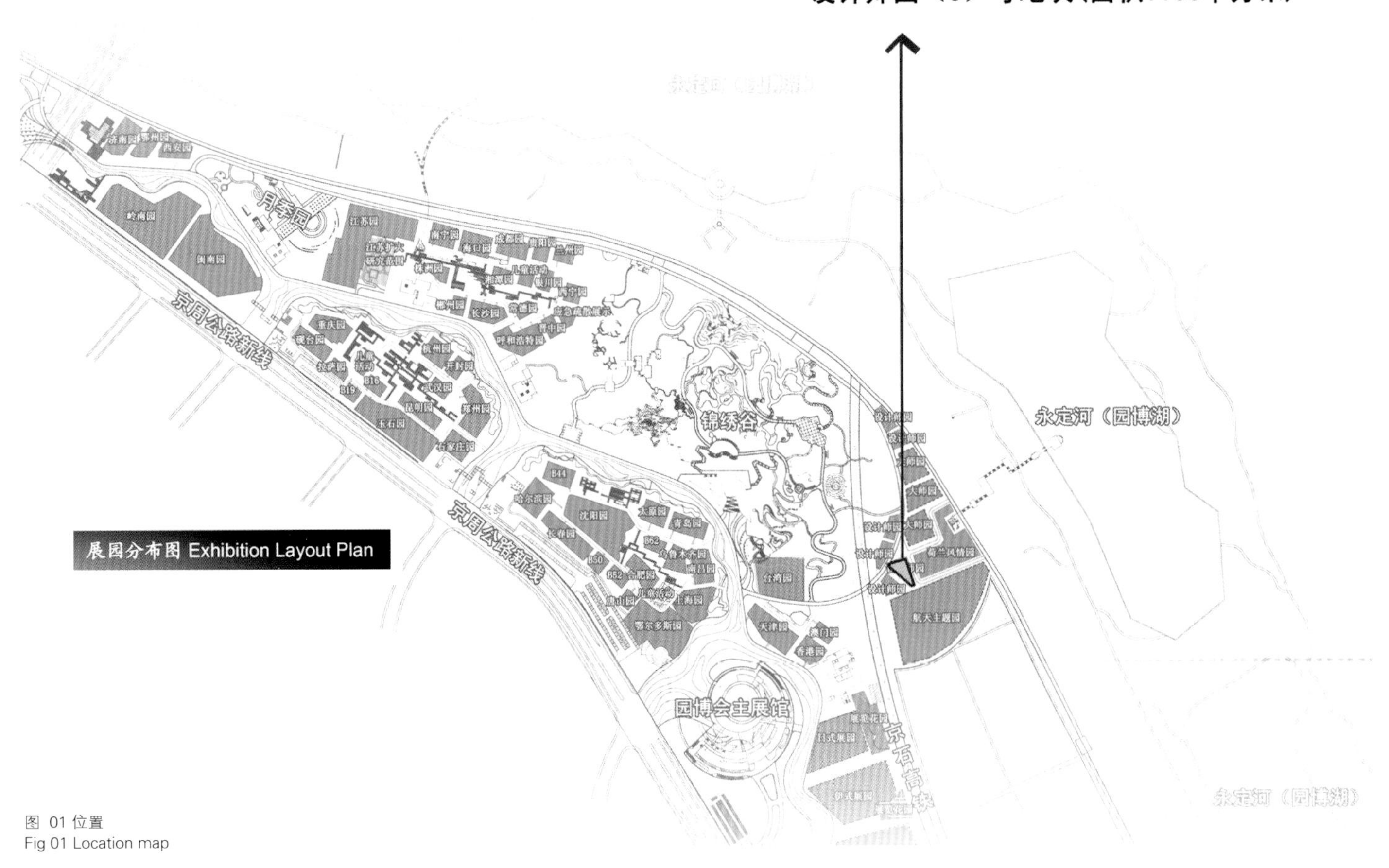

图 01 位置
Fig 01 Location map

项目位置

项目位于北京市丰台区永定河北岸，本方案设计选用第九届中国国际园林博览会组委会提供的大师园 5 号地块，面积 1,100m²（图 01）。

构思

本设计试图跳离风格化框架，不设定具体特别指向性设计语言，全部交由观者去感受。试图以对照手法激起观者的设计通感，一条廊为线索，界定南北、分隔空中和地面，形成开放景观面（图 02–03）。

空间划分和主景

亦园空间由 112m 长的曲线光影廊分隔为三个空间。由此，廊下、廊上、成南北互望；廊内、廊外，成明暗对比。光影廊将空间划分为空中和地面，地面上又分为南北西三个空间（图 04–05）。

北侧空间靠近园区主路，空间相对较大，以现代简洁的圆形水面和一株海棠为景，光影廊面向主入口的里面处理为一个半圆，与镜水面中倒影，形成另一个完整的圆形，两个圆形成一个融合的交集（图 06–08）。

南侧空间相对较小，更像传统园林中的天井，四周抱角石与类似古代屏风一般可以滑动的木格板形成围合，中心绿地中一柱瘦石（可用太湖石或现代材料）（图 09）。

西侧空间最小，作为边缘入口，曲径通幽，面竹而入。

光影廊，廊由上层空间和下层空间组成，上层空间提升了观景的高度，可南北同望两组庭院，甚至可以看到周边的几个展园；下层空间借助廊顶板镂空出的天井和透光条，同时结合屏风式可移动玻璃和木格板滑动，形成了明暗的光影变化（图 10–11）。

游线组织

园子打破了以往多数展园单一入口或者单向性游线的游览方式，

Project location

The project is located in the Fengtai District, Beijing Yongding River north shore, The program design was chosen for the Masters Park Lot No. 5 of the Ninth China International Garden Expo Organizing Committee with an area of 1100 square meters. (Fig 01)

Idea

The design attempts to jump off from a stylized framework and does not set specific special directional design language, all is for the viewers interpretation. Trying to arouse the viewer's synaesthesia by using control practices, a gallery for clues to define the north and south, separating the air and the ground, and forming an open landscape surface. (Fig 02-03)

The divided of the space and the main King

Yi Yuan is separated into three spaces by a 112 meters long curved gallery of light and shadow. This gallery is oriented from north to south. Inside the gallery and outside the gallery one can experience a chiaroscuro effect. Both the air and the ground, the ground are divided into three space. (Fig04-05)

The north side of the space near the main road of the park is a relatively large space, designed with a simple modern circular surface of water and a Begonia King. The lighting Gallery on the inside of the main entrance is shaped into a semicircle, which fuses with the reflection on the surface of the water to produce a full circle. (Fig06-08)

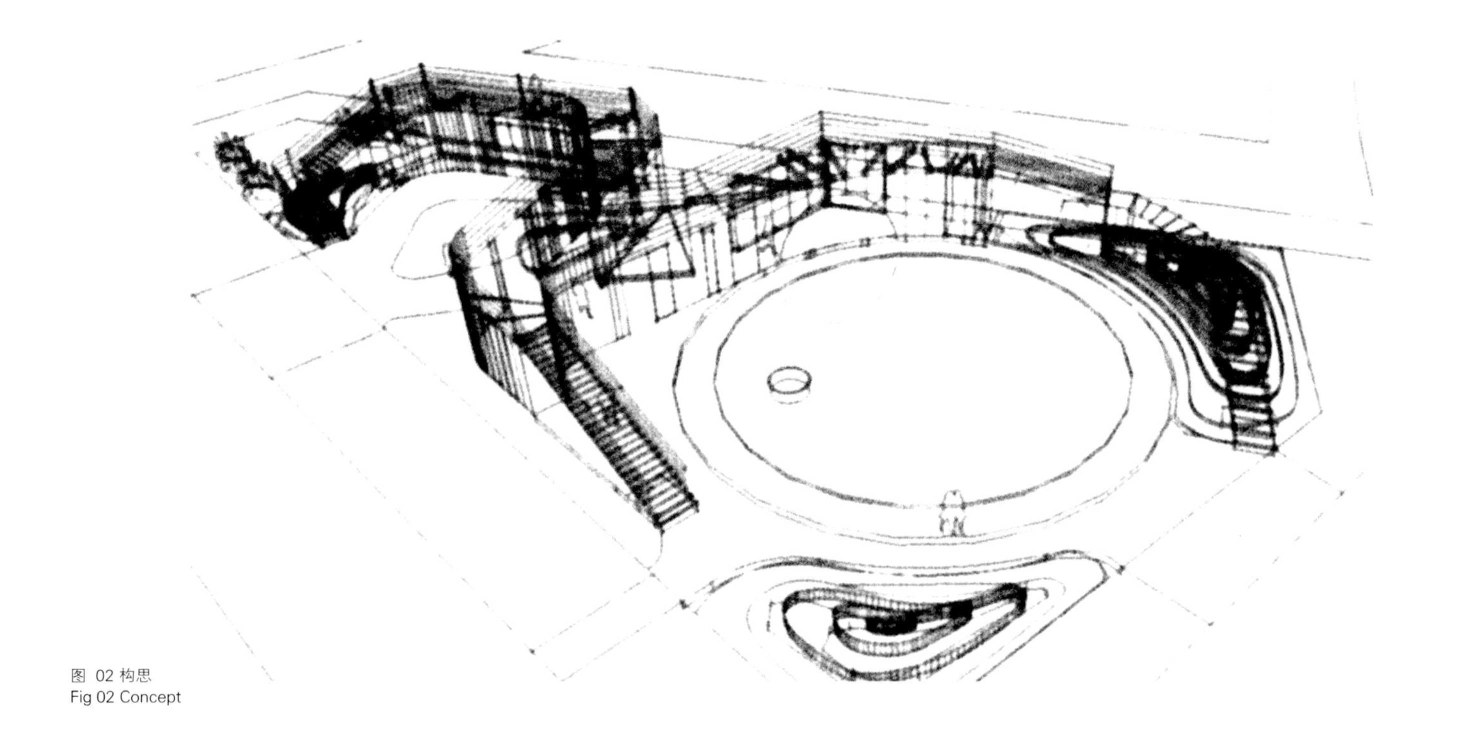

图 02 构思
Fig 02 Concept

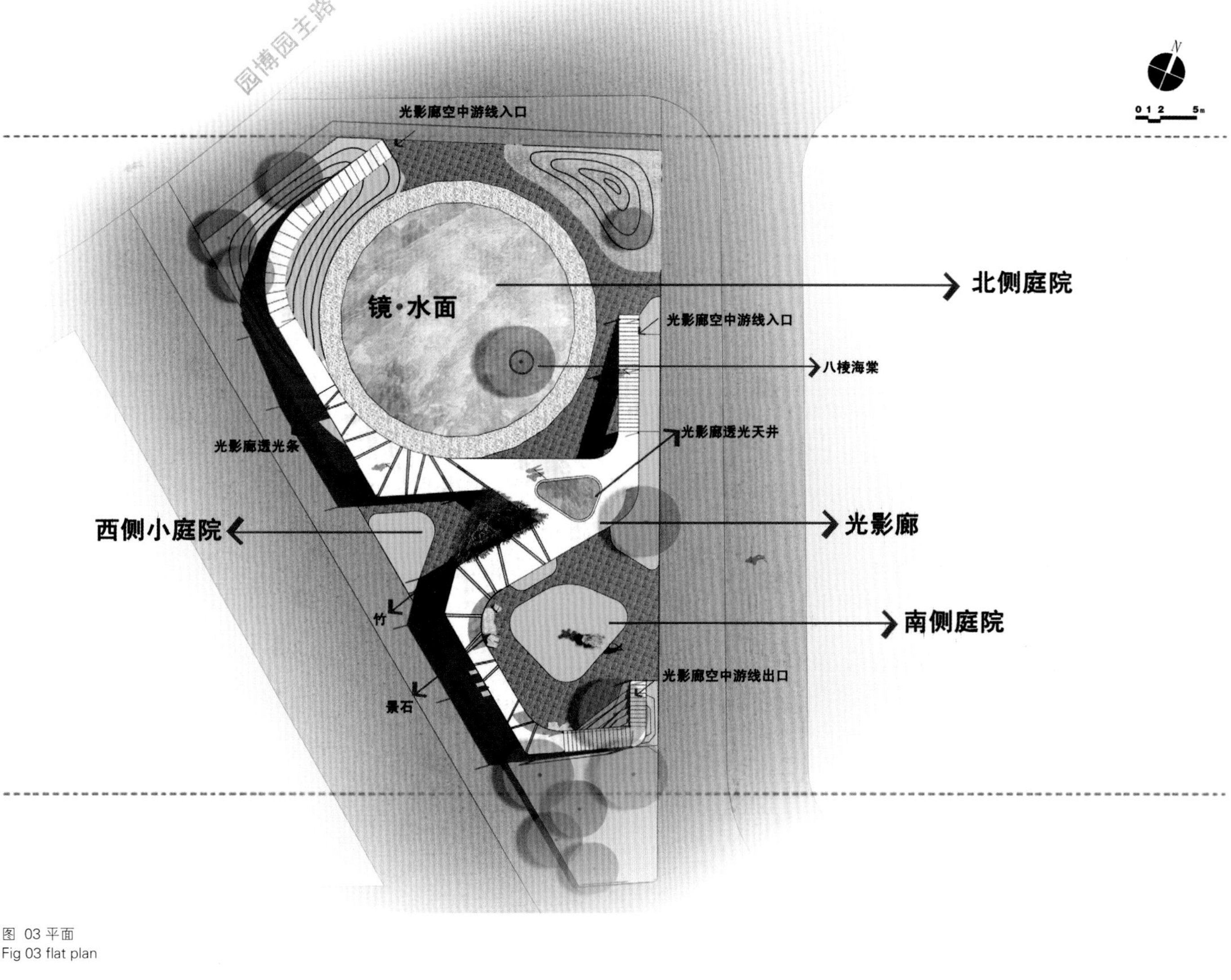

园博园主路
N
0 1 2 5m
光影廊空中游线入口
北侧庭院
镜·水面
光影廊空中游线入口
八棱海棠
光影廊透光天井
光影廊透光条
西侧小庭院
光影廊
竹
南侧庭院
光影廊空中游线出口
景石

图 03 平面
Fig 03 flat plan

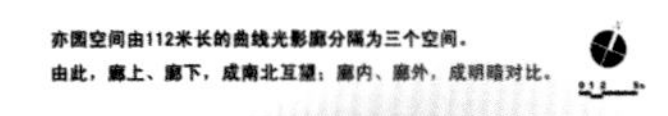

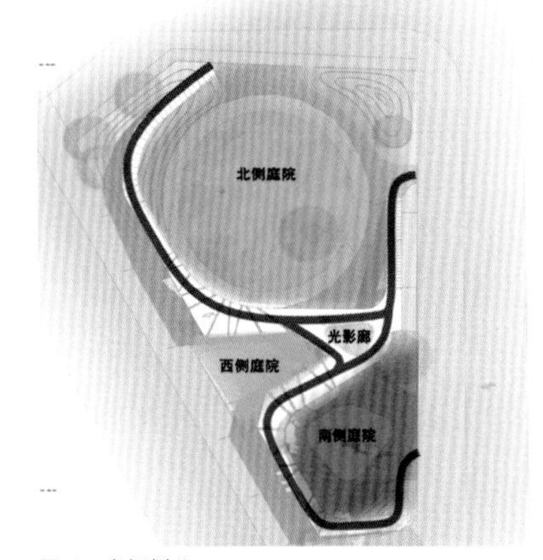

图 04 空间划分
Fig 04 Spatial Analysis

图 05 东侧鸟瞰
Fig 05 The east bird's eye view

图 06 北侧鸟瞰
Fig 06 The north bird's eye view

设置了地面和空中游线。袪除强迫性的游线设计，任何方向都可以很自由很舒适的步入，简言之：游线自由选择，出口，亦是入口（图12–13）。

种植设计

种植设计体现为三个临界面，北侧主景为一株八棱海棠，两侧绿地和台阶周边种植一些迎春、蔷薇等低矮丛生灌木；东侧种植5株元宝枫；西侧以高密度种植为主，出西侧入口对景为早园竹之外，均已修剪的桧柏篱与光影廊结合，适当留出通光通道。树种的选择均采用适合北京地区的乡土树种（图14）。

铺装及材料设计

铺装设计，材料相对简洁并试图形成明暗和质感对比。北侧、西侧庭院铺装主要为灰色和白色碎石，南侧庭院为具有一定现代感的花

The south side of the space is relatively small, more like a traditional garden patio. The corner stones and the sliding wooden grating, which resemble ancient screens, enclose the space. Thin stone pillars stand in the central green area (Fig09).

For the smaller space on the west side, an entrance faces the bamboo groves.

The gallery of light and shadows, is divided into upper and lower components. The upper space provides an enhanced viewing point, which looks into both the north and the south garden. Glimpses of the surrounding fair park can also be seen. The lower space use the hollow patio on the gallery roof and the translucent strips, combined with screen-type removable glass

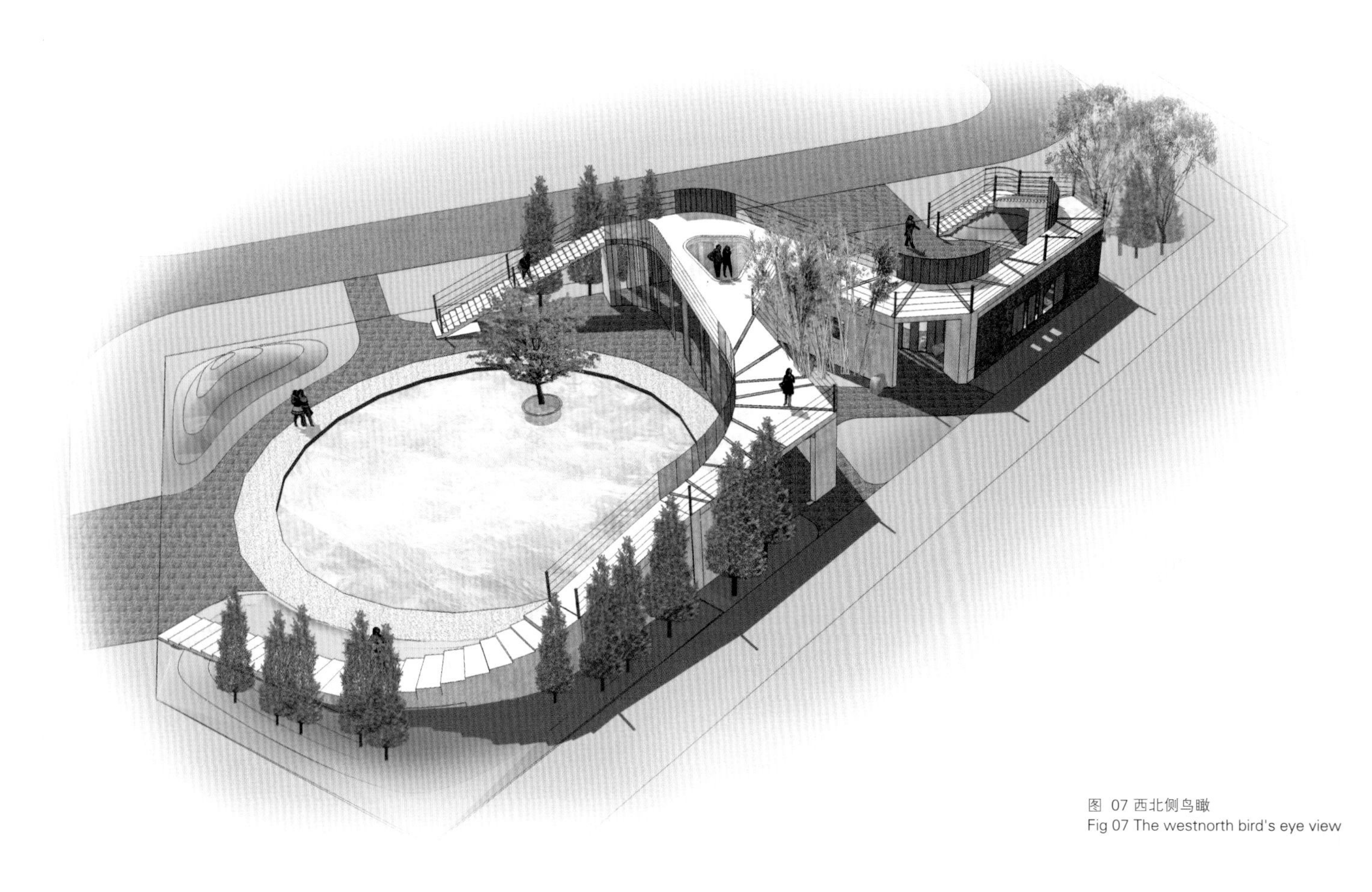

图 07 西北侧鸟瞰
Fig 07 The westnorth bird's eye view

图 08 人视效果 1
Fig 08 perspective view 1

图 10 人视效果 2
Fig 10 perspective view 2

图 09 人视效果 3
Fig 09 perspective view 3

图 11 人视效果 4
Fig 11 perspective view 4

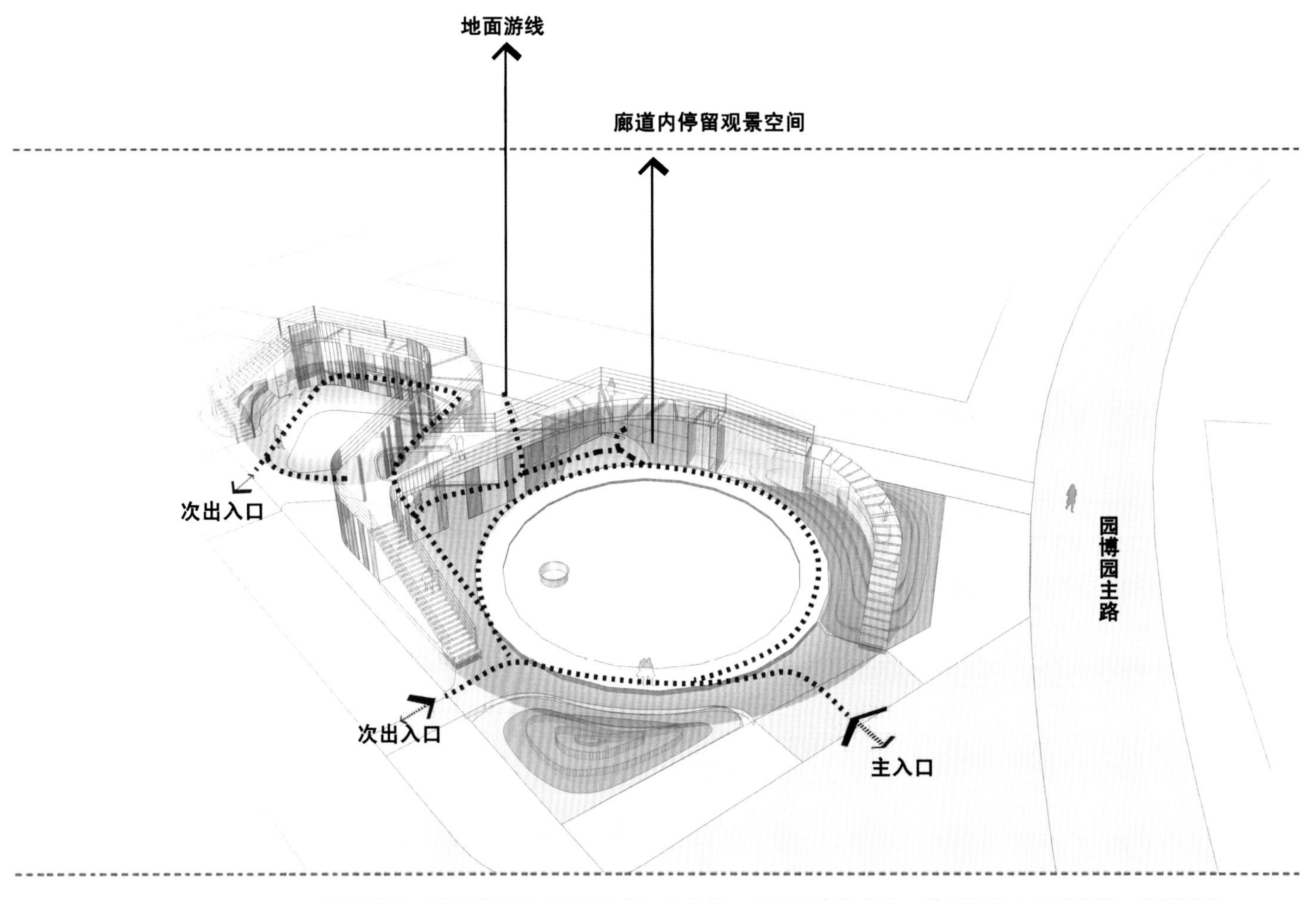

图 12 地面游线
Fig12 Tour route ground

街铺地。

光影廊的外立面材料为面北一侧为彩色玻璃和部分单透光镜面；面南一侧为屏风式木格板，面西一侧镂空为窗，以绿植填充。

光影廊顶板为料石拼接，拼接处以玻璃覆盖，即可透光。

光影廊护栏主体为玻璃，部分采用横向钢绳（图 15）。

寄语：

设计者与观者一同期待，在空中或地面，南或北，透过光影感受到对比和融合：亦上亦下，亦内亦外，亦明亦暗，亦中亦西，亦现代亦传统，殊途同归，出口，亦是入口。□

and wood lattice sliding panels, to produce variations of light and the shadow(Fig10-11).

The pathway organization

Our garden breaks from traditional pathway organization, like single entry or one-directional movement. By implementing both upper and lower pathways, we eliminate the forced adherence to a single route. One is free to enter or exit the garden from any direction (Fig12-13).

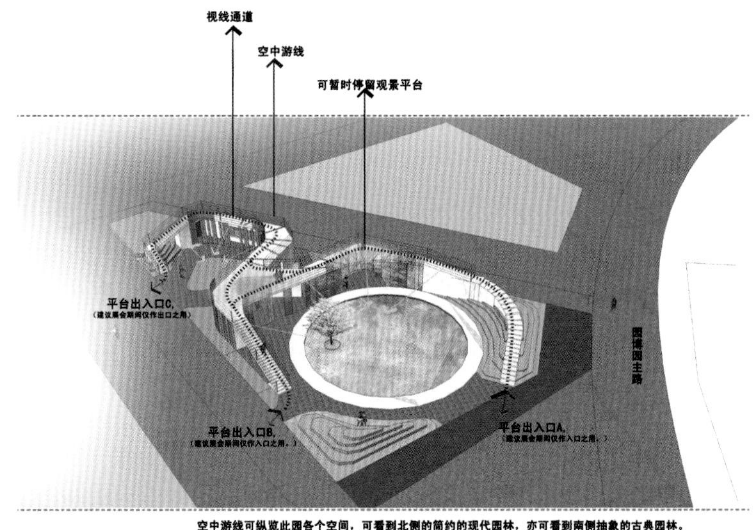

图 13 空中游线
Fig13 Tour route Upstair

种植设计

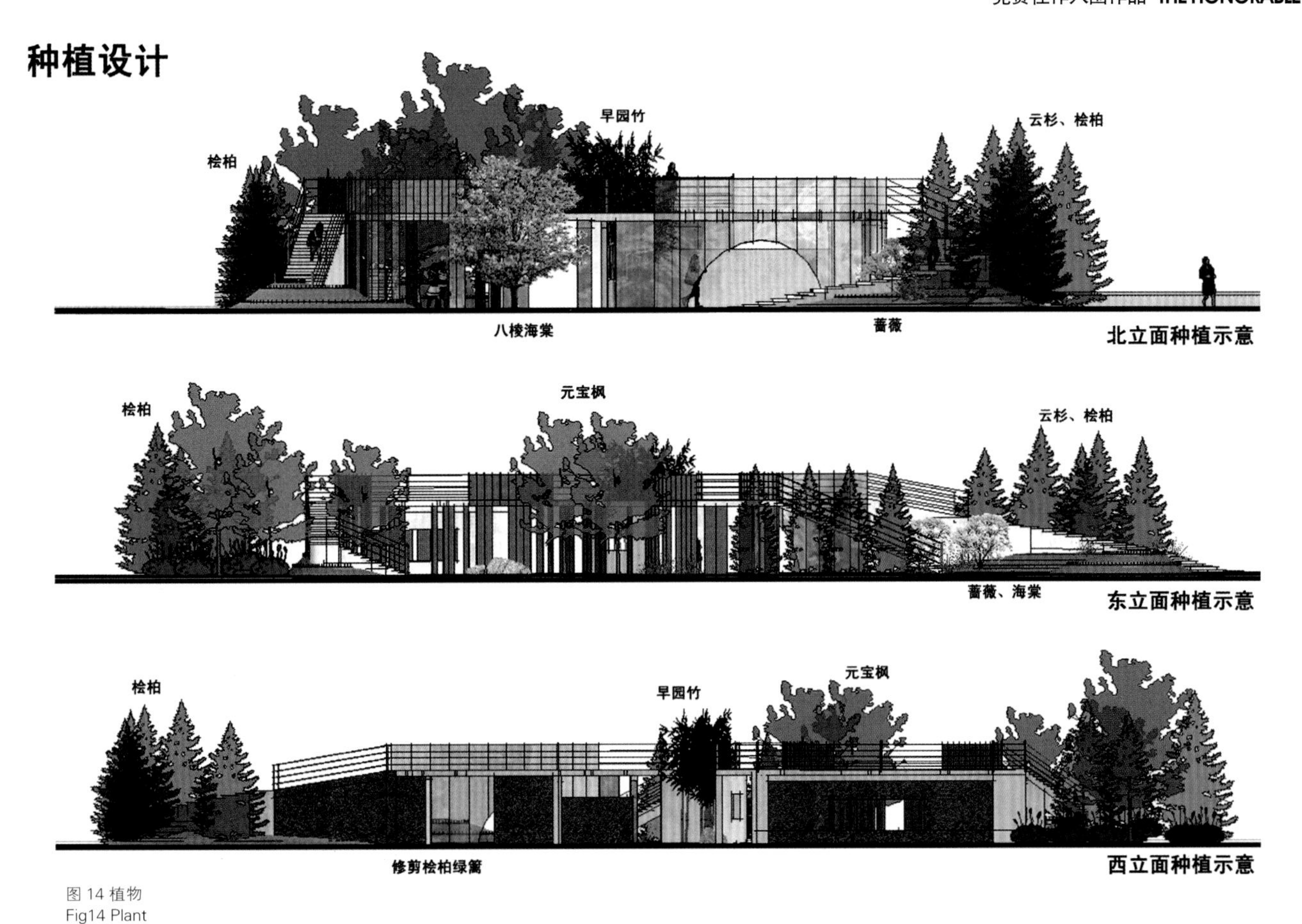

图 14 植物
Fig14 Plant

Planting Design

The planting design is based on three ornamental surfaces, The main scene on the north side is a Malus micromalus. Surrounding both sides of the green space and steps are jasmine, rose and other low profusion shrubs. There are five Shandong maple trees on the east side. The west side is comprised of high-density planting. In addition, on the west side of the entrance is a patch of bamboo, a wall of Chinese juniper and the Lighting Gallery (Fig 14).

Surface treatment and materials design

The surface treatment employs relatively simple materials and forms a pattern of light and dark tones. Most of the north side and the west side of the garden is covered with gray and white gravel. A modern flower pattern is created in the pavers on the south side.

The Lighting Gallery's facade on the north side is comprised of stained glass and a single translucent mirror, the south side is comprised of a wooden screen. On the west side there are windows with plantings.

The Lighting Gallery's roof is comprised of stone and glass, allowing light to enter.

The main part of the lighting Gallery's guardrail is glass and cables (Fig15).

Message:

Whether in the air or on the ground, north or south, light and shadow produce contrast and fusion. Whether up or down, inside or outside, bright or dark, east or west, traditional or modern, all of these have the potential to express the same meaning: exits are also points of entry.■

图 15 材料
Fig15 Material

作者简介：

毛子强 / 男 / 园林设计师 / 北京市园林古建设计研究院 / 中国北京

潘子亮 / 男 / 园林设计师 / 北京市园林古建设计研究院 / 中国北京

Biography:

Ziqiang Mao / male / landscape architect / Beijing institute of landscape and traditional architectural design and research / beijing, China

Ziliang Pan / male / landscape architect / Beijing institute of landscape and traditional architectural design and research / beijing, China

查尔斯・沙（校订）
English reviewed by Charles Sands

北京的记忆
MEMORIES OF BEIJING

中国国际园林博览会设计师广场展园方案
INTERNATIONAL GARDEN EXPOSITION

马晓伟　　　　Xiaowei Ma

项目位置：中国，北京，第九届北京园博会
项目面积：930m²
委托单位：第九届中国（北京）国际园林博览会组委会
设计单位：上海意格环境设计有限公司
景观设计：马晓伟、宣笛、黄婵
完成时间：2012 年 4 月
（图 01 ）

Location: The 9th China (Beijing) International Garden Expo, Beijing, China
Area: 930m²
Client: The 9th China (Beijing) International Garden Expo committee
Designer: Shanghai AGER Environmental Design Co. Ltd.
Landscape Design: Xiaowei Ma, Di Xuan, Chan Huang
Completion: April , 2012
(Fig 01)

图 01 区位分析图
Fig 01 Location Map

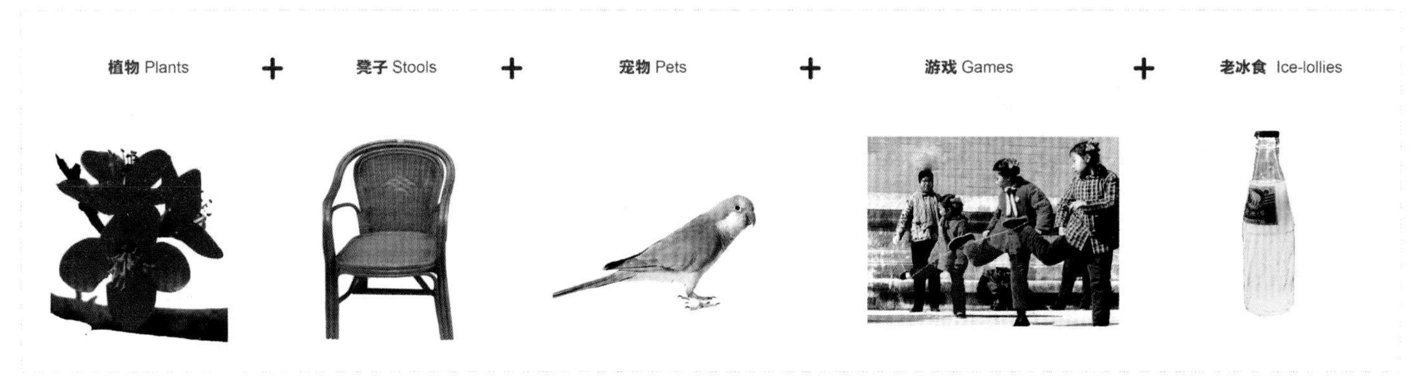

图 02 设计由来
Fig02 Design Inspiration

设计由来

用一个大院代表北京城，水绕城也穿城，五个小院代表生活化的院落，院落之间形成丰富的胡同街巷。每一个院落都是一个鲜活的生活场景，每一个小院都会有一种树、一种鸟、一种小食。让人们坐下来回忆当年北京生活中一些琐碎的片段。在这个日新月异的城市里，老北京的身影早已变得模糊。那里有过平凡生活的诗意，浓得化不开的人情味，还有浓厚的历史文化。它们曾经是那么真实，却正在慢慢消失（图 02）……

设计解读

本地块的设计以老北京的城市布局为原型，以环绕着地块的水池、砖墙抽象概括地表示了以护城河、城墙环绕的较为方正的城市布局，胡同横平竖直，四合院错落有致，并在地块中央以“水院”代表了城市中央北海、中南海等大面积水域（图 03–04）。

设计构思

唯有了解京城厚重的文化底蕴和历史渊源，才能真正领会这座城市的精髓。了解每一座城市，就从历史开始，从民俗风物开始，从市井街巷的每个剪影开始。

老北京的记忆，庙会旧京风华，饱含京腔京韵。人们一提起北京的古都风貌，总是说起“古槐、紫藤、四合院”。老北京也流传着一句俗语：“天棚、鱼缸、石榴树，先生、肥狗、胖丫头。”在设计中希望能充分体现老北京的植物造景、人文景观、旧时风物、传统美食（图 05–08、19–21）。

Design Inspiration

One large walled space representing the historical Beijing city wall, water surrounds the wall and also cutting through the space like what happened in old time Beijing. Five small courtyards representing the historical living courtyards, the in-between spaces form various alleyways. Every courtyard is a living scene; there will be on kind of tree, seat, pet, game and cold snack. People could sit down while recalling their trivial pieces of their old life in Beijing (Fig 02).

Design Interpretation

This design is inspired by the old Beijing city plan. The water body surrounding the garden reminds people of the moat; brick walls remind people of the city walls. The regular square city is bestowed of life by straight lanes and scattered quadrangles, which are in picturesque disorder. In the center of the garden lies a “Water Courtyard”, representing large water bodies Beihai and Zhongnanhai right in the center of the city (Fig 03-04).

Design Inspiration

The charms of the city only expose to those who know well of its culture deposits and rich history. To embrace the city, you need to learn its history, appreciate local traditions and customs starting

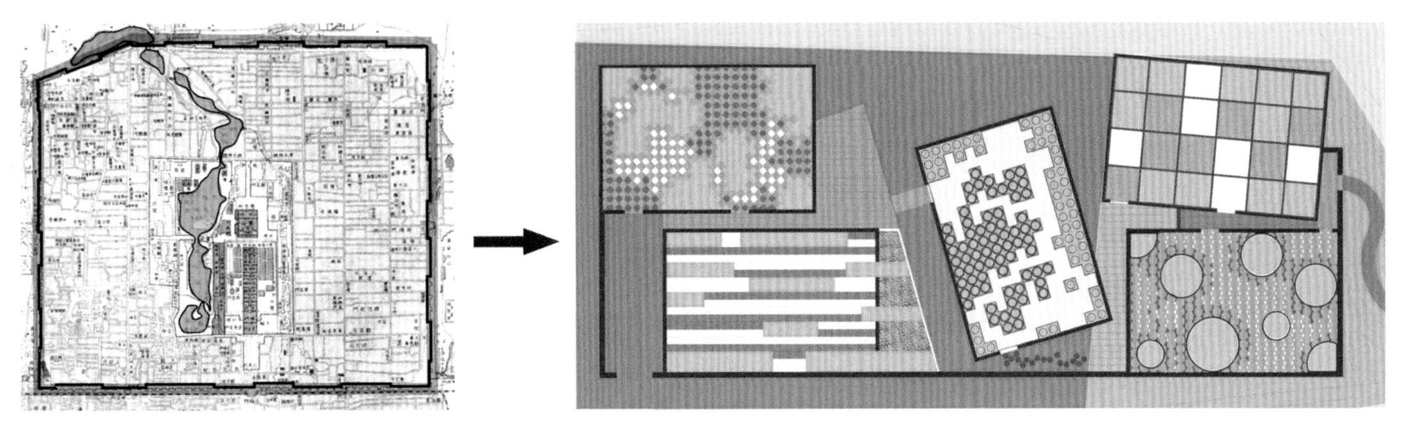

图 03 北京布局
Fig 03 Beijing Distribution

图 04 “环绕水”与“穿越水”
Fig 04 Surrounding Water and Crossing Water

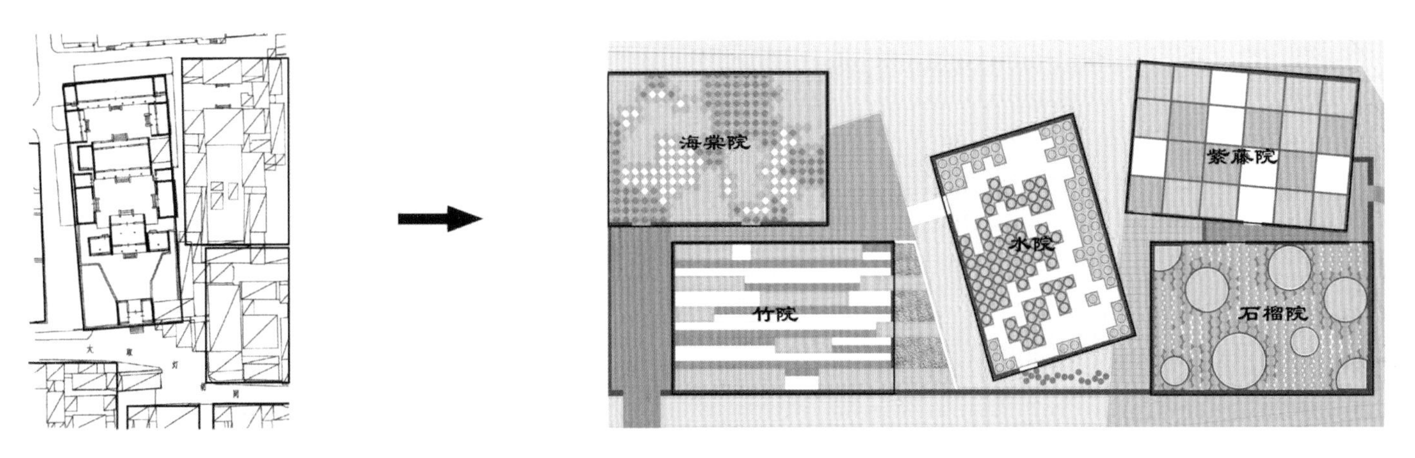

图 05 胡同格局规则的四方院子的不规则排列所创造的丰富空间体验
Fig 05 The unique spatial experience created by the irregular combinations of regular square courtyards

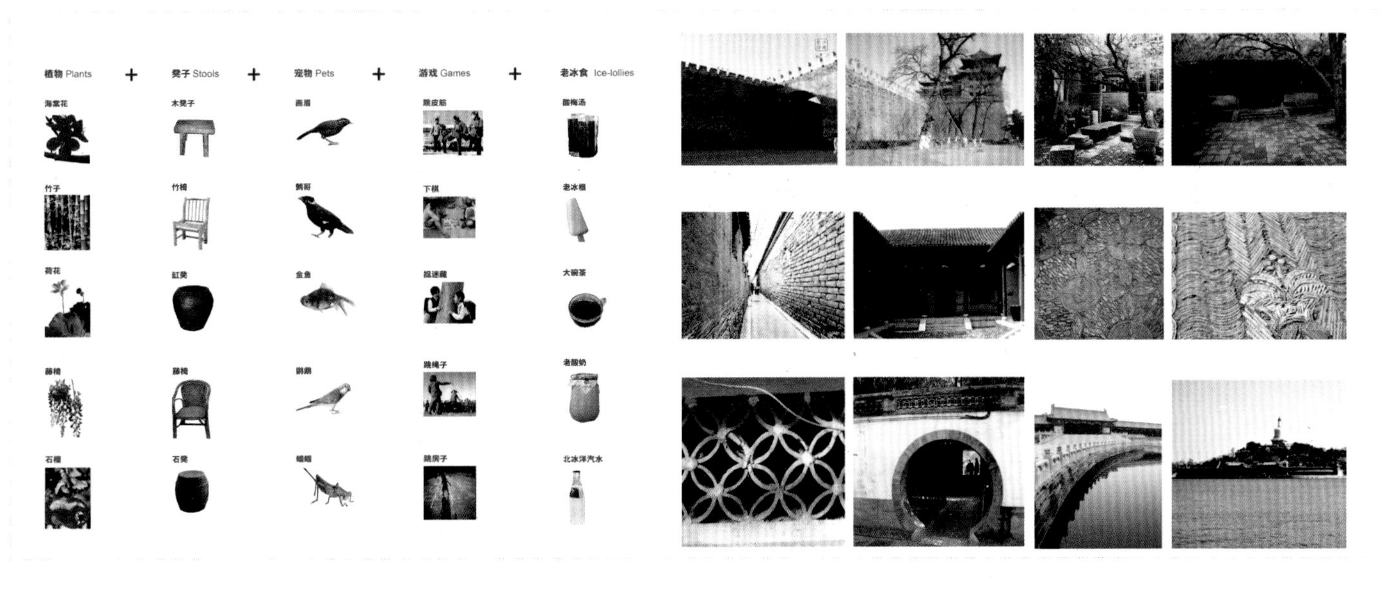

图 06-07 设计说明
Fig06-07 Design Illustration

植物 + 凳子 + 宠物 + 游戏 + 老冰食	Plants+Stools+Pets +Games+Ice-lollies
海棠院	Chinese Cherry-apple Courtyard
竹院	Bamboo Courtyard
水院	Water Courtyard
紫藤院	Chinese Wisteria Courtyard
石榴院	Pomegranate Courtyard

外墙	External Walls
内院	Interior Walls
窗与墙	Windows
树	Trees
地面	Ground
水面	Water Surface

海棠院构思

海棠——姹紫嫣红，温馨清香

北京城内从皇城、皇宫到王府、庙宇、祭典场所不胜枚举，没有哪个像样的去处不种海棠花的。

北京的民间建筑、四合院、公园均种植海棠花，处处都留有海棠花树亭亭玉立的倩影，就是最好的证明。“海棠”与“海堂”谐音，寓意美好、幸福、富贵、和谐（图 09–10）。

竹院构思

竹——四季常青，格高韵胜。

没有哪一种植物能像竹子一样对人类文明有如此深远的影响，东坡老先生一句“无竹令人俗”道出了中华文化与其他文化的迥然不同之处。竹子在北京也能生长得很好，像潭柘寺、红螺寺、紫竹禅林（紫竹院）的竹子是驰名京城的（图 11–12）。

水院构思

“在夏天，什么地方都是烫手的热，只有这口缸老那么冰凉的，而且缸肚儿以下出着一层凉汗。一摸好像摸到一条鱼似的，又凉又温”。——老舍（描述北京胡同生活）

荷花——花光人面，掩映迷离

在夏天，老北京有“夏赏荷”的习惯，比如到什刹海赏荷。在《燕京岁时记》中描述什刹海的荷花：“花开时，北岸风景最佳。绿柳垂涤，红花腻粉，花光人面，掩映迷离。真不知人之为之，花之为花矣”（图 13–14）。

紫藤院构思

紫藤——紫云垂地，文人爱藤

紫藤是我国最著名的藤荫植物。我国古代的文人爱藤，他们不但咏藤，而且在自己的四合院中植藤。所以在北京的宣南，很多古代文人故居中多有名藤。现在北京的古藤中，最著名的一棵就是纪晓岚故居门前的紫藤。这棵紫藤为纪晓岚所植，距今已近三百年。他在《阅微草堂笔记》中描述到此藤：“其荫覆院，其蔓旁引。紫云垂地，香气袭人”（图 15–16）。

石榴院构思

石榴——五月榴花耀眼明

是老北京四合院里种植最多的一种树。石榴果红似火，象征着日子红红火火，石榴的多子则象征着多子多福。老北京的有些人家常用剪纸剪成福禄寿喜字样，贴在刚成型的果实上，待成熟石榴变红时，

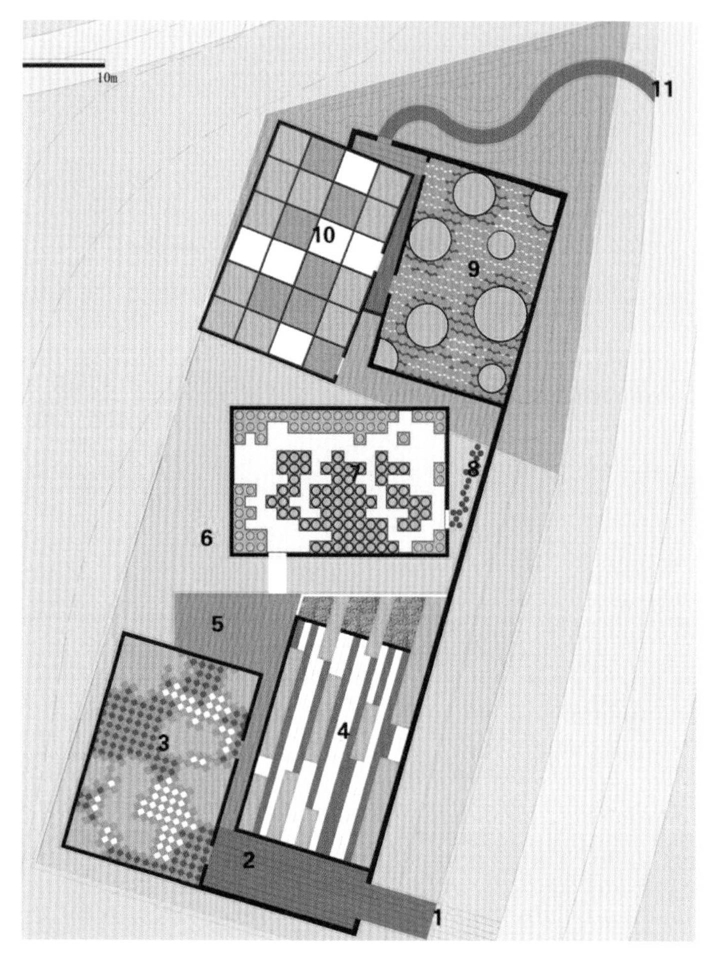

图 08 总平面图
Fig08 Master Plan

1 入口桥	Entrance Bridge		7 水院	Water Yard
2 入口巷道	Entry Lane		8 水上汀步	Steps on the Water
3 海棠院	Chinese Cherry-apple Yard		9 石榴院	Pomegranate Yard
4 竹院	Bamboo Yard		10 紫藤院	Chinese Wisteria Yard
5 休闲平台	Leisure Deck		11 出口	Exit
6 水域	Water body			

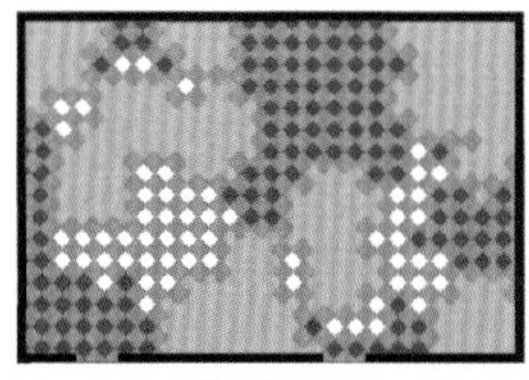

图 09 海棠院构思
Fig 09 Chinese Cherry-apple Yard

图 10 海棠院效果图
Fig10 Chinese Cherr-apple Yard Perspective

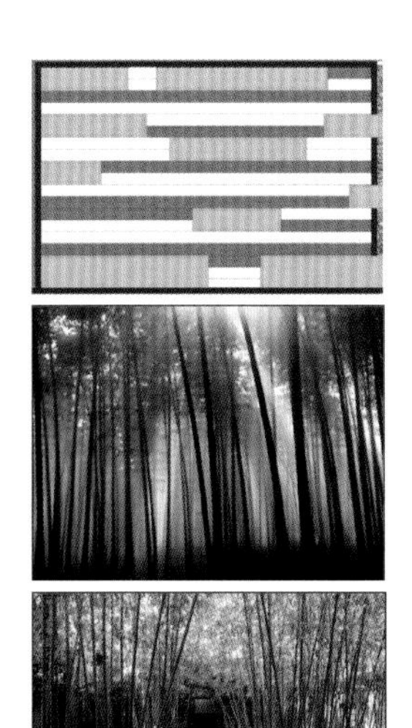

图 11 竹院构思
Fig 11 Bamboo Courtyard

图 12 竹院效果图
Fig12 Bamboo Courtyard Perspective

图 13 水院构思
Fig13 Water Courtyard

图 14 水院效果图
Fig14 Water Courtyard Perspective

揭去纸样，福禄寿喜便被拓在上面，将其献给长辈，非常喜人。满树的大红石榴成为夏秋时节四合院里的一道美丽的景观（图 17–18）。□

from a snapshot in a lane or a street. Memories of old Beijing are made up of the prosperous vistas, Beijing dialect and opera. When people think of the old Beijing, they always mention "ancient locus trees, purple wisteria trees and quadrangles". "Sunshades, fish tanks, pomegranate trees, messieurs, fat dogs and choppy girls" are prominent in old Beijing memories. We strive to show every angle of beauty in old Beijing and review the experience of traditional gardens, culture landscapes, old fashioned sceneries and traditional specialties (Fig 05-08 19-21).

Chinese Cherry-apple Yard

Chinese Cherry-apple: beautiful and luxuriant flowers with pleasant mild scent. Chinese cherry-apple trees are everywhere in Beijing, in the royal inner city, royal palace courtyard, Prince's mansions, temple courtyards or even altars, the blossoms are smiling sweetly to people passing by.

People adore Chinese cherry-apple trees, whose Chinese name "haitang" is a pun, meaning happiness, wealth, good luck and harmony fill the dwellings. That's why you can find these lucky flowers in civic spaces, quadrangles and of course public open spaces (Fig 09-10).

Bamboo Courtyard

Bamboo---An evergreen plant with the symbolic meaning of graceful and elegant nobleman. Bamboo has a profound influence on Chinese civilization. A famous poet, Sushi in from the Song Dynasty has a well-known line "if there is no bamboo in my courtyard, I would feel the space too vulgar to stand". One thing distinguishing Chinese culture from others is that plants are bestowed with anthropomorphic meanings. Bamboos

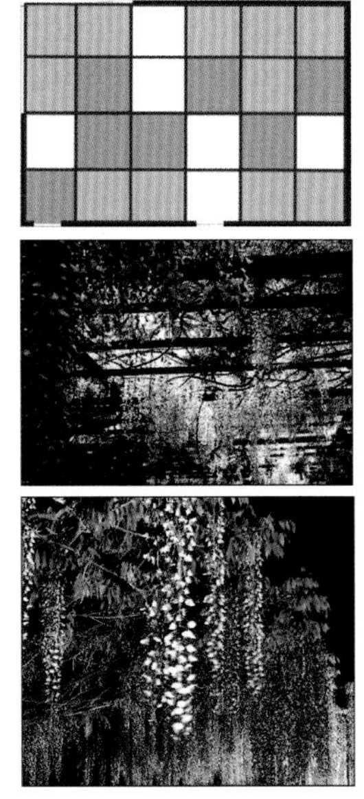
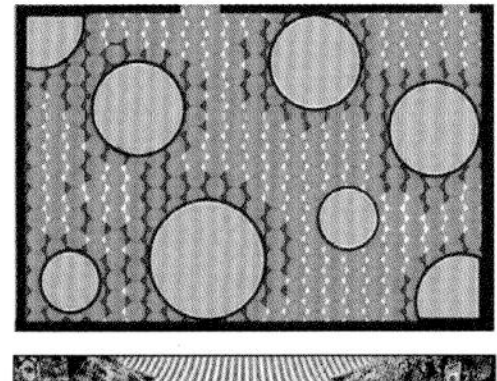

图 15 紫藤院构思
Fig15 Chinese Wisteria Courtyard

图 16 紫藤院效果图
Fig16 Chinese Wisteria Perspective

图 17 石榴院构思
Fig17 Pomegranate Courtyard

图 18 石榴院效果图
Fig 18 Pomegranate Courtyard Perspective

can grow well in Beijing. Tantuo Temple, Hongluo Temple and Purple Bamboo, Chan Temple are all famous for their bamboo courtyards (Fig11-12).

Water Courtyard

"In the summer, when everything is so hot, one is surprised to find the vat in one's courtyard is still cool. Put your head on the vat's belly to touch the hanging water beams, it's like you are touching a fish—feeling cool and wet". ——Laoshe (On Life in the Old Beijing Lanes)

Lotus——the blossom is like a beautiful lady's face. She tries to hide herself among the big leaves. The summers of old Beijing are times of lotus blossoms. People visit large lotus flower planting spaces such as Shichahai Park, just to enjoy the beautiful flowers.

In the time of "Peiking", he wrote "When the flowers are in full blossom, the scenery along the northern bank is the most beautiful. Like charming young ladies, the lotus blossoms try to hide their pink faces among green leaves. I really can't tell whether they are flowers or girls" (Fig 13-14).

Chinese Wisteria Courtyard

Purple wisteria——purple wisteria's vines hang down to the ground. This is the scholars' favorite plant. In ancient China, scholars felt especially attached to vines. They planted purple wisteria vines and made poems about them. Many quadrangles left by eminent writers are accompanied by big purple wisteria vines. The most famous one is the purple wisteria in the entrance of Prime Minister Ji XiaoLan's former residence. It was planted by Ji Xiaolan dating back to the Qianlong period in the Qing Dynasty, and it's more than 300 years old. Its owner wrote in his "Fantastic Tales by Ji Xiaolan": it covers the better half of the courtyard and its vines stretch out to embrace the fence wall. When the flowers

图 19 入口效果图
Fig19 Entry Perspective

图 20 节点效果图
Fig20 Nodes Perspective

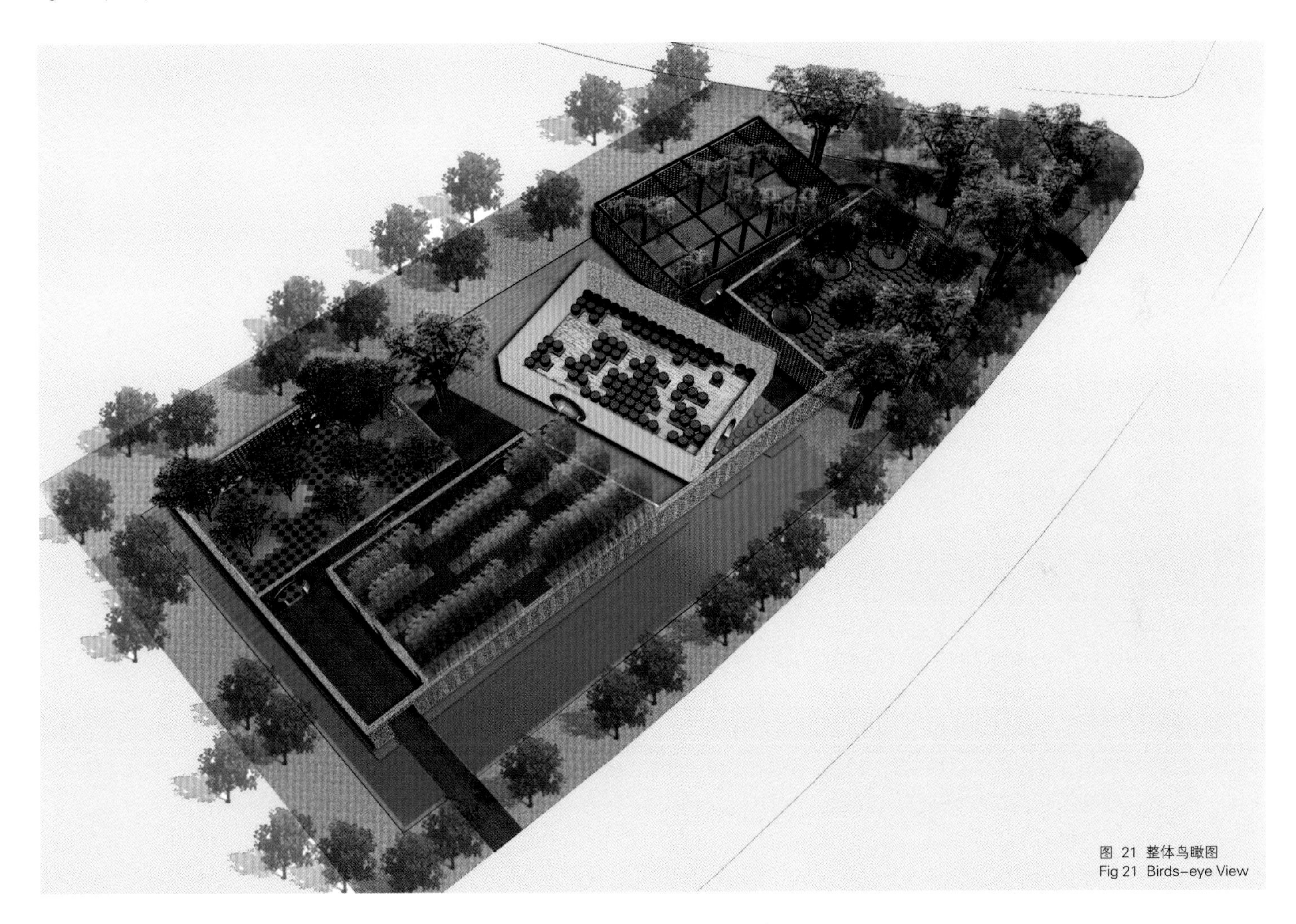

图 21 整体鸟瞰图
Fig 21 Birds-eye View

are in full blossom, it's like purple clouds hanging on the wall to the ground. The whole space is fragrant with flowers (Fig15-16).

Pomegranate Courtyard

Pomegranate——the sparkling pomegranate flowers in May. Pomegranate is one of the most frequently planted types of trees in old Beijing's quadrangles. Pomegranate seeds are red, which symbolizes prosperity and many offspring. Some families used to scissor-cut lucky Chinese characters and stick them to the fruits. When the fruit is ripe, the characters have been stamped on the peels. Those fruits with lucky characters are usually given to the old to make them happy. In early fall, pomegranate trees in the quadrangles bend with heavy red fruits. It became a beautiful scene for old Beijing (Fig 17-18).■

作者简介：

马晓伟 / 男 / 意格公司创始人 / 现任上海意格环境设计有限公司总裁兼首席设计师 / 中国北京

Biography:

Xiaowei Ma / Male / The founder of AGER Group / The President of AGER Group and Chief Designer / Beijing,China

查尔斯・沙（校订）
English reviewed by Charles Sands

跃然纸上

FLESHED OUT

2013 园博会设计师广场竞赛入选作品

2013 LANDSCAPE ARCHITECTURE DESIGN FOR THE DESIGNER'S SQUARE FOR THE 9TH CHINA (BEIJING) INTERNATIONAL GARDEN EXPOSITION COMPETITION SELECTED WORKS

曾宥榕　曾宥源　张婉钰　赵立勤　Ruth Yu-jung TSENG, Joe Yu-yuan TSENG, Wan-yu CHANG, Bonnie Lic-hin CHAO

图 01 全景 3D 模拟图
Fig 01 Panoramic 3D simulation of Fig

项目位置：中国，北京，第九届北京园博会
项目面积：1,000m²
委托单位：第九届中国（北京）国际园林博览会组委会
设计单位：米页空间设计有限公司
景观设计：曾宥榕、曾宥源、张婉钰、赵立勤
完成时间：2012 年

Location: The 9th China (Beijing) International Garden Expo, Beijing, China
Area: 1,000m²
Client: The 9th China (Beijing) International Garden Expo Committee
Designer: Millet landscape design Co., Ltd.
Landscape Design: Ruth Yu-jung TSENG, Joe Yu-yuan TSENG, Wan-yu CHANG, Bonnie Lic-hin CHAO
Completion: 2012

图 02 北京市中心位丁北纬 39 度，东经 116 度。雄踞华北大平原北端。北京的西、北和东北，群山环绕，东南是缓缓向渤海倾斜的大平原。

Fig02 Beijing Center is located at 39 degrees north latitude, 116 degrees east longitude. Dominating the northern end of the North China Plain. The west, north and northeast of Beijing, surrounded by mountains, southeast of the Bohai Sea is slowly tilt the Great Plains.

图 03 第九届园博会选址位丁北京市丰台区永定河以西地区，北至莲石西路，东临永定河新右堤，南至规划梅市口路，西南接射击场路，西至规划北宫路，总面积为 267 公顷。

Fig03 Ninth Fair Park site is located in Fengtai District, Beijing Yongding River region to the west, north to Lianshixi Road, east of Yongding River, right bank, south planning Hau Road, Plum City, southwest Shooting Range Road, west planning Kitamiya Road, with a total area of 267 hectares.

图 04 年平均气温 10~12 摄氏度，1 月 -7~-4 摄氏度，7 月 25~26 摄氏度。平均降雨量 600 多毫米，降水季节分配不均匀，全年降水的 75% 集中在夏季，7、8 月常有暴雨。冬季盛行沙尘暴，风向为西北向。

Fig04 Annual average temperature of 10 to 12 degrees Celsius, -7 to -4 degrees Celsius in January, July 25 to 26 degrees Celsius. Average annual rainfall of 600 mm, uneven seasonal distribution of precipitation, 75% of the annual precipitation is concentrated in the summer, and often heavy rain in July and August. Prevail in winter dust storms, wind direction is northwest to.

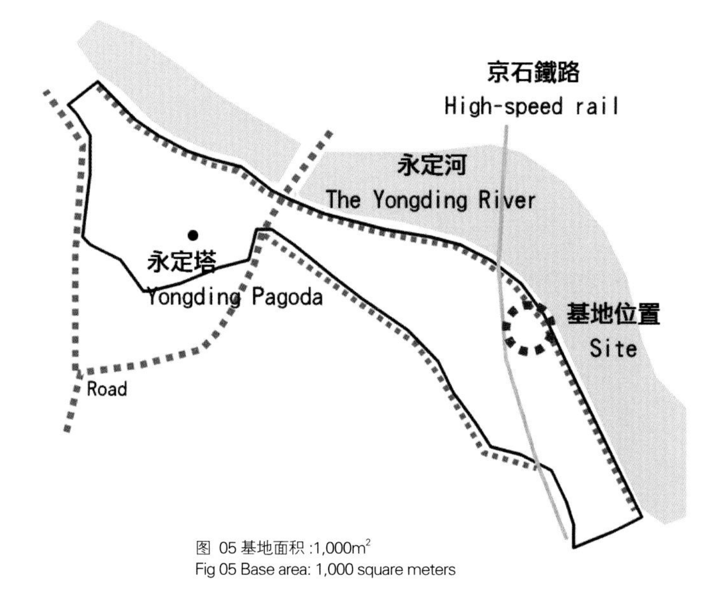

图 05 基地面积 :1,000m²

Fig 05 Base area: 1,000 square meters

设计缘起——诗、画、园三者的结合

中国现代园林（景观）朝着注重自然与传统文化的结合发展，倡导生态建园和文化建园，这与现今国际上所倡导的“永续发展”理念是相通的。

中国的古人寄情于山水，利用“诗、词、歌、赋”赞咏江山美景，并加上对景物的描绘，而有了诗与画的结合，进而达到“诗中有画，画中有诗”的境界。

然而中国山水画与中国古典园林的起源和发展有着密切联系，两者的创作一脉相承，诗、画、园三者的紧密结合成为中国传统园林的一大特色 (图 01-05)。

设计概念

本设计案最主要的概念是结合传统园林艺术、地域特色、历史与

Design origin - the combination of all three “poems, paintings, garden.”

China's modern landscape field is moving towards a focus on the integrated development of natural and traditional culture and the promotion of eco-build gardens, following the concept of sustainable development.

China's ancients abandoned themselves to nature. They used "classical Chinese poems, songs and poetic essays to praise the beauty of nature, and coupled these with painted scenes.

Chinese landscape painting is also closely linked with the origin and development of Chinese classical gardens. Thus poetry, painting and garden design were closely integrated in the classical tradition(Fig 01-05).

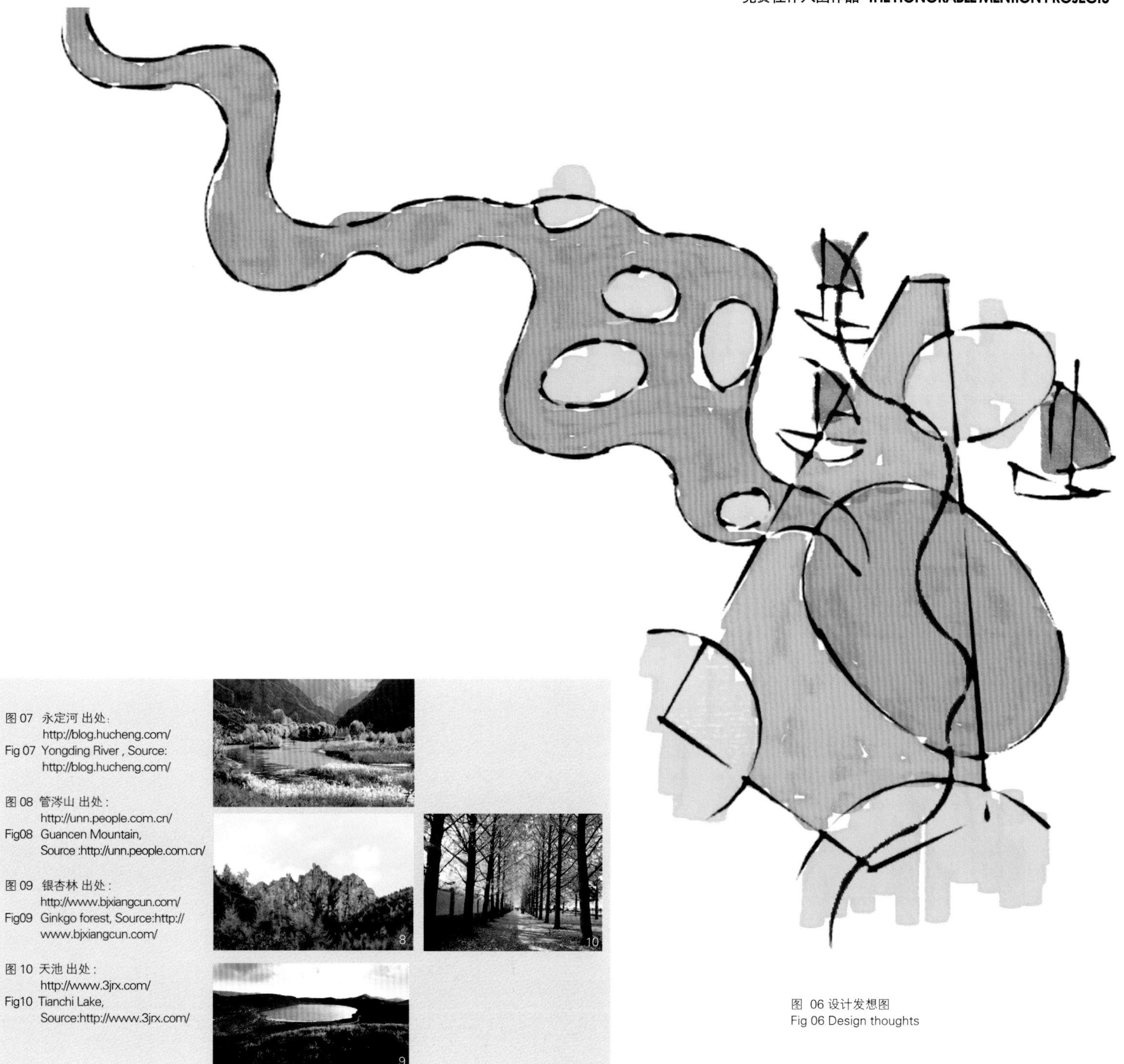

图 07 永定河 出处：http://blog.hucheng.com/
Fig 07 Yongding River , Source: http://blog.hucheng.com/

图 08 管涔山 出处：http://unn.people.com.cn/
Fig08 Guancen Mountain, Source :http://unn.people.com.cn/

图 09 银杏林 出处：http://www.bjxiangcun.com/
Fig09 Ginkgo forest, Source:http://www.bjxiangcun.com/

图 10 天池 出处：http://www.3jrx.com/
Fig10 Tianchi Lake, Source:http://www.3jrx.com/

图 06 设计发想图
Fig 06 Design thoughts

文化元素，透过设计的巧思、新旧材料的转换与创新的手法应用于现代的景观设计里（图 6)。

经由本设计案让所有人了解我国园林建筑艺术的精神，特别是文人诗画导入园林的艺术成就，以及领会欣赏中国园林的方法，将传统园林诗、画、园三者结合的独有特色做一「跃然纸上」的重现。

其中借鉴诗文来创造意境，如: 诗人叶令仪（清）“帆影都从窗隙过，溪光合向镜中看。” 另外，引用历史人文及地域特色来加深文化内涵，如山西宁武天池，古代为朝廷饲牧军马的地方，也是北京境内最大最古老的河流（永定河）的发源地。上述两者最能体现本案的设计概念及贯穿全局的精神。

山、水与树林是中国国画里的重要元素，将自然的山水美景撷取并描绘于国画里，最后成为传统园林的造园依据（图 07–10）。

Design concept

The main concept of this design case is to combine traditional garden art, geographical features and historical and cultural elements, through the conversion of old and new materials and innovative techniques used in modern landscape design (Fig 6).

The priority of the design is conveying the relationship between landscape architecture and art in China, particularly the infusion of the poetry and painting of the literati into the artistic achievements of the garden. In other words to “flesh out” the interrelationship of these arts in the tradition of garden design.

Poets such as Ye Ling-yee (Qing Dynasty) produced descriptive verse to draw inspiration from: " Sails from the window gap, River

图 11 书画中撷取的元素
Fig 11 Elements retrieved from the Chinese Painting and Calligraphy

书画中撷取的元素（图 11）

张大千－山水出处：《中国近现代名家画集 张大千》出版者：锦绣文化企业

体现地域特色

本案以传统园林为基础结合地域特色发展出新的创作元素，具有传承人文历史的功能。水是生命的泉源和文明的摇篮，每一座城市的兴起几乎都伴随着一条大河的流淌，北京也不例外。北京是享誉世界的历史名城及国家首批的园林城市，而永定河是北京境内最大也是最古老的河流，发源于山西宁武县管涔山天池。

新理念的运用导入

A 取其“形”——在形体上做简化、变形、抽象，以新的材料取代，并做空间重组，使原有的传统形象失去真实性，却又保留了历史的延续，成为隐喻性的新语言。

photosynthesis seen in the mirror." In addition, reference to the history and geographical features in garden design deepened the cultural connotations, one finds ancient references to Yongding River, and Ningwu Tianchi in Shanxi which embody the design concept and spirit of the present case(Fig 07-10).

Mountains, water and forests are important elements in Chinese painting and are also the basic elements of traditional garden design.

Capture the elements of painting and calligraphy(Fig 11)

Chang Dai-chien-Landscape , Source: "Masters of Modern Chinese Art Collections- Chang Dai-chien " Publisher: Kam Running Cultural Enterprises

图 12 设计概念图
Fig12 Conceptual Design

图 13 树的剪影
Fig 13 The silhouette of the tree
图 14 卵石
Fig 14 Pebble
图 15 屋瓦 出处 :http://www.nipic.com/
Fig 15 Roof tiles
图 16 再生纸
Fig 16 Recycled paper

图 17 竹编
Fig17 Bamboo weaving

B 承其“意”——对于传统与现代的过渡，既能强调统一、协调、融合、逻辑上的因果关系，也能体现出传统元素在现代景观中的新发展。

C 借其“势”——需延续传统对形体的所蕴藏的气韵及其表现于外的气势及氛围。

D 传其“神”——是一种摆脱传统符号，并提高领域及发展的新民族精神。

E 蕴其“情”——通过景观符号把感情客观化，并把符号作为情感的寄托（图 12）。

新旧材料的应用

1. 将传统的旧材料重新组合，取代传统的工艺却保留历史的精神。

本案利用传统园林建筑的旧瓦片，垂直镶嵌于地面，既成为地坪的装饰线也能传达河面波光粼粼的意向（图 13–14）。

2. 赋予现代建筑新材料对于传承传统园林与文化的新使命，如：钢、瓦片、防水再生纸及竹编等。

本案以型钢、防水再生纸及竹编等新材料构筑墙体，取代传统的版筑墙、乱石墙、磨砖墙、白粉墙等构造（图 15–17）。

Reflect the regional characteristics

Identify the basis of the geographical features of a traditional garden to develop new creative elements, that contain the heritage of cultural and historical features. Water is the source of life and the cradle of civilization. The rise of almost every city is accompanied by a flowing river and Beijing is no exception. Beijing is a world-renowned historic city and the first national garden city, while the Yong-ding River in Beijing is the largest and oldest river originating in Guan-cen Tian-chi Ning-wu County.

The criteria of referenced elements

(A) Retrieve the shape - Simplify the physical into an abstraction and replace it with new material. Reorganization the original the image of lost authenticity but retain a continuation of the historical, metaphorical language.

(B) Order the meaning - For the transition of the traditional and the modern, both stressed unity, coordination, integration, logic

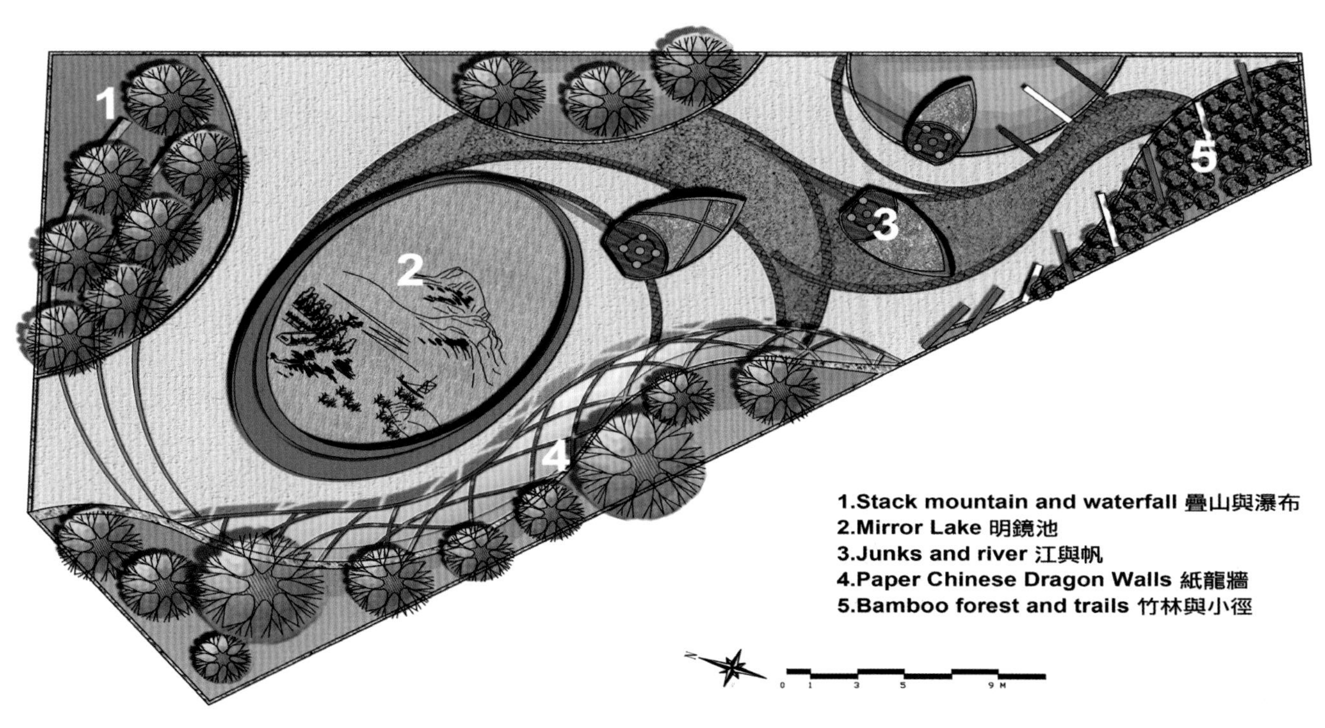

图 18 全区平面配置图
Fig 18 Site Plan

图 19 迭山与瀑布
Fig 19 Stack mountains and waterfalls

图 20 明镜池
Fig 20 Mirror Lake

图 24 全区活动图
Fig 24 Activity diagram in the region

图 21 江与帆
Fig 21 Jiang and sails

图 23 竹林与小径
Fig23 Bamboo forest trails

图 22 纸龙墙
Fig 22 Paper Chinese Dragon Walls

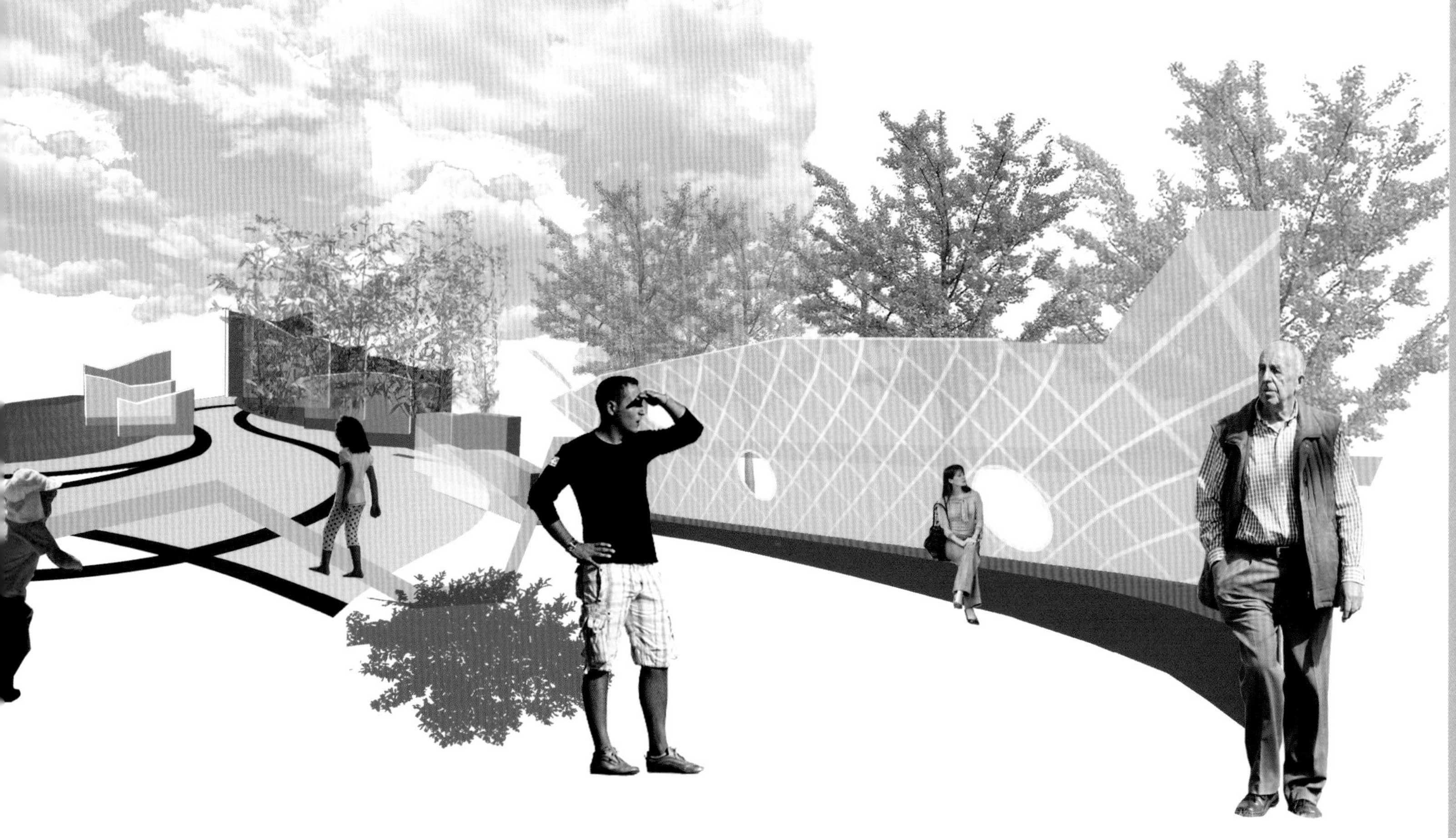

新的设计方法

将中国传统的书画与园林山水的创作手法与精神加以运用，例如留白、虚实相生、自然观、禅意、诗情画意、虽由人作宛自天开、小中见大、移步换景等。

本案利用帆船在水池中形成的倒影凭镜借景，使景映镜中，化实为虚，这就是所谓“镜借”了。

1. 将大自然的山水景色浓缩并撷取于本园中，达到“自然与意境”的结合。

2. 本案中撷取永定河的发源地山西宁武天池、北京常见的银杏树林及竹林、以禅意的手法重现古代永定河湿地及河上的帆船等（图 18)。

分区设计说明

叠山与瀑布

运用自然写意式的叠山手法，模仿并撷取宁武管涔山区内高山耸峙、峰峦奇拔、山脉绵延的自然山景，力求体现自然山峦的形态和神韵，并将顺流而下的瀑布一并表现在该区域内（图 19)。

明镜池

本区运用简约的线条撷取及写意位于山西宁武县管涔山上的天池，用以表现静态的水景，水面平静如镜可供人们观赏园中山水景物在水中的倒影、天上的浮云以及夜晚皎洁的明月（图 20)

宁武天池古称祁莲池，形成于新生代第四纪冰川期，距今有 300 万年的历史了，它是我国三大高山天池之一，世所罕见的高山湖泊湖。唐代曾在此设立天池牧监，为朝廷饲牧军马，故又称马营海。是一处高山群湖。

江与帆

永定河是北京境内最大也是最古老的河流，对于北京的发展与人文历史有着绝对性的影响，因此本区运用枯山水的写意手法带出禅意，更用诗词入画与当时的人文与历史背景相结合，加深了本案所要传达的意境。在细节上除了给予船身现代艺术的绿化外，风帆可随风向而转动并带动下方的座椅旋转，增添游客的互动性及趣味性。刘长卿（唐）“吾亦从此去，扁舟何所之。迢迢江上帆，千里东风吹。”达到“会移动的山水画”新境界（图 21)。

永定河发源于山西宁武县管涔山天池，流经山西、河北、北京、天津，全长 680km。其中，流经北京市门头沟、石景山、房山、丰台、大兴五区的河段长 159.5km，流域面积 3168km^2。

纸龙墙

利用纸片及画轴之意象作为全案最主要的构件，并以现代建材钢、防水油纸及竹编等，构成贯穿古今及围合空间的“园墙”，造型很前卫的“园墙”也具有廊道遮风避雨的功能。它将空间巧妙的分割成若干区域，并利用墙上竹编纹理留设漏窗、洞门、空窗等，形成虚实对比和光影的变化效果，使墙面丰富多彩，又可作为取景的画框，达到“小中见大”、“移步换景”的精神。

走廊既可在交通上连通自如，将园林串通一气；又可让游人仔细品味周围景色。在酷暑风雨之时，仍然可以观赏不同季节和气象时的园林之美，还可以欣赏书法字画，领略历史文化。可于隔墙上开设众多花窗，两边可对视成景，既移步换景增添景色，进一步又扩展了园区内空间感（图 22)。

竹林与小径

利用山的造型数组排列出犹如舞台背景之效果，增加层次与延伸透视之感，山型与穿插其中的竹林，交错出永定河蜿蜒曲折的古河道印象，人们步行其间除了赏景更可乘凉，进一步体会古代文人雅士之思古幽情（图 23)。

娱乐与趣味性

融合园林的多元性功能，即除了观赏性外亦增加了游戏与互动的实用性。例如：大人们坐在廊下聊天休憩，儿童们在池上玩水或池边溜冰、玩耍或上风帆操作，提供风与生活上的乐趣与交流性（图 24–32)。

四季变化的植栽设计

1. 在植栽选择上以春夏观花、秋季观果、冬季观花枝为原则。人文意涵的传神：将古典园林中植物所赋予的人文意涵传达到设计上。例如：

of causality, thereby reflecting the traditional elements of a new development in the modern landscape.

(C) Potential - The need to continue artistic tradition in a new form

(D) God - Get rid of traditional religious symbols, and replace them with symbols of national spirit.

(E)Love - through the landscape develop symbols of emotional sustenance (Fig 12).

The application of new and old materials

1. Use traditional materials but replace the traditional process.

For example, old tile inlaid vertically in the ground, conveying the effect of the river waves sparkling (Fig13-14).

2. Give new modern building materials new purpose in relation to the traditional garden, such as: steel, tiles, waterproof recycled paper and wood.

For example, steel, waterproof recycled paper and wood to build walls, to replace traditional versions of construction, chaotic stone walls, brick walls, white walls and structures (Fig15-17).

New design methods

Use a creative approach that maintains the spirit of traditional Chinese painting and calligraphy in garden landscapes, such as the void.

1.The case of the a sailboat in the water:

Natural landscape scenery concentrated to achieve a combination of natural and artistic conception captured in the park.

2. Retrieve the birthplace of the Yongding River and Ningwu Tianchi with common ginkgo trees and bamboo groves, use a Zen-inspired approach to reproduce ancient Yongding River wetlands and sailing on the river.

Partition design description

Stack mountain and waterfall

Strive to reflect the shape and charm of natural mountains and waterfalls working together in the region (Fig 19).

Mirror Lake

This area uses simple lines to capture Shanxi Ningwu County Guancen mountains, Tianchi. The water is as calm as a mirror for people to observe the garden landscapes in reflection, the clouds of heaven, and the moon (Fig 20).

Ningwu Tianchi called Qi Lianchi, formed in the Cenozoic Quaternary Ice Age, dating back 300 million years, is one of China's three major alpine lakes.

Junks and river

Yong-ding River is Beijing's largest and oldest river, human history has transformed it into a dry landscape. We attempt to convey this transformation through the poetry of Liu Changqing. (Fig 21).

Yong-ding River rises in the the Shanxi Ningwu County Guancen mountains, Tianchi, flows through Shanxi, Hebei, Beijing, Tianjin, at a total length of 680 km. The basin area is 3168 square kilometers.

Paper Chinese Dragon Walls

Paper and scrolls are the main components, and modern building materials such as steel, waterproof and greaseproof paper and bamboo, constitute the "garden wall". Through ancient and modern spaces, the style is very avant-garde. The garden wall and gallery road, act as shelter. This clever space is divided into a number of regional bamboo wall textures leaving in the grilles, Portals, and windows. The effect of the wall changes with the lighting.

Corridors can create connectivity with ease but also allowing

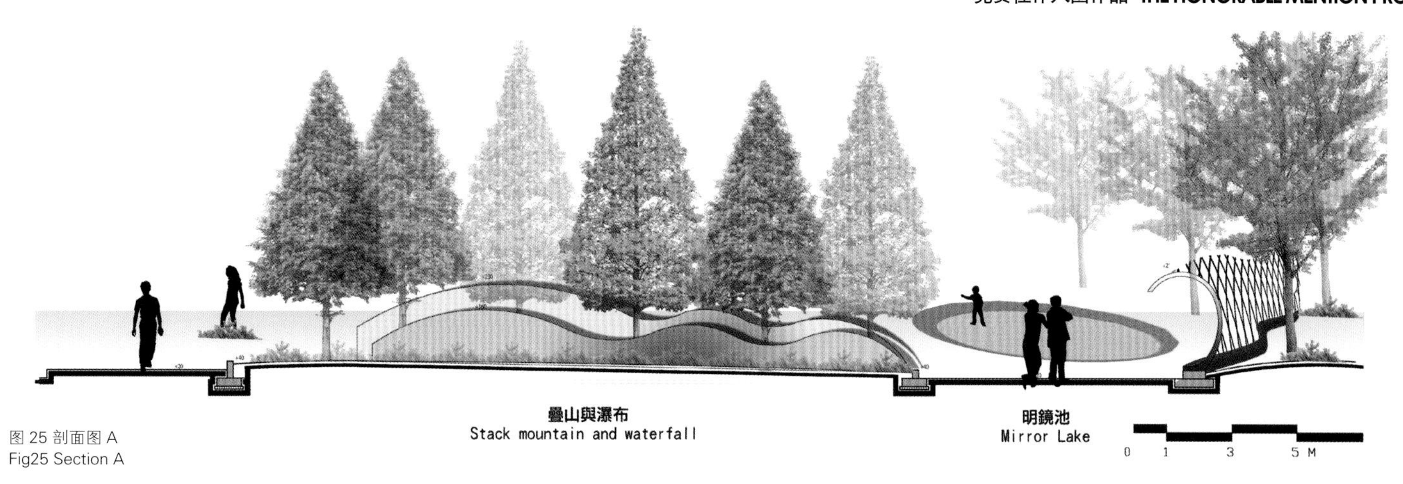

图 25 剖面图 A
Fig25 Section A

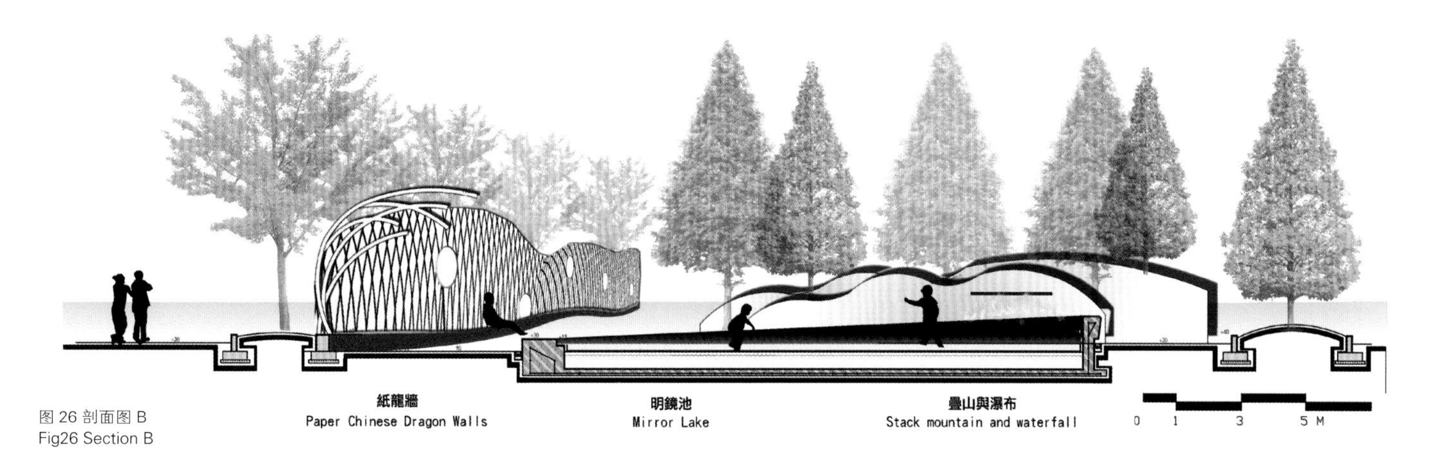

图 26 剖面图 B
Fig26 Section B

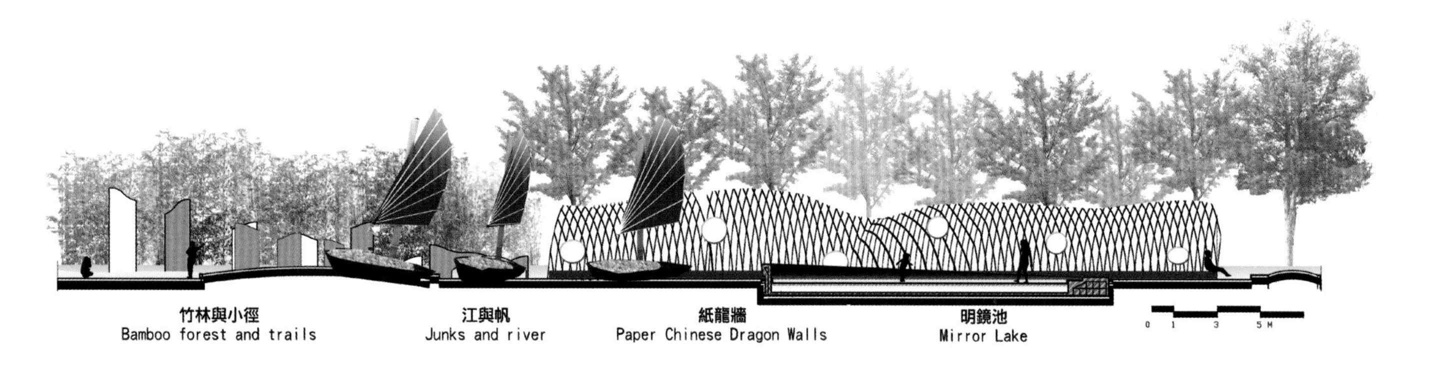

图 27 剖面图 C
Fig27 Section C

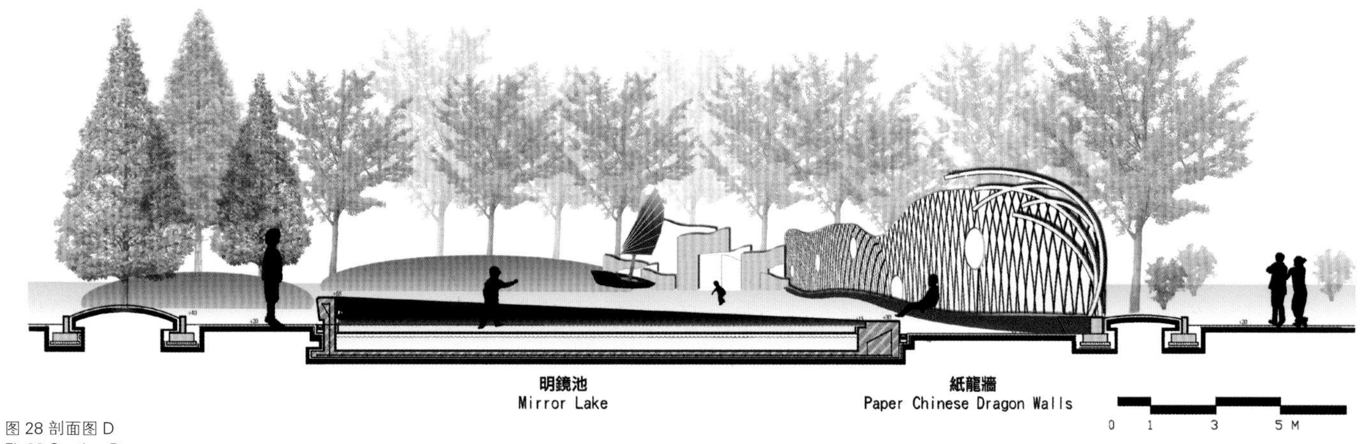

图 28 剖面图 D
Fig28 Section D

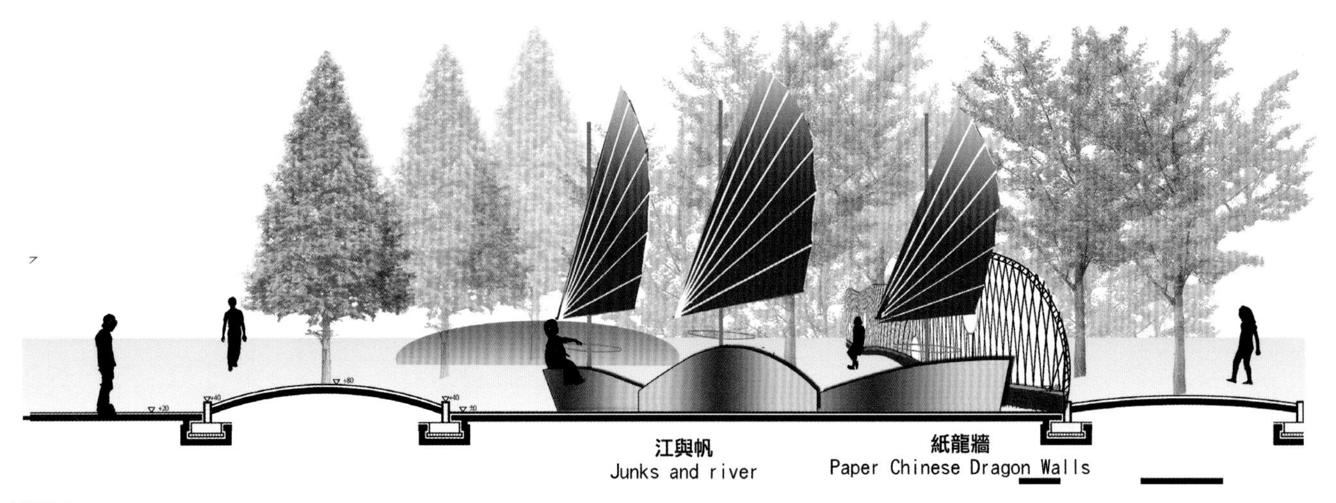

图 29 剖面图 E
Fig29 Section E

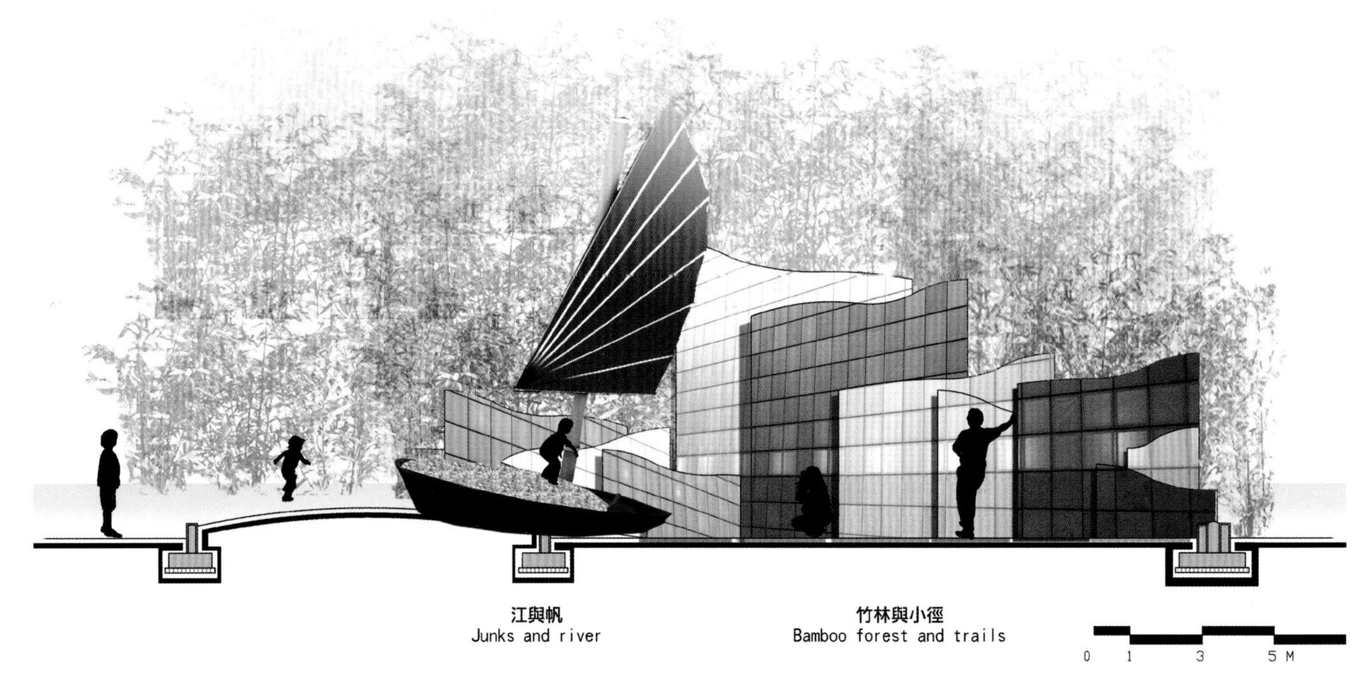

图 30 剖面图 F
Fig 30 Section F

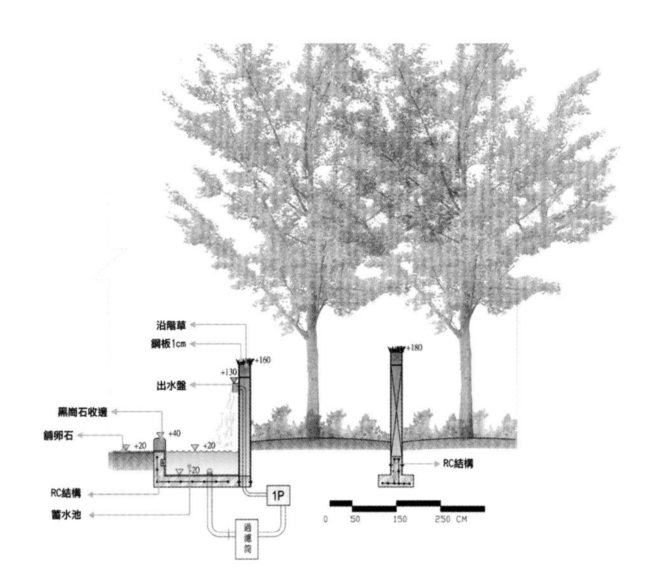

图 31 剖面图 G
Fig31 Section G

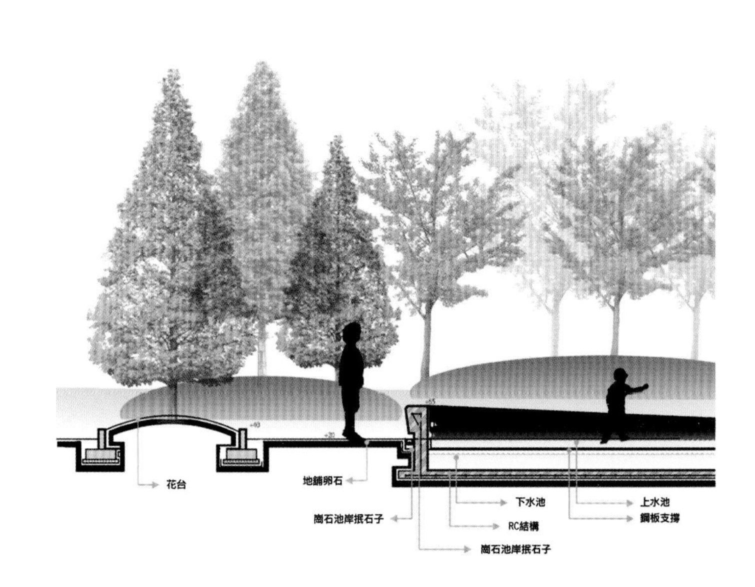

图 32 剖面图 H
Fig32 Section H

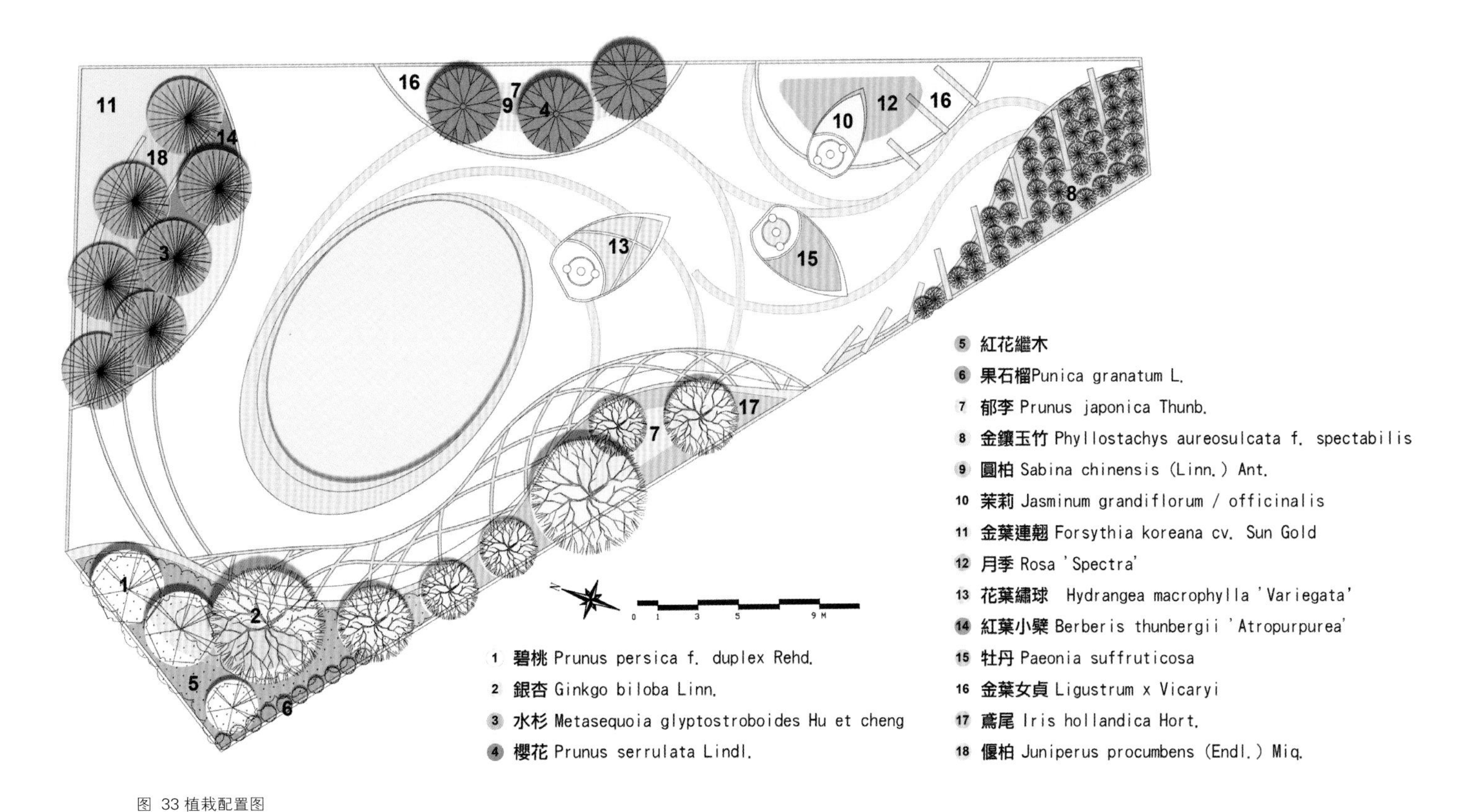

图 33 植栽配置图
Fig 33 Planting configuration

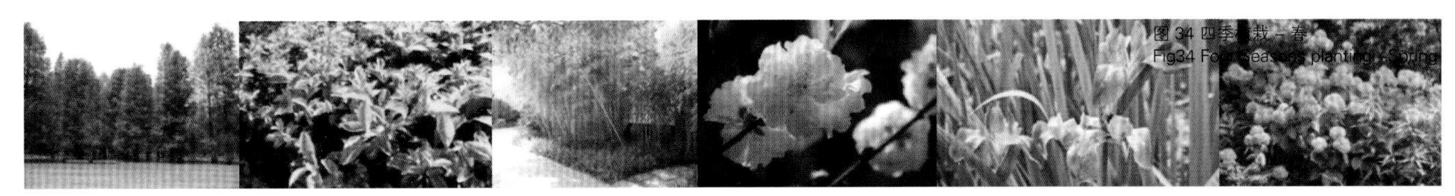

图 34 四季植栽 – 春
Fig 34 Four Seasons planting –Spring

图 36 四季植栽 – 秋
Fig36 Four Seasons planting – fall

图 35 四季植栽 – 夏
Fig35 Four Seasons planting – Summer

图 37 四季植栽 – 冬
Fig 37 Four Seasons planting – Winter

38

40

39

图 38 全区灯光计划图
Fig 38 Lighting plan map in the region

图 39 迭山与瀑布
Fig 39 Stack of mountains and waterfalls

图 40 明镜池
Fig 40 Mirror Lake

图 41 明镜池照明细部图
Fig 41 Mirror Lake lighting detail drawings

图 42 江与帆
Fig 42 Jiang and sails

图 43 竹林与小径
Fig 43 bamboo forest trails

图 44 纸龙墙
Fig 44 paper dragons wall

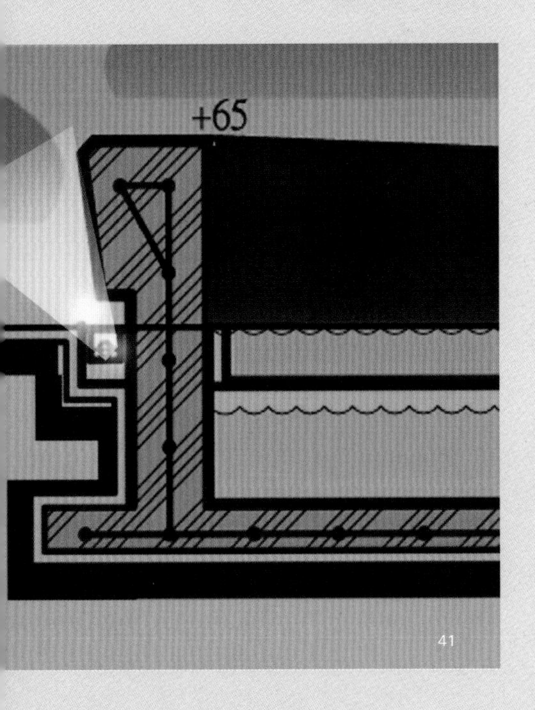

41

43

42

44

银杏为最古老的化石树，代表着传承之意；金镶玉竹，代表着文人的气节等。

2. 生态原则：选择最能吸引鸟、昆虫的桃、李、樱花等，除了乔木种植外，灌木采复层式植栽配置手法，呈现园林植栽的多样性与提高观赏的层次（图 33）。

3. 因地制宜：考虑北京当地气候，选择适合的树种、挑选适合西南风水气与温带的树种，如水杉、桃花、鸢尾等，既可观果又可为生物的食物（图 34–37）。

照明计划 – 节能

1. 夜间的大明镜：夜晚除了星空的繁星外，更加入了树林的投影效果，借由对象元素本身的发光映像于"镜池"的水体中，于黑暗中创造了美丽的倒影。呈现出隐入自然山林中，水墨丹青的内敛，让人陶醉沉浸其中。

2. 动线导引：运用琥珀色与暖色调、太阳光的 LED 灯投射于步道动线，引导游客夜间的行进与安全。

3. 蕴含诗意的画布：在纸墙内设置太阳光的 LED，透过再生纸呈现出如薄雾清透的效果，佐以树木枝叶的剪影效果创造出充满想象且蕴含诗意的景观画面。

4. 公园的地标：在全区最高的 8M 高处风帆杆上，装设明亮的 LED 投射灯，可为夜晚带来地标性的指引（图 38–45）。□

visitors to carefully taste the surrounding scenery. In the heat of the wind and rain, you can still watch the beauty of the garden in different seasons and weather. As with calligraphy and paintings, you can also enjoy a taste of history and culture. The many flower shaped windows on both sides further extends the sense of space in the park (Fig 22).

Bamboo forest and trails

Mountain shapes arrayed like a stage background, increase the sense of space and extend the sense of perspective. These mountains create the impression of the meandering Yongding River, which people appreciate today as in the past (Fig 23).

Entertainment and fun

Integration of the diversity of landscape features creates the possibility for play and social interaction. For example: people chat at leisure on the porch and children splash or skate on the pool. (Fig 24-32).

The changing seasons of planting design

1. In the choice of planting flowers in the spring and produce autumn fruit and winter views. The cultural implications are vivid: the humanistic meaning given by classical garden plants are conveyed in the design. For example: Ginkgo the oldest fossil Tree.

2. cological principles: select plant species which attract birds and insects, such as peach, plum and cherry, In addition, shrubs using stratified planting configurations, show the diversity of planting and improve the level of viewing (Fig 33).

3. Local conditions: Consider the local climate, select appropriate species suitable to the southwest feng shui, for example, Metasequoia, peach, and iris which are ornamental but also provide food (Fig34-37).

Lighting plan - energy-saving

1. Night mirror: In addition to the stars, the light of the garden elements themselves are reflected in the mirror pool. Introverted, intoxicated, immersive.

2. Moving line guide: Warm white LED lights on the path, guide visitors traveling in the night.

3. LED lights glow through the recycled paper producing a misty effect, accompanied by the silhouette of the tree branches and leaves to creating imaginative poetic landscape screens.

4. LED projection lights, landmark guidelines for the night(Fig 38-45).■

图 45 透视图
Fig 45 a perspective view

作者简介：

曾宥榕 / 女 / 东海大学景观学系讲师、米页空间设计有限公司设计总监 / 台湾
曾宥源 / 男 / 米页空间设计有限公司设计部经理 / 台湾
张婉钰 / 女 / 米页空间设计有限公司 景观设计师 / 台湾
赵立勤 / 女 / 米页空间设计有限公司 景观设计师 / 台湾

Biography:

Ruth Yu-jung TSENG / Female/ Department of landscape architecture, Tung-Hai university, Master degree. / Millet landscape design Co., Ltd. , Landscape design director / Taiwan
Joe Yu-yuan TSENG / MaleMillet Design Co., Ltd. , Manager of designing department / Taiwan
Wan-yu CHANG / Female Millet landscape design Co., Ltd. Landscape Designer / Taiwan
Bonnie Lic-hin CHAO / Female / Millet landscape design Co., Ltd. Landscape Designer / Taiwan

查尔斯・沙（校订）
English reviewed by Charles Sands

未来的人，未来的景观
FUTURE PEOPLE, FUTURE LANDSCAPE

俎志峰 Zhifeng Zu

项目位置：中国，北京，第九届北京园博会
项目面积：1,000~1,200m²
委托单位：第九届中国（北京）国际园林博览会组委会
设计单位：重庆天开锦城园林景观设计有限公司
景观设计：俎志峰
完成时间：2012 年

Location: The 9th China (Beijing) International Garden Expo, Beijing, China
Area: 1,000 ~1,200m²
Client: The 9th China (Beijing) International Garden Expo committee
Designer: Chongqing Tiankai Jincheng Landscape Design Co.,Ltd
Landscape Design: Zhifeng Zu
Completion: 2012

图 01 鸟瞰图
Fig01 Aerial View

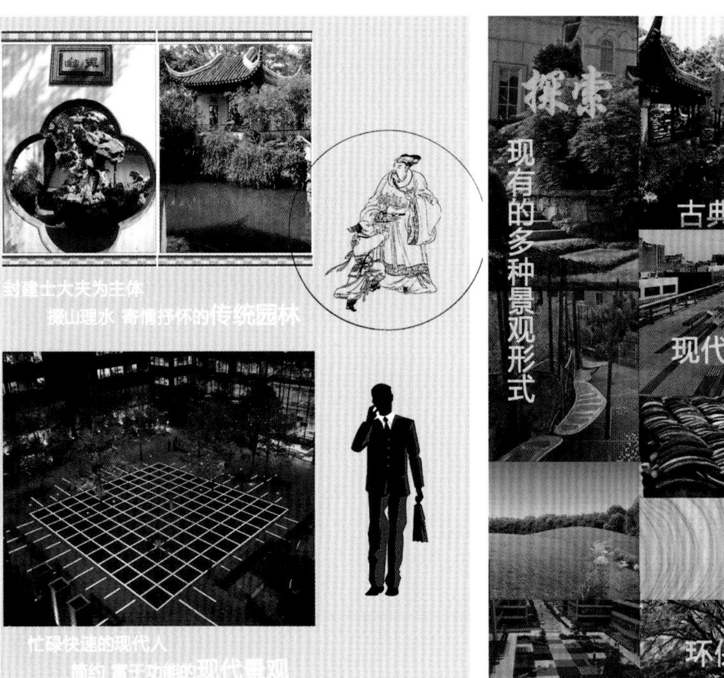

衍生—未来的景观

我们希望通过采用新的设计观念、手法与材料，重新诠释中国古典园林中的传统审美情趣。营造出富有前瞻性的未来中式之园。通过新思维与传统观念的碰撞衍生出未来景观发展之路一隅。同时以此一种未来景观的可能性引发观者对景观发展之路的臆想。在参与中交流，在交流中迸发启示的火花。使建者与观者共同参与到对未来景观发展之路的探索中来。

图 02 设计理念
Fig 02 Landscape design

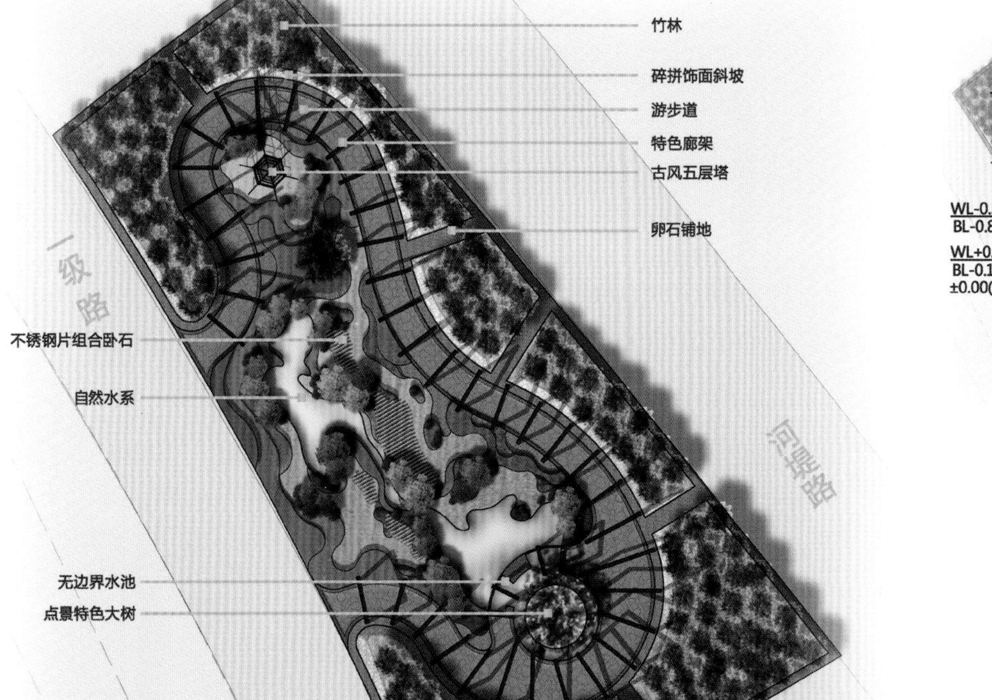

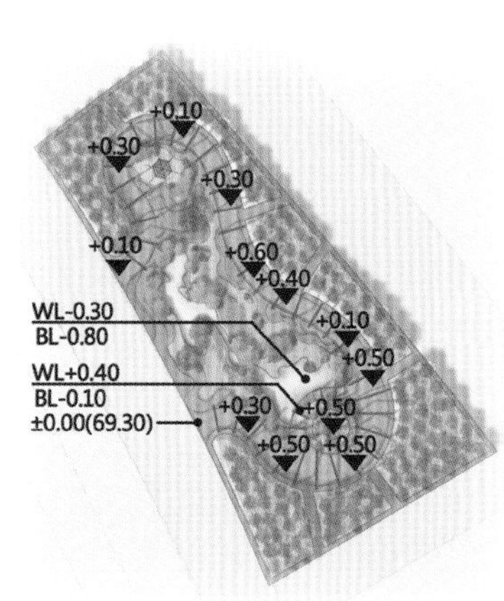

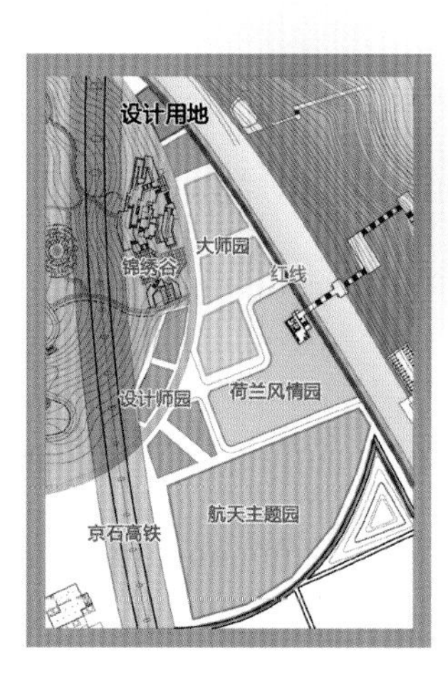

图 03 总平面
Fig03 General Layout

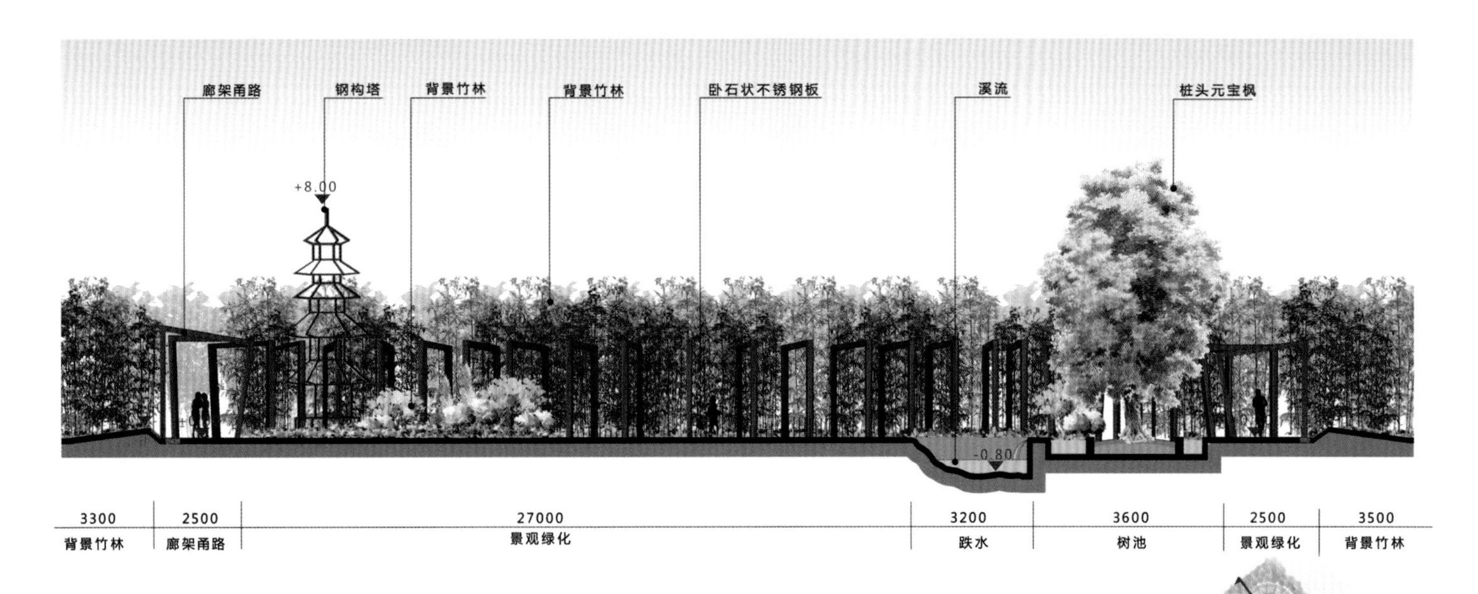

图 04 剖面图 01
Fig 04 Cutaway View01

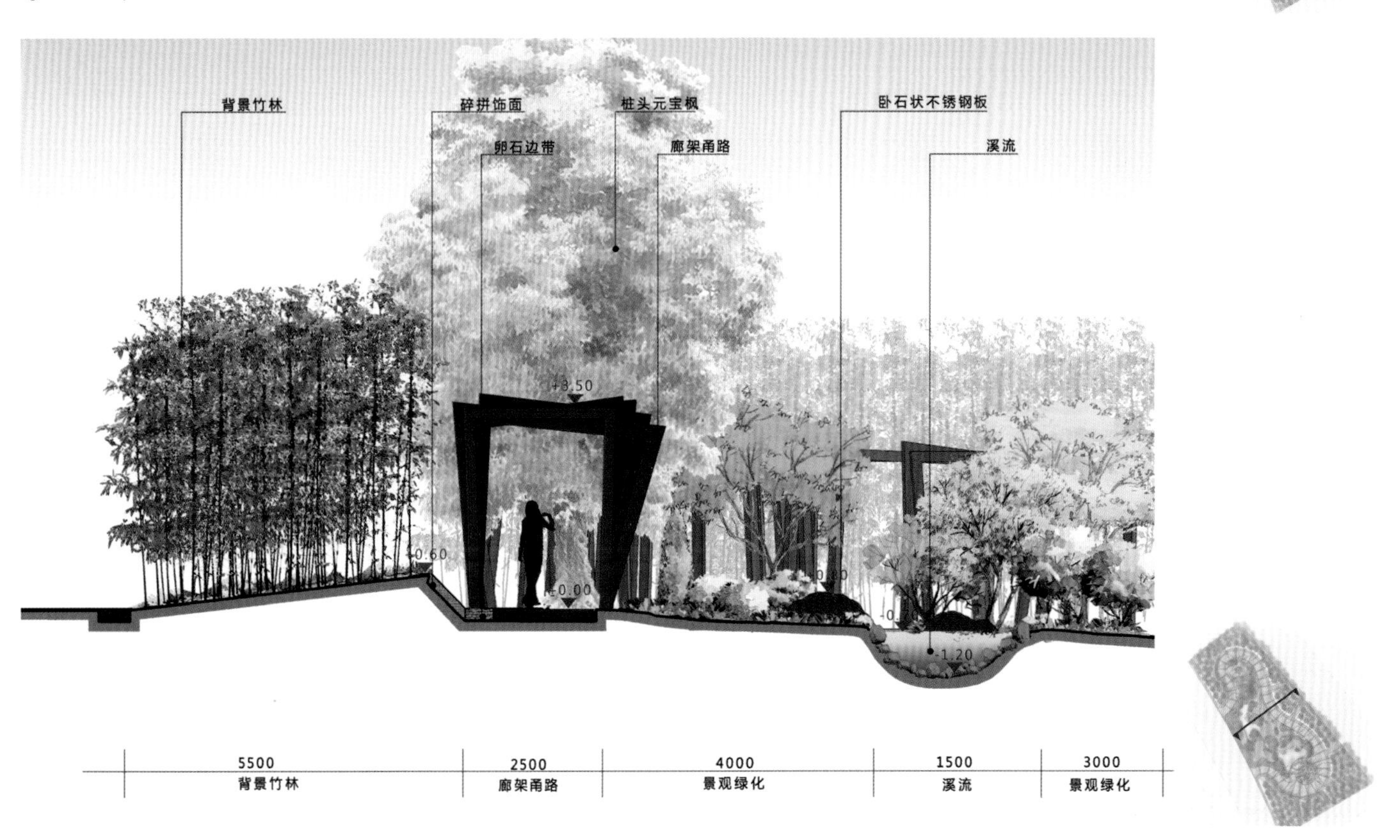

图 05 剖面图 02
Fig 05 Cutaway View02

我们希望通过采用新的设计理念、手法与材料，重新诠释中国古典园林的传统审美情趣，营造出富有前瞻性的未来中式之园（图 01–03）。通过新思维与传统观念的碰撞衍生出未来景观发展之路一隅。同时以此一种未来观景的可能性引发观者对景观发展之路的臆想（图 04–10）。在参与中交流，在交流中迸发启示的火花。使建者与观者共同参与到对未来景观发展之路的探索中来（图 11–13）。

We hope to interpret traditional aesthetic tastes of Chinese classic gardens and build future perspectives on Chinese Gardens with methods, materials and design philosophy(Fig01-03). We hope to find a way to develop future landscapes by blending new thinking with traditional views. Meanwhile, we hope to lead visitors to think about the developing trends in landscape(Fig04-10). The audience will communicate when they participate and obtain ideas in the communication; in this way, the designer and audience can explore the development of the future of landscape together(Fig11-13).

图 06 效果图 1
Fig06 Effect Picture 1

图 07 效果图 2
Fig07 Effect Picture 2

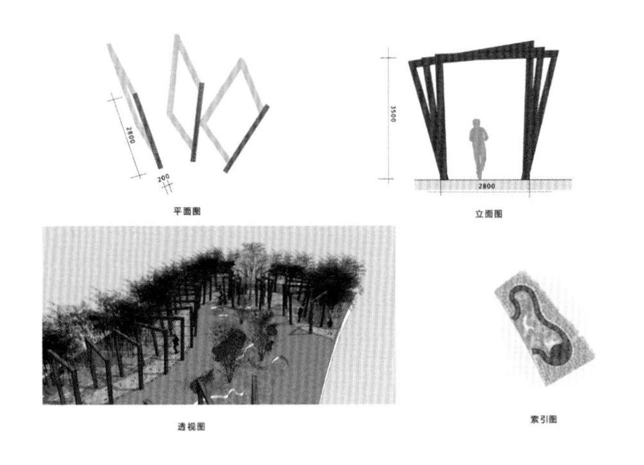

图 08 小品图 1
Fig 08 Sketch1

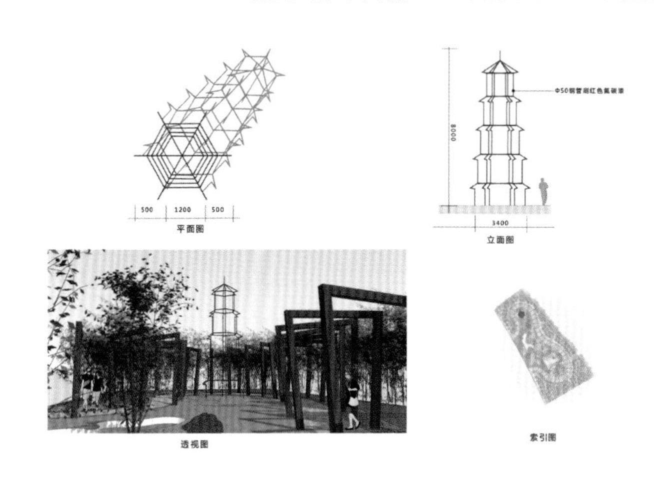

图 09 小品图 2
Fig 09 Sketch2

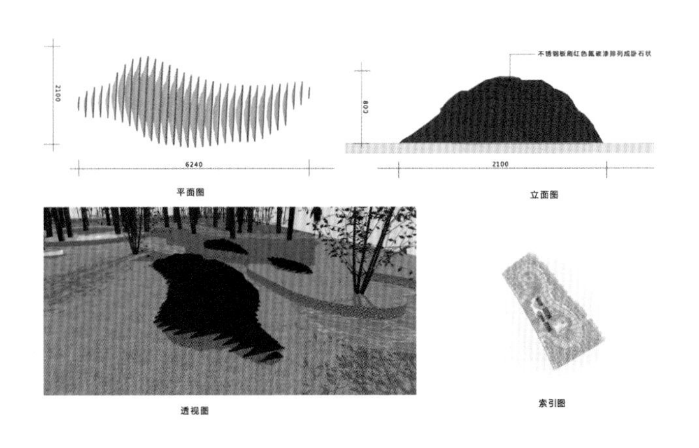

图 10 小品图 3
Fig 10 Sketch 3

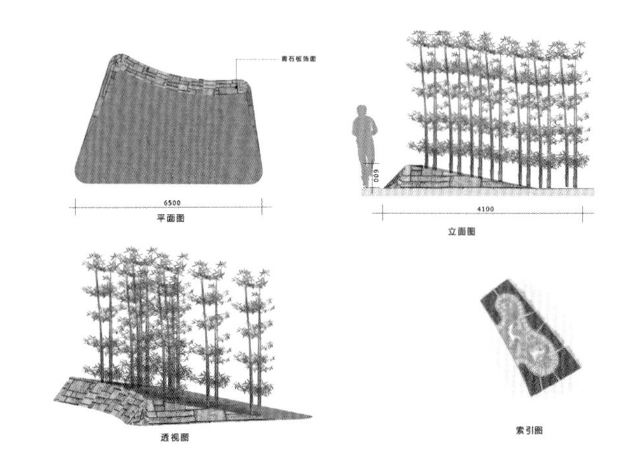

图 11 小品图 4
Fig 11 Sketch 4

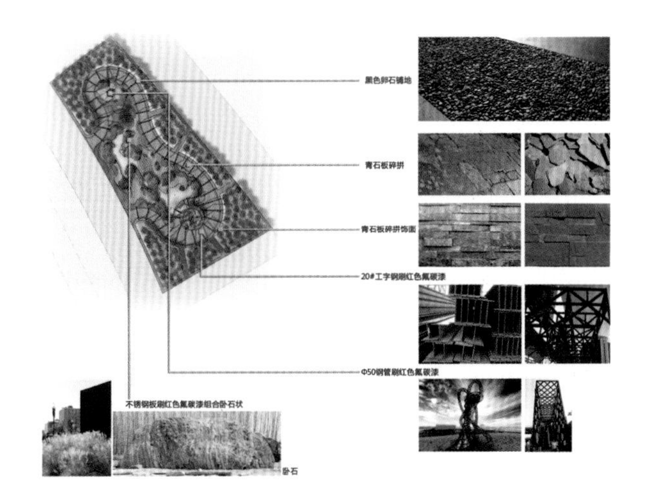

图 12 材料运用
Fig12 Material Application

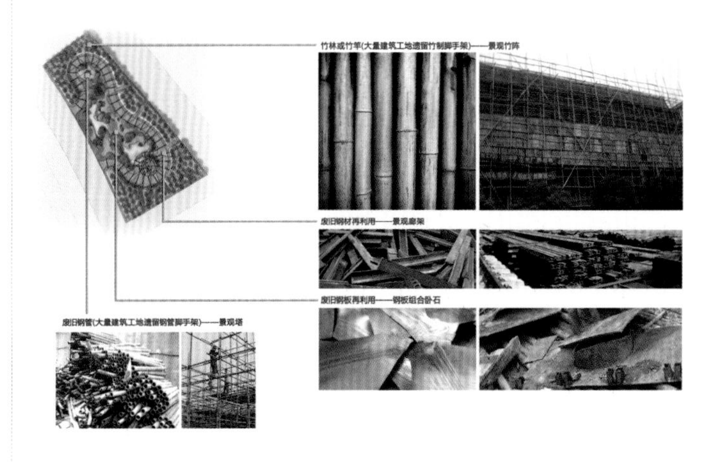

图 13 环保运用
Fig13 Environmental Protection Application

作者简介：

俎志峰 / 男 / 风景园林师 / 重庆天开锦城园林景观设计有限公司 / 中国重庆

Biography:

Zu Zhifeng / Male / Landscape Architect / Chongqing Tiankai Jincheng Landscape Design Co.,Ltd / Chongqing China

查尔斯 • 沙（校订）
English reviewed by Charles Sands

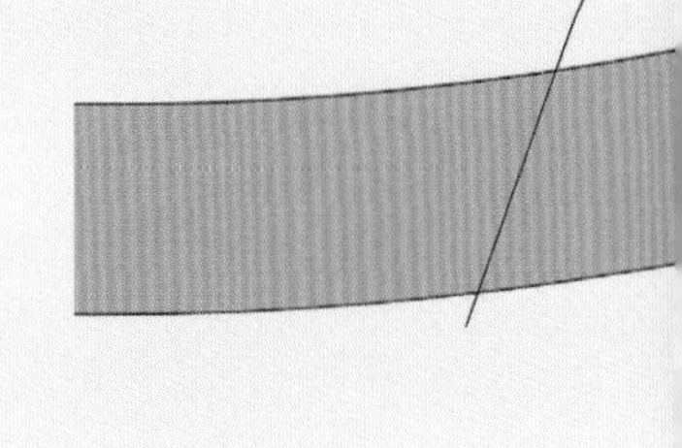

流动的窗

THE FLOWING WINDOWS

第九届中国（北京）国际园林博览会设计师广场设计获奖作品

A WINNING PROJECT FOR DESIGNER PLAZA OF THE 9TH CHINA (BEIJING) INTERNATIONAL GARDEN EXPO.

顾志凌　王伟　　Zhiling Gu　Wei Wang

项目位置：中国，北京，第九届北京园博会
项目面积：1,000m²
委托单位：第九届中国（北京）国际园林博览会组委会
设计单位：北京海韵天成景观规划设计有限公司
景观设计：顾志凌、王伟
完成时间：2012 年 4 月

Location: The 9th China (Beijing) International Garden Expo, Beijing, China
Area: 1,000m²
Client: The 9th China (Beijing) International Garden Expo committee
Designer: Macromind Landscape Architectural Landscape Planning & Design
Landscape Design: Zhiling Gu, Wei Wang
Completion: April, 2012

White Birch 白桦

Cherry Plum 紫叶李

Poplar 新疆杨

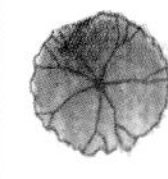
C.Plum Feature 紫叶李

Roof 廊架顶棚

Paving 花岗岩

Window 窗雕塑

Balcony 平台

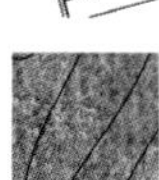
Grasses 草地

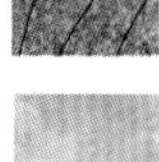
Water 水

Pebbles 卵石

Walls 景墙

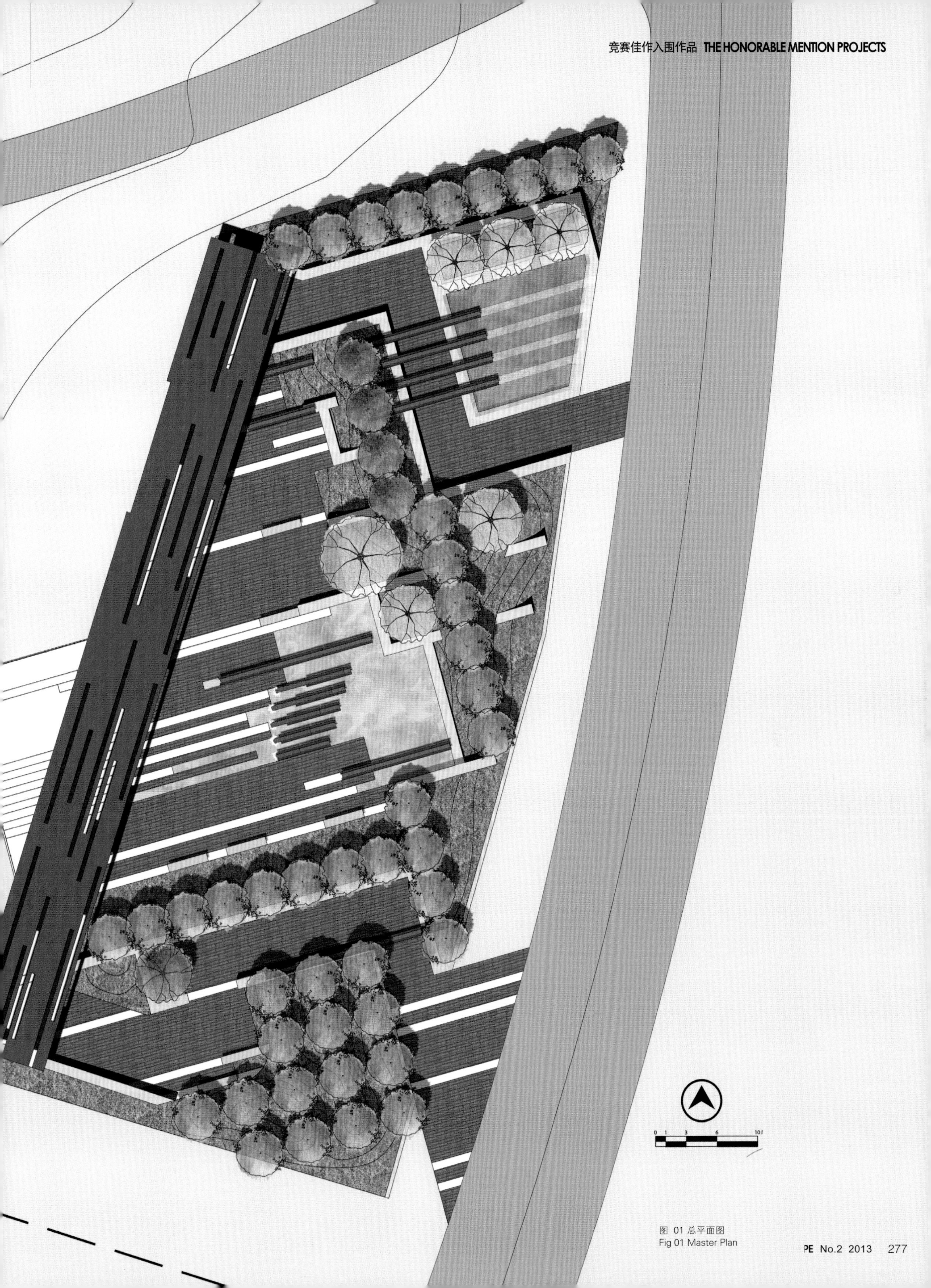

图 01 总平面图
Fig 01 Master Plan

图 02 北入口
Fig02 North Entrance

图 03 “窗”套“窗”
Fig03 “window” cover “window”

图 04 南入口
Fig04 South Entrance

第九届中国（北京）国际园林博览会旨在建设当代园林建设最高科技水平和艺术成就的示范区。应园博会组委会邀请，2012 年 2 月顾志凌先生应邀参加了本届园博会的设计师广场方案征集。

本案自命名为“流动的窗”，它以“窗”为设计主题，注重视觉上的愉悦性和体验者的参与性，通过简单、时尚的元素在不同场所与不同材料间的使用来展示丰富空间，运用各具形态的窗框有机组合，并与水景相间构成完美画面，展示了“窗”的魅力。

窗，自从诞生那刻起，就作为一种独立的元素而存在，它联通空间，沟通内外；透过窗，能看见与墙内不一样的另一世界，感受另一风景。本园设计以“流动的窗”为设计主题，漫步园中，去感受窗景之间形成的流动空间、窗景与水景的结合，以及窗与光产生的流动光影。设计从根本上脱离传统意义上的“窗”，以不同形式和存在方式来诠释“窗”的内涵，体验“窗”带来的乐趣。

方案注重视觉上的愉悦性和游客的参与性，强调视觉光影的美轮美奂与游历者的互动，通过简单、时尚的元素在不同场所与不同材料间的使用来展示空间的多样性与丰富性，设计师运用各具形态的窗框有机组合，并将水景融入其中，构成一幅和谐自然的画面，展示了“窗”的魅力（图 01）。

园区的南北两侧各设有一入口。北入口采用无边界水池和金属窗框相结合的手法，以干练的线条与净溢的水景吸引路上的游客，在有限的空间内，延长入口路径的长度（图 02）。游客从北口进入后，首先看到水景，穿过序列窗框，经过较封闭的墙体，进入设计师广场主景区（图 03）。每一个转折点都有一处与周边环境相结合的对景，在引导游客的同时，也产生空间的“障与引”。丰富的设计空间给游客目不暇接的感受。

相对北侧入口来说，南侧入口在设计上相对“封闭”一些（图 04）。阵列式的白杨林简洁有力，将游客引到近前（图 05），运用人们的猎奇心理，来回转折的景墙围合成一条百转千回的入园通道，各种高度的“窗”洞让各类人群都能对园景“略见一斑”，欲扬先抑的

the 9th China (Beijing) International Garden Expo is intended to be a demonstration plot of the highest technological and artistic achievements in modern landscape architecture. At the invitation of expo headquarters, Mr. Gu Zhiling participated the project collection of Designer Plaza in February, 2012.

Entitled “Flowing Windows”, the creation of design “windows” is aimed to produce joyful visual effects and enhance participation, using simple but fashionable elements in different areas and with different materials to demonstrate the diversity and richness of space. The designer presents the charm of “window” by combining a variety of window frame with distinctive shapes and bringing water into the site, creating a harmonious and natural picture.

From the moment of birth, windows exist in our lives as independent links between spaces—a connection between outside and inside. Through the window, we can see a world different from where we stand and experience another atmosphere. The theme of the garden is "Flowing Windows", as viewers are led to stroll and experience the flow of space between the various windows, the spaces are integrated by the flow of water as well as changing shadows created by windows of light.

The design is aimed to create joyful visual effects and to enhance participation of people, highlighting the interaction between attractive shadow effects and visitors. Using simple but fashionable elements in different sites and materials, it demonstrates the diversity and richness of space. The designer presents the charm of “window” by combining a variety of window frame with distinctive shape together and bringing water

图 05 条窗
Fig05 Strip Windows

图 06 曲折道路的焦点
Fig06 The focus of winding road

图 07 简约景观要素材质的对比
Fig07 Material contrast in simple elements

图 08 虚实辉映的长廊
Fig08 Virtual-real synthesis corridor

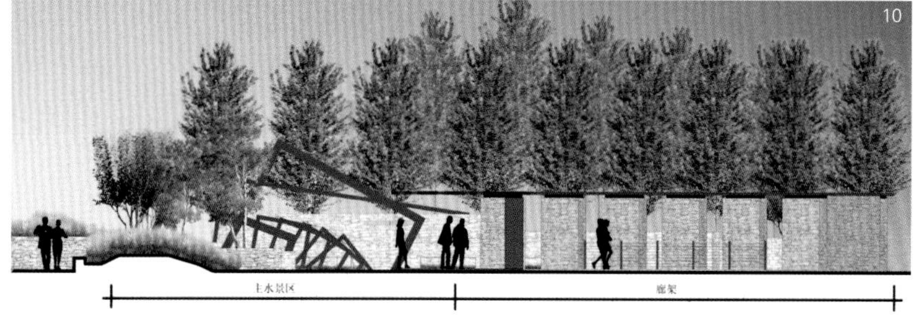

图 09 翻腾的窗框
Fig09 Rolling window frame

图 10 剖面图
Fig10 Section draw

撩拨着人们的心境，颇具趣味性。入口空间以一颗树形优美的紫叶李作为收尾，同时作为对景，加强空间色彩效果（图 06）。小空间的设计与草坡、树林为伴，阴凉舒适，供游客驻足休憩（图 07）。

进入主空间前的小空间起到缓冲作用，镶有三棱镜的景墙作为对景，投下彩虹般的阳光，充满趣味性。转身即可看到极具韵律感的长廊（图 08）。

隐映于树影之间的，就是主景——的窗。一个个仿佛刚从大地翻腾而出铁框，似乎还带着泥土的新鲜，以裂帛般的气势撕裂着大地的宁静，滚滚而来，一往无前。磅礴的气势震撼着每一个人心——仅只是一个片段，深陷入绿化中黑乎乎的洞口似乎随时都可能翻腾出另一个“窗”（图 09–10）；与地面呈鲜明对比的白色条带，就像是窗流动的轨迹。一系列带着韵律的水景窗让这生硬的家伙有了一丝温存，透过不同窗的角度与轮廓，我们可以看到不同的风景。

阳光透过廊架窄隙投射在地面上，与地面铺装虚实相交，产生奇幻的光影效果。长廊的延伸，同时对游客起到引导作用，配合条形开口的廊架顶棚，宽厚敦实的柱墙有韵律的向远方延展。（图 11）

漫步前行，西向视野豁然开朗，锦绣谷优美的景色映入眼帘。观景平台的设计强化游客的感受，站在平台上眺望，锦绣谷的景色尽收眼底，一览无余。远处高铁架桥排列与廊架的韵律遥相呼应，相映成彰（图 12–13）。回看园内，柱墙构成的框景如同一幅幅精美的图片，窗框和水景结合的雕塑群突出“窗”这一主题。

组委会给每块园区设定的面积为 1,000m^2，设计师并非将视野局限在此范围内，通过分析周边关系，结合自身特点，借用周边锦绣谷

in the scenery, creating a harmonious and natural picture(Fig.01).

Entrances are sited on the north and south sides of the garden. Borderless pools and metal window frames constitute the north entrance. It attracts visitors by showing simple lines and ample water pools to lengthen the entrance road in a limited space (Fig.02). When entering into the north entrance, visitors can see the waterscape immediately, then walk through a series of window frames, passing less opened landscape walls and finally going into the main square (Fig.03). Every turning point is bonded with an accompanying scenery, it becomes a guide to visitors as well as creating a "hinder and leader" effect. People can enjoy dizzying views with abundant spatial experience.

Compared with the north entrance, the south entrance is designed more "closed" (Fig.04). Planted in a grid, the aspen grove is simple but expressive, attracting people to come close. (Fig.05)With curious psychology, the back and forth landscape walls enclose an entry road full of twists and turns. Windows are installed at different levels, creating stimulating views. The entry space is capped off with a beautiful Prunus ceraifera. It is also used to enhance the color effects of the site. (Fig.06)Small spaces are provided with lawns and groves creating a cool and comfortable environment, providing people with a place to stay and rest(Fig.07) .

Before entering the main space, there is a small buffer area with a playful atmosphere. Landscape walls beset with triple prisms bring rainbow light into the area(Fig.08).

The key windows are hidden in tree shadow. The frames seem to billow from the earth which still retains the smell of fresh soil. It

图 11 富于几何张力的线条
Fig11 Expressive geometric line

图 12 远眺的平台
Fig12 View Platform

图 13 剖面图
Fig13 Section draw

的景色，充分利用场地独特的区位优势和周围优美的自然条件，在有限的空间内创造无限的景观。方案设计通过场地与锦绣谷之间的高差优势，进入观景平台似是打开园区一扇无形的窗，让锦绣谷优美景色成为园区的景观之一，并与之前封闭的入口空间形成鲜明对比，这种先抑后扬和借景的设计手法给游客带来无胜惊喜（图 14）。

gives out a powerful vigor breaking the peace of the area and coming in a roll, and never turning back. The majestic momentum shocks every visitor there, it seems like each fragrance, the black hole falling into the green, can stimulate another "window" (Fig.09-10). White paved lines present distinctive contrast like flowing traces of windows. A series of rhythmic waterscapes soften the hard scenery. People can sense different scenery through transitioning between different viewpoints.

Light permeates the corridor onto the ground, integrating with the pavement to create magical effects. The porch stretches itself in rhythm with the slatted ceiling, and the stocky column wall acts as a guide to visitors(Fig.11) .

Strolling to the west, the view is suddenly enlightened by the graceful scenery of Jinxiu Valley. A viewing platform enhances the visiting experience by presenting the whole landscape of Jinxiu Valley. The range of the high-speed bridge in the distance responds to the corridor's rhythm in the garden (Fig.12-13). Looking back to the garden, the framed scenery displays elegant pictures. The composition of the window frame and the water scenery highlights the theme of "window".

Although the expo headquarters set the area of each garden 1,000m^2, the designer did not limit their vision within the garden but took full advantage of the site characteristics to bring in the surrounding natural scenery of Jinxiu Valley by analyzing the neighboring relationship and responding to its characteristics, creating an infinite scenery in a limited space. Using the height difference between the site and Jinxiu Valley, the viewing platform acts as an invisible window taking Jinxiu Valley's natural scenery into the garden and creating a distinctive contrast to the closed scenery of the entry space. This sequence of foreshadowing and presenting brings visitors an exhilarant experience (Fig.14) .

图 14 园区全景
Fig14 Full view of garden

作者简介：

顾志凌 / 男 / 北京海韵天成总裁兼首席设计师 / 中国昆明

王　伟 / 男 / 北京海韵天成主任设计师 / 中国内蒙古

Biography:

Zhiling Gu/male/President and chief designer of Beijing Macromind Architectural Landscape Planning & Design/Kunming,China

Wei Wang/male/Director of designer of Beijing Macromind Architectural Landscape Planning & Design/ Inner Mongolia,China

查尔斯·沙（校订）
English reviewed by Charles Sands

图 01 全园鸟瞰图 1
Fig 01 Bird' s eye view 1

点滴园
WATERDROP — SHAPED GARDEN
第九届园博会设计师园 1 号地环境设计
THE 9TH CHINA (BEIJING) INTERNATIONAL GARDEN EXPO — DESIGNER GARDEN 1 ENVIRONMENTAL DESIGN STATEMENT

郭明

Ming Guo

项目位置：中国，北京，第九届北京园博会
项目面积：1,000m^2
委托单位：第九届中国（北京）国际园林博览会组委会
设计单位：中外园林建设公司
景观设计：郭明
完成时间：2012 年

Location: The 9th China (Beijing) International Garden Expo, Beijing, China
Area: 1,000m^2
client: The 9th China (Beijing) International Garden Expo committee
designer: Landscape Architecture Company of China
Landscape Design: Ming Guo
Completion: 2012

图 02 总平面图
Fig 02 Master Plan

图 03 全园鸟瞰图 2
Fig 03 Bird' s eye view 2

园博会已成为风景园林界最大规模和最有影响力的博览会，对于引导社会宜居生态环境建设的关注，推动社会经济人口资源环境协调发展起到了积极的作用。作为历史名城，北京在古城保护的同时，又要打造现代国际城市，对于景观设计者来说是难得的机遇和挑战。园博会依托永定河道，与卢沟古桥遥相呼应，历史文化气氛浓郁，山水相依。该地原为建筑垃圾填埋场，在原有垃圾填埋场建设一座生态、环保的园博园，有利于修复永定河生态环境，打造生态修复新亮点，并拉动北京西南地区经济社会发展。本届园博会以“绿色交响·盛世园林”为主题，以“园林城市·美好家园”为口号，设计理念是“化腐朽为神奇”，展现地域文化，展示多彩魅力，展示先进理念，促进区域发展，打造中国传统园林和国内外现代园林的艺术精品，建成生态和谐、景观优美、可持续发展的园博园（图 01–03）。

对于我们这些有志的本土设计师来讲，“继往开来”一直是我们设计师的原动力，这就需要我们继承几千年的传统园林文化，也需要我们的设计作品追赶现代设计的潮流与时俱进。

一个一千平方米的设计师园无法彻底改变生态环境，但是它却可以给人们一种启示。地质学家已经研究证明人类目前移动的岩石和沉积物比地球上所有河流加在一起移动的岩石和沉积物都多，人类对地球的影响与一个冰川时期相等，以至于科学家必须宣布全新纪的结束和人类纪的到来。改善环境给人类自己一个生存空间，是全体人类的当务之急，把保护生态环境可持续发展的理念及现代生活方式传播给大众也是我们设计师的任务，只要把这些点点滴滴的对历史、对文化、对记忆、对生活及对生命的理解交织在一起形成一个整体，给人们以启迪，我们的目的就达到了。

把这些废弃材料、可回收材料通过一定的组织让它承载一定的文化内涵，表现新时代的文化精神是这一作品的初衷，水与生命永远是人类灵感的来源，可回收资源代替不可回收资源也是生态环境的主流，

China international Garden expo has become the largest and one of the most influential exhibitions in the field of landscape architecture. It plays a significant role in attracting attention to ecological environmental construction and the coordination of the development of society, the economy, the population, and the environment. For landscape designers, the reconstruction of Beijing is an uncommon opportunity and challenge，On the one hand, they must make this city into an international and modern city, but on the other hand, they also have the responsibility of protecting the historic achitecture. The geographical position of the Garden expo is superior, as it sits along Yongding river and echoes Lu gou bridge. The historical and traditional atmosphere is also pervasive. As the area was a refuse landfill in the past, building an ecological and environmental expo park on such a site has many advantages: it is not only suited for creating an ecological environment for the Yong ding river, but also accelerates the economic development of the southwest area of Beijing. The theme of this year's expo is Green Symphony, Flourishing Gardens；the slogan is Garden City，Beautiful Homes；the design philosophy is Make The Ugly Beautiful. This expo aims to show the regional culture，colorful charm and advanced ideas. In addition, it combines the traditional Chinese garden with modern garden style in order to build a Garden expo that creates beautiful landscapes(Fig.01-03), a harmonious ecology and sustainable development.

For our ambitious designers,the motivation is to carry on the past and open a way to the future. Therefore, we have to inherit

图 04 花园入口
Fig 04 Entrance of the garden

图 06 蹈深
Fig 06 Dao Shen

图 05 缘理
Fig 05 Yuan Li

图 07 障清
Fig 07 Zhang Qing

从现代符号学的角度来分析一个简单的图形，即有“能指”又有“可指”，它可以表现出不同层次和不同深度的含义，八个简单的圆形水滴是皇家三山五园的暗指，三个圆水池也是皇家园林中一池三山的暗指。从整个园子的布局来讲，正合皇家古典园林布局形式即“西北高，东南低”，如避暑山庄、圆明园等。《赤霞经》中天下地理形势的论述中讲到“天下山脉发于昆仑，以西北为首，东南为尾……此天下之势。”同时 1 号园又是设计师园的入口（图 04），西北起地形也为整个设计师园园区增加了竖向变化，起到了障景的作用，而八个园的景名分别取自《大唐新语隐逸》及汉韩婴《韩诗外传》中对山水的描写，分别为 ①“嘉遁”，②“幽栖”，③“超逸”，④“缘理”（图 05），⑤“动下”，⑥“蹈深”（图 06），⑦“障清”（图 07），⑧“致远”。

从现代设计层面上讲，方案构图也是表现水与生命相互交织在一起的主题，同时也表现岁月的涟漪不断扩散而逝的主题。所以一个水滴符号其实有多重含义，正如音乐中的“胡麦”或“卡农”的表现手法，前一种是一个人同时唱两个声部，即一个作品表现两种内涵；后一种是复调音乐，后部重复前部，即一个作品不断的衍生。两种手法在设计中重复使用，交织在一起，在简单中表现多重复式含义，同时它也符合中国人的思维模式——“言外之意”。

一个好作品离不开细部材料的支撑，用可回收材料表现生态主题，用废弃材料表现场所精神。把汉白玉弃料给了历史与文化主题；把钢筋、锈钢板、水泥涵管给了回忆主题；把鱼缸、啤酒瓶、玻璃球给了生活主题；把植物和水给了生命主题。让简单材料有了多层含义，用可回收资源的生态利用来围绕主题和加深主题，从材料的选择角度来讲确保了材料和场所精神的相似性，把材料单独提出来解释了场所的过去历史与现代生态，同时也是一种新的组景尝试，而这种尝试又是具有可操作性的，设计师园仿佛是一个容器，把简单元素放在一起来表达深刻的内涵，同时也是一种现代设计中异质同构的手法运用（图 08）。

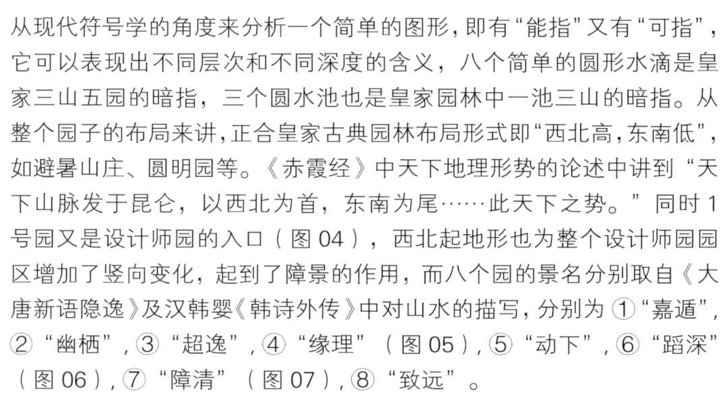

traditional garden culture and advance with the times.

This garden is not able to change the ecological environment greatly, however, it is a kind of inspiration for people. Geologists have proved that the rocks and areas of land moved by human beings are much more than that moved by all rivers in the globe. The influence on the world from human beings is as much as that of the whole glacial period so that scientists have to announce the end of Holocene and the coming of Anthropogen. The changing environment and habitat for us is a crisis. It is designers' responsibility to spread the philosophy of sustainable development and sustainable lifestyle to more people. As long as we can provide people some inspiration, our goals have been reached.

The original intention of this work was to express the new age's cultural spirit by reorganizing waste and recycled resources and instilling them with meaning. To analyze a simple graphic from the perspective of modern semiology, the terms "can refer to" and "refer to", can show different meanings from different levels. Eight simple circular water droplets implies royalty "three mountains and five gardens", the three round pools imply "a pool and three mountains". According to the layout of the whole garden (Fig.02-03), it is a typical royal classical garden layout, which is "high in the northwest and low in the southeast." The famous summer palace and Yuanmingyuan are good examples. In terms of 'Red Nepheline', Kunlun Mountain is the beginning of all other mountains in the world, starting from the northwest, and ending in the southeast. Meanwhile, No.1 garden is the entrance of this park(Fig.04); the terrain increases vertically and creates a scenic barrier effect. The names of the eight gardens are

事实上，传统与现代从来不矛盾，而只是延续与扩散。传统园林庭园或曰“一壶”，或曰“残粒”，或曰“半亩”，“点滴园”正合此意，小中见大从来都是庭园设计的准则，点滴春水，吞吐朝晖，接纳夕霞，永定河水逝者如斯，承载着过去和未来，点滴记忆只是一种苍然和感怀，水滴如澄澈的心境，圣人以此洗心，凡人以此隔尘。一个小水滴也是一个小祝愿，“半亩方塘一鉴开，天光云影共徘徊，问渠那得清如许，唯有源头活水来”。“大象无形”，“真水无香”。北京的母亲河，北京的故有文化精髓只有通过那些有志于此的设计师勤勤恳恳点点滴滴的努力，才能够让点滴细流汇聚成河。

from 'Datang hermit' and 'HanYing poetry' respectively: (1) jia dun (2) You qi (3)Chao Yi (4) Yuan li(Fig.05)(5) Dong Xia (6) DaoShen(Fig.06) (7) ZhangQing (Fig.07)(8) ZhiYuan.

"Drop" is a symbol with multiple meanings. Water and life are eternal Castalia for people, Water and life intertwine together, the ripple is in continuous but fleeting diffusion and, and so is life. "Choral" and "Canon" are two constructions of music, the former one means one person singing two part, namely one work revealing two different meanings; the latter is a kind of polyphonic music, with the background echoing the foreground, namely, a work that is constantly derived from itself. These two techniques are used in my design here, repeated and intertwined, revealing complex meaning in simple performance.

Details and materials are important support for a good piece of work. Recyclable resources replacing unrecyclable resources has been a mainstream trend. Waste and recycled materials are used here to reveal the theme of ecology and represent the spirit of the site. Discarded white marble represents history and culture; Steel bars, rusted steel, and cement culvert pipes represent memory; Fish tanks, beer bottles, and glass balls represent life. These reclaimed materials have been grouped and designed carefully in this garden. They do not only intensify the theme but also become the major elements of the scenery. This can be seen as an operable and optimistic trail, which helps us to have

图 08 夜景效果图
Fig.08 Nightscape

some new perspectives on material in landscape design. This garden is a container, simple elements are gathered together to express complex and profound connotations. This is also a kind of heterogeneous isomorphism(Fig.08).

Actually, the tradition and the modern are never contradictory. Tradition is always continuous and diffused. Chinese traditional gardens are named as" one pot" or "half-grain", " to see a world in a grain of sand" has always been our goal. In my work the "drop" also concentrates on this point. Yongding river is continuously flowing, passing history and moving towards the future, every drop in it recorded every laugh and tear, departing together. A droplet always lives together with us, from the sunrise to the sunset, it not only rinses the dust but also clears our mind. A "drop" is also a blessing. Designers always concentrate on different parts of life, and are dedicated to making our lives better. My concern and work on both environmental protection and Chinese traditional design is only a little drop in this massive project, nevertheless, drops converge into a vast ocean, I believe, if every designer could make this small effort, we can make our lives more pleasant.

作者简介：

郭明 / 男 / 风景园林师 / 中外园林建设有限公司设计院设计总监 / 中国北京

Biography:

Guo Ming/Male/Landscape Architect/Landscape Architecture Corporation of China Chef Designer/Beijing, China

查尔斯・沙（校订）
English reviewed by Charles Sands

青草地园林市政

——营建城市新型绿地　探寻花卉发展模式

浙江青草地园林市政建设发展有限公司是具有国家壹级城市园林绿化企业资质、市政公用工程施工总承包贰级、绿化造林施工资质乙级、绿化造林设计资质乙级、园林古建筑工程专业承包叁级、河湖整治工程专业承包叁级、城市及道路照明工程专业承包叁级、体育场地设施工程专业承包叁级、机电设备安装工程专业承包叁级，集园林绿化设计施工、园林植物科学研究、花卉生产销售、园林信息咨询和鲜花礼仪服务一体的综合型园林市政企业。

湖州潜山公园景观

海天公园

海天公园假山景观

湖州潜山公园景观

中华树艺苑

公司确定以"质量立业"为发展定位，辨证地处理量的跨越和质的提高的关系。先后承接了海天公园（包括海天高尔夫球场）整体绿化工程、杭州樱花小筑室外景观工程、萧山区风情大道北伸绿化工程、丽水市滨江景观带工程、红谷滩新区总体绿化工程、南昌市象湖公园景观工程、湖州潜山公园景观工程、湖州南浔嘉沁园室外总布景观绿化工程、湖州大剧院景观工程、江苏省太仓港口开发区管理委员会、绿城•桂花园一起景观绿化工程、上海大上海会德丰广场硬景工期、上海新华路一号景观工程、银亿海尚广场、乐清东山公园一期建设工程、上海三甲港江畔御庭别墅绿化工程、温州广汇景园绿化工程等400余项，多项工程获得过"杜鹃花奖"、"百花奖"、"茶花杯"、"最佳人居住环境奖"、"飞英杯"、浙江省优秀园林"金奖"等奖项，绿化市场逐渐向全国拓展。

和景中豪城一期景观

温州广汇景观

国色之颜，天香之醉
——世界芍药新品介绍

AWE-STRUCK APPEARANCE, INTOXICATING FRAGRANCE
——THE INTRODUCTION OF NEW HERBACEOUS PEONY CULTIVARS

王 琪 陆光沛 于晓南　　Qi Wang　Guangpei Lu　Xiaonan Yu

图 01 ‘落日珊瑚’
Fig 01 ‘Coral Sunset’

芍药（*Paeonia lactiflora*），历来被认为是我国的传统名花，并常以“花相”之名与牡丹并称。在我国，牡丹耀眼的光芒常常将芍药花的特质隐去，甚至很多花农和百姓，误将芍药称为牡丹。但如果细细留意中国的花文化史，还是能够发现，在我国历史上，芍药并不是完全被牡丹的阴影所笼罩，其自身娇艳的色彩、婀娜的风姿，也被众人所喜爱，并留下了浓重一笔，尤其是其“爱情之花”的寓意，流传甚广。如早在《诗经·郑风·溱洧》中便有关于芍药传递爱情的记载：“维士与女，伊其相谑，赠之以芍药”。

在西方，芍药远比牡丹享誉盛名。希腊神话中，她是药神（Paeon）的化身，具有驱除病魔、治愈疾病的神奇功能；在14世纪的英国，她是厨师的宠爱，其种子制作的调味品一直活跃在英国贵族的厨房里；在美国，她的圣洁和高贵，既被用作Memorial Day的指定用花，更是5月著名的新娘花，有着“Wedding flower”的美誉。

十九世纪初，芍药被英国著名植物猎人约瑟夫·班克斯发现，‘芳香’（‘Fragrans’）、‘慧氏’（‘Whitleyi’）、‘福美’（‘Humei’）、‘波特西’（‘Pottsii’）等品种，被更名改姓之后，相继远离故土，来到大不列颠的花园，并在异国他乡，杂交繁育，产生了一系列优秀的后代。

时值今日，全世界的芍药品种已逾500个。依据其野生种源的不同，国际上将芍药品种群分为三类：中国芍药品种群，欧洲芍药品种群，以及杂种芍药品种群。其中，中国芍药品种群作为优良的亲本，得到了世界育种家的充分重视和发掘，欧美国家亦纷纷以此进行杂交育种工作，得到了该品种群的许多优良品种；而杂种芍药品种群则是多种起源的、由多个种参加杂交组合而形成的品种系列，该品种群具有花头直立、茎杆粗壮、花色明亮新奇、适应性强等优点，逐渐成为世界芍药中备受瞩目的新星。北京林业大学于晓南芍药课题组于2006 ~ 2010年陆续从美国引进了大量国外培育的观赏芍药新品种，经多年生物性特性、生态适应性的观察以及分株繁殖试验，各品种均

图 02 ‘夏威夷珊瑚粉’
Fig 02 ‘Pink Hawaiian Coral’

图 03 ‘红色魅力’
Fig 03 ‘Red Charm’

图 04 ‘巴克艾美人’
Fig 04 ‘Buckeye Belle’

图 05 ‘堪萨斯’
Fig 05 ‘Kansas’

Paeonia lactiflora cvs is one of China's most famous garden ornamentals, with the high reputation of 'prime minister of flowers'. Although Chinese people respect tree peony as 'king of flowers', if we take a birdview of China's floriculture history, it is not too difficult to find that, actually, herbaceous peony was introduced to garden cultivation much earlier than tree peony. Herbaceous peony is appreciated by many people for its beautiful colors and graceful flower types, what's more, as 'China's love flower', it is popular with many young lovers. The record that herbaceous peony used to convey a message of love could be found in *the Book of Poetry • Songs Collected in Zheng • Riverside Rendezvous:* The lovely lad and lass played together and had a happy hour, and then each gave the other herbaceous peony flower.

In the West, the herbaceous peony is more renowned than the tree peony. In Greek mythology, the peony is named after Paeon (also spelled Paean), a student of Asclepius, the Greek god of medicine and healing. Asclepius became jealous of his pupil; Zeus saved Paeon from the wrath of Asclepius by turning him into the herbaceous peony flower. In 14th century, the seed of herbaceous peony is one of British chefs' favorite spices--only British aristocracy could afford that. In the United States, the herbaceous peony is used as a particular flower not only in Memorial Day, but also in May—as 'Wedding flower', she gains favor by many brides.

At the beginning of the 19th century, *P. lactiflora cvs* was found by the famous British plant hunter, Joseph Banks, in China. Some cultivars, such as 'Fragrans', 'Whitleyi', 'Humei', 'Pottsii' were introduced from their homeland to Britain. Even though many

图 06 '公爵夫人'
Fig 06 'Duchesse de Nemours'

表现良好，适合在我国华北大部分地区的园林绿化及切花生产中推广应用。

提到杂种芍药品种群，就不得不提其中的经典代表——'珊瑚'（'Coral'）系列。自从该系列中的'珊瑚魅力'（'Coral Charm'）于1986年首次获得美国芍药牡丹协会（APS）金牌奖之后，'珊瑚'在美国人心目中就成了一个经久不衰的品系，受到人们的热烈追捧，历久弥新。这个品种系列共育有5个品种，其中3个都曾获得金奖，这在其他品种系列中是不多见的。尤为值得一提的是，'珊瑚魅力'和2003年获APS金奖的'落日珊瑚'（'Coral Sunset'）还是会变色的品种。

'落日珊瑚'（图01）由美国人Samuel Wissing-Roy G. Klehm于1965年育成，是*P. officinalis* 'Otto Froebel'和*P. lactiflora*白色半重瓣品种杂交而成的品种。其茎秆粗壮，小叶平展，花瓣圆润可人。初开时呈现娇艳欲滴的珊瑚红色，随着花的盛开，颜色慢慢褪淡，宛若落日西沉，充满诗意。有如荷花的花型，赋予了它同样"出淤泥而不染"的高贵品质；恬淡的微香，则给予了它俯瞰众花的顾盼生姿。集万千优点于一身，得到APS评委和欧美芍药爱好者的青睐也就不足为奇了。

'夏威夷珊瑚粉'（'Pink Hawaiian Coral'）（图02），这个由美国人Roy G. Klehm于1981年育成的另一个'Coral'系列品种，是由*P. officinalis* 'Otto Froebel'和*P. lactiflora* 'Charlie's White'杂交而成。花初开时呈现可爱的圆球状，远观有如耀眼的火种；完全开放后，金黄色的雄蕊在珊瑚红色花瓣的映衬下，二者相得益彰，使整朵花显得格外艳丽夺目。与'落日珊瑚'相比，'夏威夷珊瑚粉'的生长势更旺盛，重瓣性更高。而其名字的由来，主要是因为它的花色如同夏威夷群岛的珊瑚礁一般光彩照人，获得2000年APS金奖也是实至名归。

'红色魅力'（'Red Charm'）（图03），隶属于杂种芍药品种群，由美国人Glasscock于1944年育成，是P. officinalis 'Rubra Plena'和P. lactiflora杂交而成的品种。它的花呈少有的鲜红色，花色细腻、纯粹，如丝绒般质感。花外瓣平伸舒展，瓣缘稍内卷；内瓣轻褶紧凑，狭长坚挺层叠，仿佛鸟羽般迷人。'红色魅力'花开时热烈、奔放却又不失高洁的气质，使得东西方的审美观在它的身上得到了完美的诠释和统一，难怪它不仅在美国芍药界的展览上多次夺冠，更有甚者封其为"世界第一芍药"。

'巴克艾美人'('Buckeye Belle')（图04），这个同样隶属于杂种芍药品种群的品种，由美国人Mains于1956年育成，是*P. officinalis*和*P. lactiflora*杂交而成的品种。花初开时，充满了神秘感的深红色外瓣欲

cultivars were renamed later on, this beautiful oriental flower did contribute a lot for spring scenery in British gardens. From the beginning of 20th century, in the United States, the herbaceous peony from China started to be used in distant hybridization, together with other herbaceous peony species and cultivars from Europe and Caucasus, pioneer breeders such as A.P. Saunders, the Klehms, and E.Jr.Auten etc. created a series of interspecific cultivars, whose charming appearance pushed the development of herbaceous peony to a new peak.

Until now, there are more than 500 herbaceous peony cultivars in the world. According to different origination of parent, these cultivars are divided into three cultivar groups:

Chinese herbaceous peony group, which is single species originated, *P.lactiflora*; European herbaceous peony group, which is single species originated, main parents including *P. officinalis and P.tenuifolia;* Hybrid herbaceous peony group, which is two or more species originated, main parents including: *P.officinalis, P.tenuifolia, P.peregrina, and P. mlokosewitschii etc.*

As main parents, the cultivars from the Lactiflora group are widely used by peony breeders. Some offsprings from this group are very classic and popular, like: 'Sarah Bernhardt', 'Kansas', 'Duchess de Nemours'. The Hybrid group, the youngest group, with only a hundred years of breeding history, contains promising cultivars with good qualities: the flower stem is upright and strong; the flower color is bright and novel; the leaf color is deep green, which looks more vigorous; the disease-resistance is stronger than the Lactiflora group. From 2006 to 2010, sixteen cultivars were introduced by Yu Xiaonan research group (Beijing Forestry University) from Holland, and these cultivars covered all three cultivar groups. After several years field cultivation, some cultivars show good characters and can be successfully reproduced by dividing the root. They are suitable for both landscaping use and as cut flower.

When mentioning the Hybrid group, the 'coral' cultivar series is well worth of attention. For all the five cultivars in this series, they share the same color – 'coral', which distinguishes them well from the other cultivars. In 1986, 'Coral Charm' won the gold medal

拒还迎地露出内瓣一隅；花逐渐打开后，狭长的小碎花瓣搭配着黄色的花药及红色的花丝组成的雄蕊，让人赏心悦目、心旷神怡。纯正优雅的深红花色，大方且细腻的荷花花型，浓郁芬芳的醉人花香，是‘巴克艾美人’跻身成为顶级芍药切花品种的奥秘。

‘堪萨斯’（‘Kansas’）（图 05）是由前 APS 主席 Bigger 于 1940 年育成的品种，隶属中国芍药品种群，它的得名即源于 Bigger 的家乡堪萨斯州。其叶色深绿，花朵硕大，成花率高。花含苞待放时，好像一枝枝热情洋溢的月季，侧面坛状的花型仿佛斟满了沁人心脾的美酒；浪漫的玫瑰红色外瓣与绮丽的深粉色内瓣随着时间的推移渐次打开，些许金黄色的雄蕊掩映其中，令整朵花顿时活泼了起来。‘堪萨斯’曾获 1942 年美国国内成就奖章（The American Home Achievement Medal）、1957 年 APS 金奖等多个奖项，至今仍为世界各地的芍药爱好者所津津乐道。

‘公爵夫人’（‘Duchesse de Nemours’）（图 06），这个由法国人 Calot 于 1856 年育成的品种，隶属中国芍药品种群，是一个畅销百年、仍广受欢迎的芍药品种。其生长强健，茎秆粗壮。花初开时，呈现圆润可爱的酒杯状；随着花朵的盛放，外瓣逐渐打开，露出奶油黄色的花心，为纯净的白色增添了几许灵动。圣洁的花色、馥郁的芳香、强健的茎杆、深绿的枝叶，是这个得名于法国贵族的芍药品种，在芍药界经久不衰的秘诀。

图 07 ‘埃里先生’
Fig 07 ‘Monsieur Jules Elie’

‘埃里先生’（‘Monsieur Jules Elie’）（图 07），隶属于中国芍药品种群，虽然早在 1888 年就由法国人 Félix Crousse 育成，但历经一个多世纪，仍是一个熠熠生辉、活跃在芍药切花舞台的品种。其生长势旺，复叶宽大。近圆形的外花瓣宽阔整齐、平伸舒展，浅玫瑰粉的颜色令人如痴如醉；狭长的内花瓣虽有部分扭曲，但向内逐渐过渡的粉白花色却使花心部分呈现出和谐的稳重感。整朵花像闪耀的玉盘一般，与淡绿的叶子一起，向观赏者诉说着自己的清新超凡、弥足珍贵。

‘莎拉小姐’（‘Sarah Bernhardt’）（图 08）则由法国著名育种家 Lemoine 于 1906 年育成，同样隶属于中国芍药品种群，该品种是欧美市场弥久不息的经典品种。它得名于法国早期电影女演员 Sarah Bernhardt（1844–1923）——这位曾被冠以“世界所知最著名的女演员”。‘Sarah Bernhardt’ 的花期较晚，当大部分芍药品种都卸下红妆时，它才气定神闲地粉墨登场。花瓣未完全打开时，可爱的花心呈现圆球状，宛如一颗颗蚌上的珍珠，展现出迷人的魅力；粉红色的花瓣中间有几片不显眼的红边，好似飞舞翩跹的绸带，为整朵花平添了些许灵气。娇滴可爱的花色，清新淡雅的芳香，以及强壮挺拔的茎秆，使得这个品种成为欧美婚礼时新娘的不二选择。

以上各芍药品种引种以来的成活率均在 80% 以上，且自第三年开始，形态特征与原产地基本一致，表现出了较高的观赏性。繁殖方式仍以分株繁殖为主：选择株龄大于 3 年、生长健壮的母株，于秋季地下芽形成、地上部分叶片开始枯黄进行分株。这八个芍药品种同绝大数芍药一样，都喜欢阳光充足、高燥凉爽的气候；耐寒，耐旱，宜湿润及排水良好的土壤或沙质土壤，忌盐碱地和低洼地。它们在较肥沃的土壤中生长较好，开花期之前可适当增施磷钾肥，以促使枝叶生长茁壮，开花美丽。

随着这些观赏芍药新品种的引进与推广普及，相信在不久的将来，我们身边定会出现它们娉婷袅娜、风姿绰约的身影，同时也希望更多的园艺爱好者参与到芍药的育种中来，将我们身边的花园打扮地更加绚丽多彩。□

from American Peony Society (APS) -- this was the first time that a ‘coral’ cultivar received a big reward. Since then, the ‘coral’ series won more and more affection and were honored with gold medals for many times. Many peony books, when mentioned the Hybrid group, have recommended the ‘coral’ series. ‘Coral Sunset’ (Fig 01), 2003 APS gold medal cultivar, was bred by Samuel Wissing and Roy G. Klehm in 1965. It is an offspring of *P. officinalis* ‘Otto Froebel’ and a white semi-double cultivar from the Lactiflora group. It has strong stems, flat leaves and round petals. When the flowers open early in the spring, they present tender and beautiful coral red color, then the color slowly fades to creamy yellow along with the blooming, just like the sunset.

‘Pink Hawaiian Coral’ (Fig 02) is another cultivar of ‘Coral’ series. It was bred by Roy G. Klehm in 1981, and it is an offspring of *P. officinalis* ‘Otto Froebel’ and *P. lactiflora* ‘Charlie's White’. The flowers are like dazzling fire when they start to give blossom. After fully blooming, the golden stamens and coral red petals make the flower color look more attractive. ‘Pink Hawaiian Coral’ has more robust growth and more plump flower than ‘Coral Sunset’. It won the APS Gold Medal 2000 and Award of Landscape Merit 2009.

‘Red Charm’ (Fig 03), also belonging to the Hybrids group, was bred by Glasscock in 1944, and it is the offsping of *P. officinalis* ‘Rubra Plena’ and *P. lactiflora*. Its color is bright red and the texture feels like velvet. The flower type is bomb-type double: The outer petals are large and flat while the inner petals are narrow and compact. ‘Red Charm’ is one of the most popular herbaceous peony cultivars in the US, winning gold medals from different peony shows for many times.

‘Buckeye Belle’ (Fig 04), which also belongs to the Hybrids group, was bred by Mains in 1956, and it is an offspring of *P. officinalis* and *P. lactiflora*. Very dark chocolate-red petals are sprinkled with golden stamens. The combination of large outer petals surrounding narrow center petals adds depth to the bloom. Besides, the dark green leaflets on reddish stems create the finishing touch. It won Award of Landscape Merit 2009 and APS Gold Medal 2010.

‘Kansas’ (Fig 05) was bred by former APS President Bigger in 1940. It belongs to the Lactiflora group and it names after Bigger's hometown--Kansas. It has dark green leaves and large rose-type flowers. It is easy to grow and blooms freely every year. The purple-red color fades little while blooming. ‘Kansas’ had won the American Home Achievement Medal in 1942, APS Gold Medal in 1957, and many other awards.

‘Duchesse de Nemours’ (Fig 06), bred by Calot in 1856, is a cultivar belonging to the Lactiflora group. It is a classic French cultivar, named after an aristocracy. ‘Duchesse de Nemours’ has strong growth and erect stems. The flower is large, cupped shape,

图 08 '莎拉小姐'
Fig 08 'Sarah Bernhardt'

with white guards and a moderately full center of light canary-yellow deepening to a pale green at the base of the petals. The same as 'Kansas', 'Duchesse de Nemours' is also one of the most popular cut flower cultivar.

'Monsieur Jules Elie' (Fig 07), another old cultivar of the Lactiflora group, was bred by Félix Crousse in 1888. Since it is born, time has passed more than a century, but it is still a brilliant and popular cultivar. The plant is tall and very free-flowering. The flower has broad, smooth guard petals; center is incurved and silvered with light grayish pink. The pink crown-type flowers look like large chrysanthemums, very spectacular. And it is moderately fragrant.

'Sarah Bernhardt' (Fig 08), belonging to the Lactiflora group, was bred by the famous French breeder Lemoine in 1906. It is a classic cultivar and very popular. In Europe and America, when people can find peony cut flower in supermarket or garden center, for most time, they will find this cultivar – one of the most successfully commercialized cultivar. It is named after an early French film star, Sarah Bernhardt (1844-1923), who is honored as 'one of the most famous actresses in the world'. It is floriferous and has dark rose-pink flower, with inconspicuous red edges on a few central petals, medium fragrant.

All the cultivars introduced above can grow and flower well in Beijing. Some of the cultivars need a year to get settled and give blossom, but their characteristics will be the same as where they are from originally. Crown division is the technique most used to propagate herbaceous peonies. It should be done only in the early fall. The technique is as follows:

In late August, early September cut down foliage of peony to be divided. If ground is very dry, water the plant the day before dividing. Dig plant up with a strong spade, taking as much of the root system as possible and being careful not to damage the crown. Wash soil from roots with a strong jet of water. Leave cleaned root system in the shade for several hours to soften it (this makes it less brittle and gives you more precision when cutting) .Examine the plant carefully and plan your cuts. Each division must have at least one visible crown bud to grow. It is best however to have at least 3 buds per division. With a clean, strong knife and working on a rough work bench, make your first cut through the top of the crown being careful to cut down toward the work bench and away from yourself. If the plant has long tuberous roots it may make the process easier if the roots are trimmed back to between 6-8 inches from the crown. Continue to divide the plant until you have the desired number of divisions.

The introduction and acclimatization of these new ornamental herbaceous peony cultivars in Beijing inspires confidence that they will appear in many gardens in the near future. At the same time, more and more gardeners can be encouraged to participate in the breeding of herbaceous peonies and make many gardens more colorful.■

作者简介：
王琪 / 男 / 硕士研究生 / 北京林业大学园林学院 / 中国北京
陆光沛 / 女 / 硕士研究生 / 北京林业大学园林学院 / 中国北京
于晓南 / 女 / 副教授 / 博士 / 北京林业大学园林学院 国家花卉工程技术中心 / 中国北京

Biography:
Qi Wang / Male / Postgraduate Student / School of Landscape Architecture, Beijing Forestry University / Beijing, China.
Guangpei Lu / Female / Postgraduate Student / School of Landscape Architecture, Beijing Forestry University / Beijing, China.
Xiaonan Yu / Female / Associate Professor / Ph.D / School of Landscape Architecture, Beijing Forestry University, China National Center for Flower Engineering Technique / Beijing, China.

《世界园林》征稿启事
Notes to Worldscape Contributors

1 本刊是面向国际发行的主题性双语（中英文）期刊。设有作品实录、专题文章、人物 / 公司专栏、热点评论、构造、工法与材料（含植物）5 个主要专栏。与主题相关的国内外优秀作品和文章均可投稿。稿件中所有文字均为中英文对照。所有投稿稿件文字均为 Word 文件。作品类投稿文字中英文均以 1000–2500 字为宜，专题文章投稿文字的中英文均以 2500–4000 字为宜。

2 来稿书写结构顺序为：文题（20 字以内，含英文标题）、作者姓名（中国作者含汉语拼音，外国作者含中文翻译）、文章主体、作者简介（包括姓名、性别、籍贯、最高学历、职称或职务、从事学科或研究方向，现供职单位、所在城市、邮编、电子信箱、联系电话）。作者两人以上的，请注明顺序。

3 文中涉及的人名、地名、学名、公式、符号应核实无误；外文字母的文种、正斜体、大小写、上下标等应清楚注明：计量单位、符号、号字用法、专业名词术语一律采用相应的国家标准。植物应配上准确的拉丁学名。扫描或计算机绘制的图要求清晰、色彩饱和，尺寸不小于 15cm*20cm；线条图一般以 A4 幅面为宜，图片电子文件分辨率不应小于 300dpi（可提供多幅备选）。数码相机、数码单反相机拍摄的照片，要求不少于 1000 万像素（分辨率 3872*2592），优先使用 jpg 格式。附表采用 " 三线表 "，必要时可适当添加辅助线，表格上方写明表序和中英文表名，表序应于内文相应处标明。

4 作品类稿件应包含项目信息：项目位置 / 项目面积 / 委托单位 / 设计单位 / 设计师（限景观设计）/ 完成时间。

5 介绍作品的图片（有关设计构思、设计过程及建造情况和实景等均可）及专题文章插图均为 jpg 格式。图片请勿直接插在文字文件中，文字稿里插入配图编号，文末列入图题（须含中英对照的图号及简要说明）。图片文件请单独提供，编号与文字文件中图号一致。图题格式为：图 01 xxx/Fig 01xxx。图片数量 15–20 张为宜。可标明排版时对图片大小的建议。

6 文稿一经录用，即每篇赠送期刊 2 本，抽印本 10 本。作者为 2 人以上，每人每篇赠送期刊 1 本，抽印本 5 本。

7 投稿邮箱 :Worldscape_c@chla.com.cn 联系电话 :86–10–88364851

---

1 Worldscape is an international thematic bilingual journal printed in dual Chinese and English. It covers five main columns including Projects, Articles, Masters / Ateliers, Comments, and Construction & Materials (including plants). The editors encourage the authors to contribute projects or articles related to the theme of each issue in both Chinese and English. All submissions should be submitted in Microsoft Word (.doc) format. Chinese articles should be 1000-2500 characters long. English articles should be 2500-4000 words long.

2 All the submitted articles should be organized in the following sequence: title (no more than 20 characters and the English title should be contained); author's name (for Chinese authors, pin yin of the name should be accompanied; for foreign authors, the Chinese translation of the name should be accompanied if applicable); main body; introduction to the author (including name, gender, native place, official academic credentials, position/title, discipline/research orientation, current employer, city of residence, postal code, E-mail, telephone number). For articles written by two or more authors, please list the names in sequence.

3 All persons, places, scientific names, formulas and symbols should be verified. The English submissions should be word-processed and carefully checked. Measuring units, symbols, and terminology should be used in accordance with corresponding national standards. Plants should be accompanied with correct Latin names. Scanned or computer-generated pictures should be sharp and saturated, and the size should be not less than 15cmx20cm. Diagrams and charts should be A4-sized. The resolution of digital images should be not less than 300dpi (authors are encouraged to provide a selection of images for the editors to choose from). The resolution of pictures generated by digital camera and digital SLR camera should be not less than 3872x2592, and .jpg formatted pictures are preferred. Annexed tables should be three-lined, and if necessary, auxiliary lines may be used. All tables should be sequenced and correspond to the text. Chinese-English captions should be contained.

4 All the submitted materials should be accompanied with short project information: site, area, client, design studio (atelier or company name), landscape designers (landscape architects) and completion date (year).

5 All project images (to illustrate the concept, design process, construction and built form) should be .jpg format. The images should be sent separately and not integrated in the text. All images should be numbered, and the numbers should be represented in the main body of the text. At the end of the text, captions and introductions to the images should be attached (Chinese-English bilingual text). The caption should be formatted as Fig 01 xxx. No more than 20 images should be submitted. Suggestions to image typeset may be attached.

6 The author of each accepted article will be sent 2 copies of the journal and 10 copies of the offprint. In the case of articles with 2 or more authors, each author will be sent 1 copy of the journal and 5 offprints.

7 Articles should be submitted to: Worldscape_c@chla.com.cn Tel: 86-10-88364851

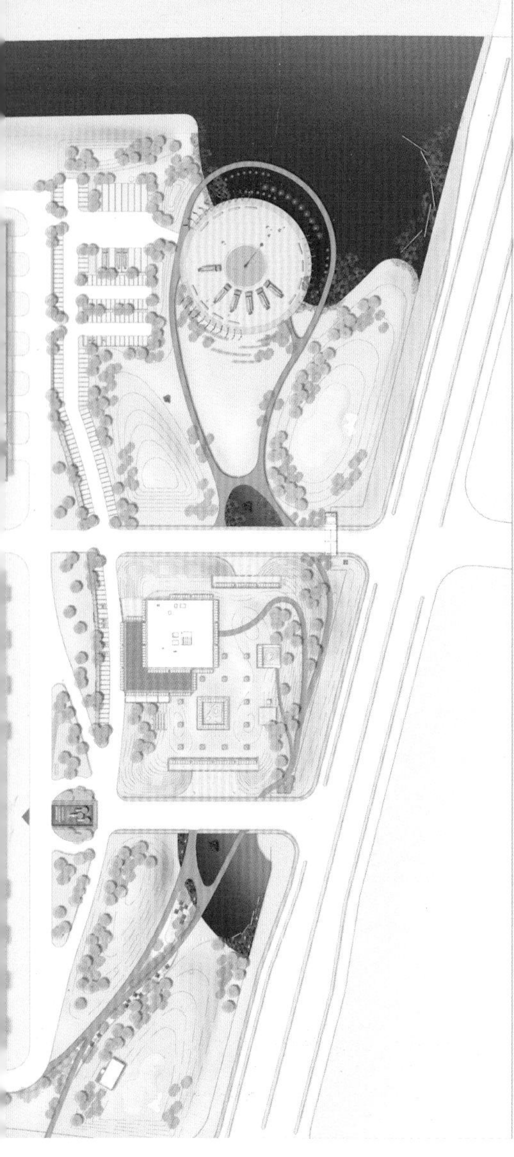